MW00849959

THE NEW GEOPOLITICS OF
NATURAL GAS

THE NEW GEOPOLITICS OF NATURAL GAS

AGNIA GRIGAS

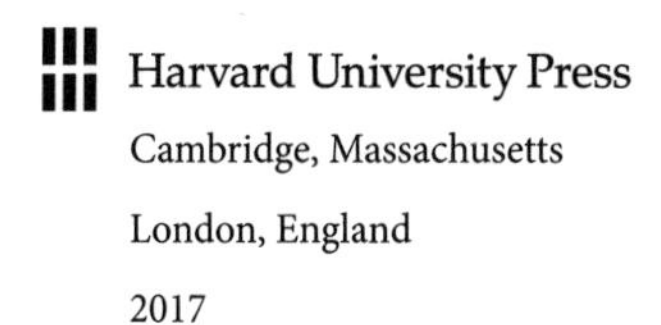

Cambridge, Massachusetts

London, England

2017

Printed in the United States of America
First printing

Library of Congress Cataloging-in-Publication Data
Names: Grigas, Agnia, 1979– author.
Title: The new geopolitics of natural gas / Agnia Grigas.
Description: Cambridge, Massachusetts : Harvard University Press, 2017. |
Includes bibliographical references and index.
Identifiers: LCCN 2016045023 | ISBN 9780674971837 (alk. paper)
Subjects: LCSH: Gas industry. | Geopolitics. | Primary commodities. | International trade.
Classification: LCC HD9581.U52 G75 2017 | DDC 382/.42285—dc23
LC record available at https://lccn.loc.gov/2016045023

For my husband, Paulius

Contents

THE NEW GEOPOLITICS OF NATURAL GAS

Introduction: A New Era of Gas

"Are we entering the golden age of gas?" "Are we on the edge of a truly global gas market?" Are we "breaking Russia's natural-gas chokehold?" Or "are low oil [and gas] prices killing the US shale boom?"[1] Headlines like these have been spinning every bullish and bearish story on natural gas since the early 2010s, when it became increasingly evident that something extraordinary was taking place as a result of America's shale gas boom and that the long-awaited global gas market could finally be on the horizon. Newfound abundant resources, new producer states, the growth of trade in liquefied natural gas (LNG), the buildup of infrastructure, and the rise of global gas trade are indeed exciting developments for businesses, entrepreneurs, and financiers.

However, it may very well be that the most profound implications of the transforming global gas markets will be for governments and policymakers. The shifts in gas markets are upsetting the half-century-long status quo of global gas relations and carry profound geopolitical implications. If the optimistic projections of the gas revolution come to fruition, American gas could secure and diversify Europe's supplies, reign in Russia's energy influence, woo energy-hungry Asia, and ensure that the twenty-first century once again remains in the firm grasp of the United States and its allies. Other countries have also made it clear that they will not remain on the sidelines. Canada, Australia, China, Argentina, and some European states have either reinvigorated their own gas development or considered shale exploration. Gas-hungry states like China, India, Japan, and South Korea look to import American LNG, while the Central and Eastern European states have likewise lobbied for American LNG exports to reduce their dependence on increasingly revanchist and aggressive Russia. For some, as a cleaner fossil fuel, gas holds promise as a means for reducing carbon dioxide emissions in the face of climate change concerns. If a new era of gas is arriving across the globe, then the new geopolitics of gas is coming on its heels.

What will the new geopolitics of gas mean for the United States, Europe, Russia, China, and beyond? This book explores this question by assessing the political implications of the transforming global gas markets. Gas is no longer a scarce, localized, difficult-to-transport resource doled out by energy monopolists (and their affiliated states), not infrequently in expectation of commercial and political concessions under the threat of price hikes or supply cuts. Instead, gas is becoming a freely traded and increasingly available commodity worldwide, and with abundant gas and rising global gas trade, the established modus operandi between importing states and traditional suppliers is changing—strengthening the bargaining position of the former and weakening the leverage of the latter. Markets are increasingly setting the terms of trade, and gas monopolies look poised to lose some of their geopolitical clout. This analysis will focus on the main players in the new geopolitics of natural gas: the new gas leader, the United States; the largest gas importer, Europe; the historic gas powerhouse, Russia; key gas transit states such as Ukraine; isolated gas producers such as Azerbaijan, Turkmenistan, and beyond; and new centers of gas demand such as China, India, and other Asian nations.

The strategic and geopolitical role of energy has long colored interstate relations. As a result, energy security has consistently remained a top concern for most governments, even as its definition greatly expanded.[2] In the past, policymakers and scholars focused mainly on oil rather than natural gas, which played second fiddle to oil on the energy markets. Initially, gas was little more than a waste product of oil production because of its lower energy density. Historically, it has also been mainly a consumer-oriented commodity, while oil has been crucial for the military and industry. Indeed, it was oil rather than gas that won the First and Second World Wars for the Allies, and it was American oil supplies that established the country as the leading power of the twentieth century.

At the same time, gas was more susceptible to political variables than oil. Difficulties in transporting gas long distances over land or across seas have made it a regional fuel rather than a fungible global commodity. In addition, gas transportation has often been operated by monopolies. Consequently, gas-producing and gas-importing countries have had to forge direct and lasting links with each other via long-term gas contracts and codeveloped pipeline infrastructure.[3] For instance, the centerpiece of twentieth-century gas geopolitics—the gas supply relationship between European states and Russia—emerged in the 1960s, with the Soviet buildup of pipeline infrastructure and establishment of long-term contracts. This relationship has remained largely unchanged until the present day. In contrast, in the international oil markets, middlemen freely trade oil. Moreover, because gas historically had been supplied primarily by land-based pipeline monopolies rather than by competing ship tankers moving in any direction across the seas, gas-importing countries have had to rely on a limited number of suppliers. As a result, until recently, many countries, especially in Central and Eastern Europe and the Caucasus, have been 100 percent dependent on a single gas pipeline, a single gas-producing country, and even a single company, such as the Russian gas giant Gazprom. Meanwhile, gas-producing countries such as Russia have likewise been dependent on a fixed set of pipeline export routes and consumers. Then, too, with other states serving as transit territories for the pipelines carrying gas from producing to importing countries, these territorially driven gas trade relationships have impacted national, regional, and international politics. At the start of the twenty-first century, however, changes in the global gas sector are upsetting the status quo and rewriting the rules of the game in the new geopolitics of gas.

The Prospect of a Global Gas Market

The recent exuberance stemming from the American shale boom and the prospect of a global gas market has also been mixed with skepticism as market conditions have seemed to shift with the wind, exceeding even the rapid developments of foreign affairs. Still, since the early 2010s, and especially since the fall in energy prices of 2014, the overall prospect has been one of energy abundance, accompanied by optimism and even rhetoric about a gas glut, which differs sharply from the pessimistic mind-set surrounding energy scarcity that dominated as late as the mid-2000s. At that time, eminent think tanks such as the Council on Foreign Relations and agencies such as the US Department of Energy's Energy Information Administration (EIA), along with other scholars and analysts, predicted that America's energy imports would only continue to rise.[4] Political pundits worried about the vulnerability stemming from the dependence of the United States and its allies on unfriendly or unstable oil- and gas-exporting states, while gas companies and entrepreneurs planned to import significant quantities of LNG and invested some $100 billion worth of LNG import terminals along the US coasts.[5] However, almost none of these developments came to pass because by then the American shale boom, or the rapid production of gas and oil from shale deposits, was already underway. Shale gas swiftly and consistently increased as a percentage of American natural gas production, from 1 percent in 2000, to over 20 percent in 2010, to about 48 percent in 2014, and to 50 percent in 2015.[6] While the United States had been competing neck and neck over gas production with Russia for three decades, in 2011 it overtook Russia as the world's greatest producer of natural gas, a position it held briefly in 2009 upon the start of the shale revolution and before that in 2001.[7] As a result, in 2016 the United States commenced LNG shipments to Latin and South America, Asia, Europe, and the Middle East and looked poised to eventually emerge as one of the world's leading LNG producers, joining the ranks of Australia and Qatar. By 2020, the United States is expected to add the equivalent of approximately 20 percent of total LNG volumes that were traded worldwide in 2014 into the international gas market.[8] The global superpower is poised to become an energy superpower.

These newfound energy resources will bring many domestic benefits and economic opportunities as well as improve America's energy security, or the availability of sufficient supplies at affordable prices. These advantages, however, are only part of the story.[9] Energy insecurity—or constrained access to supplies or acute dependence on undiversified imports—has not been an acute dilemma for the United States. Before the shale boom, US net imports of gas accounted for only about 16 percent of total domestic consumption in 2005; by 2015 this figure fell to just over 3 percent. More than 90 percent of imported gas came from friendly neighboring states such as Canada and Mexico, which were not perceived to pose a threat to American energy security. US oil imports have also declined from their peak in 2005 of 60 percent of US consumption, with net imports contributing to some 40 percent of total consumption in 2012 and only 24 percent in 2015, the lowest level since 1970. Oil imports were also well diversified across a number of states with Canada, Saudi Arabia, and Venezuela being the top suppliers in 2015.[10] Thus, the US shale revolution can only relatively improve America's energy security, especially its gas security.

The greatest strategic potential of American gas lies overseas—to support America's existing allies and win over new ones, to contain its foes, and to bring its rivals into the fold. However, many of these geopolitical developments will be driven by markets rather than by policymakers. Unlike the energy sector of other energy-producing countries such as Russia or those in the Middle East, the American gas industry is made up of and driven by private corporations and entrepreneurs rather than state-owned companies or state interests. Nonetheless, market developments will be both consequential for and somewhat dependent upon politics. The administration of US president Donald J. Trump will support robust domestic energy production. The availability of US LNG for export will strengthen Washington's leadership on the world stage, while the appetite for US LNG in foreign markets will depend to a degree on whether states such as the United States or Russia are perceived as trustworthy gas suppliers, potential threats, or potential allies.

With its booming shale oil and gas production and LNG exports, the United States is well positioned to challenge Russia's position of energy influence over Europe and its efforts to become a dominant supplier to Asia. Even if the United States' LNG export volumes were modest, its domestic gas production would meet America's demand, which otherwise would have been met by imports. At minimum, this would free up gas in other markets, thereby increasing liquidity of global gas markets and theoretically driving global gas trade in a convergence toward a global gas market. Since the mid-2010s, the so-called gas glut has emerged as a popular refrain among market watchers. This surplus of gas in the global market has driven down gas prices and impacted pricing and contract schemes—a key development both for traditional suppliers such as Russia and for import-dependent states in Europe and Asia. Moreover, American LNG exports may eventually make their way to energy-vulnerable US allies. If so, American LNG imports could help meet Europe's and Asia's demand for diversification, reduce reliance on long-term contracts with Gazprom and dampen its price gouging, and limit the influence of other potentially hostile or unstable energy-producing states. The result: a shift of the current geopolitical landscape in America's favor.

These gains are not a given, however. Like stock markets, energy markets are cyclical, with boom years alternating with downturns. Oil markets have demonstrated such decade-long cycles in which low prices lead to high demand but result in low investment in production, which, in turn, lead to a constriction of supplies and a rise in prices.[11] With gas prices having been traditionally linked to those of oil, natural gas has been described as "a manic-depressive industry that is prone to wild swings in mood."[12] Thus, changing market factors such as economic slowdowns, which reduce global energy demand, or falling global oil prices since 2014, which create less incentive for US shale producers, could dampen gas production and reduce the global gas glut. For now, American shale gas producers have shown remarkable resilience, increasingly their productivity and efficiency by cutting their costs and raising their yields. Regardless of any short- or medium-term cyclical market downturns, it is increasingly evident that a gas revolution is in the making.

Indeed, the story of the changing global gas sector is not solely about the boom or bust of American shale gas or even global shale gas development. This book will focus on three key transformations in the gas markets that together contribute to the integration of regional markets into a global gas market: the growth of LNG trade, the shale boom, and the buildup of inter-connective gas infrastructure, all of which, in turn, have consequences for gas pricing, contracts, and trade.

There is considerable debate among leading energy scholars and analysts about what a global gas market would entail and even more debate about when it would emerge, with some cautioning that we are still a long way from its arrival. In a truly global gas market, gas trade would take place on short-term and long-term contracts offered by a large array of gas sellers and traders; the European, Asian, and American markets would become more interconnected while regional price disparities ceased to exist; gas prices would reflect supply and demand economics as opposed to indexation to the price of oil; and regional monopolies would theoretically be forced to compete with plentiful gas volumes transported from various pipelines and various LNG sources in a more integrated and connected region.[13] The gas markets are already seeing many of these developments, which point to an increasing interconnection between different regional markets and a rise in global gas trade.

The original and possibly most important catalyst for a global gas market was the boom of LNG trade that started in 1997 when Qatar created the world's largest LNG export plant and took its business global.[14] The availability of sizable volumes of LNG for export and the ability to ship LNG by tankers across the globe first opened the theoretical possibility of gas prices converging across the three main gas markets of Asia, Europe, and North America. By 2012, when Japan's nuclear power plants were shut down and LNG filled the energy gap, Daniel Yergin, a highly influential energy expert, said that gas had become a global commodity.[15] This trend has only continued with the increased supply and demand for LNG from gas producing and importing countries at the expense of dependence on piped gas. In 2015 global LNG trade rose to a highest ever of 244.8 million tons (MT) in the LNG trade history, surpassing the previous high of 241.5 MT in 2011.[16]

The subsequent shale boom, which unlocked previously unavailable shale gas resources primarily in the United States and Canada, has been the second powerful factor in the transformation of global gas markets. Together with shale newcomers of Argentina and China, these four are the only countries in the world to produce commercial volumes of shale gas. Coupled with LNG liquefaction, regasification, shipping technologies, and the rising global LNG market, it is now possible to leverage the shale revolution outside of North America with American LNG exports. Moreover, a number of other countries also possess shale gas resources. The top ten in order of greatest to least are: China, Argentina, Algeria, the United States, Canada, Mexico, Australia, South Africa, Russia, and Brazil. The United States has a unique advantage over most other countries due to the confluence of know-how, technology, business environment, financing availability, legislation such as landowners' mineral rights, and entrepreneurial drive. However, it is possibly just a matter of time before other nations like China will overcome existing obstacles and eventually emerge as notable producers of shale gas. The economic and strategic implications could be similar to or even greater to those of the US shale revolution.

The third key development is the buildup of gas transport infrastructure such as gas pipelines, pipeline interconnectors, and LNG liquefaction and regasification terminals, all of which help integrate different local and regional gas markets in North and South America, Europe, Asia, and beyond. This growth of interconnectivity also comes from new uses of existing pipeline infrastructure, such as for reverse flows, which enable greater flexibility and diversity in gas suppliers. The rise in traditional pipeline infrastructure should not go unnoticed in light of the more dramatic breakthroughs and technologies of shale and LNG. In fact, the results of pipeline interconnections and reverse flow may come sooner in many parts of the world than progress from shale development or LNG trade, especially for landlocked countries that always have to rely on traditional pipeline infrastructure rather than LNG terminals, which require coastal access. Furthermore, other game-changing gas market developments pertain to distinct regions or countries. These include the rise of energy market regulation in the European Union (EU), as well as calls for reduction of carbon emissions and thus increased use of cleaner forms of energy such as gas in the United States, the EU, and China.

Shale abundance, LNG flows, and new gas infrastructure are in turn altering the global gas trade. New gas-exporting states and gas-importing states are forming new relationships. Gas trade is poised to be less dependent on traditional east-west pipelines for Russian gas exports to Europe, while across the Atlantic, it is likely to become less dominated by north-south flows through Canada, the United States, and Mexico. These shifting gas trade flows (especially of LNG) are having an impact on gas prices as well as gas pricing models and contract schemes. Price differentials in different parts of the world are converging, long-term contracts are being challenged by increased trade on the spot markets, and pressure for removing destination clauses for gas flows and de-linking the price of gas from the price of oil is building. All this favors the emergence of a global gas market and shakes up the status quo of the geopolitics of gas.

The Study of Geopolitics, Energy, and Eurasia

The term *geopolitics* comes from the Greek *gê*, meaning earth or land, *politikē* and refers to the interplay of geography and international politics, particularly the former's effects on the later.[17] Although many prominent political thinkers, including Montesquieu, Kant, and Hegel, had an implicit understanding of the concept, geopolitics as we know it today developed in the 1970s, when prominent American statesmen and national security advisors such as Henry Kissinger and Zbigniew Brzezinski combined notions of power politics with regional dimensions.[18]

For simplicity's sake, geopolitics can be distilled into two key elements—namely, territory and power. Together, these form the foundations for understanding energy because natural resources are derived from delineated territories of states, and their endowment or lack thereof has to a considerable extent determined the power and influence of nations. Access to and command of transit territories and the seas for the transportation of energy has also been a critical factor.

The study of energy emerged in no small part due to energy security considerations of states as they became increasingly dependent on energy imports. Concerns first emerged among government officials in the early twentieth century in the context of oil supplies for armies and navies. Subsequently, scholars started studying energy security in the 1960s and particularly after the oil crises of the 1970s,[19] while conversely, the stabilization of oil prices and the receding threat of political energy embargoes in the late 1980s and the 1990s dampened interest in the subject. In the 2000s, however, the issue once again came to the fore due to rising energy demand in Asia, tensions over Russian gas supply in Europe, and concerns over climate change.[20] Then, in the 2010s, during the early years of the US energy renaissance, the links among energy, security, and international relations drew attention, with oil markets once again garnering most of it.[21] The fundamental shift in the global gas sector likewise requires attention and assessment—something this book will tackle for the first time.

If the study of the geopolitics of energy is relatively novel, that of the geopolitics of gas is even more so. While there is no agreed to, or precise, definition of the geopolitics of gas, David G. Victor, Amy M. Jaffe, and Mark H. Hayes, who are among the original theorizers on the subject, view it through the prism of "the immensely political actions of governments, investors, and other key actors who decide which gas trade projects will be built, how the gains will be allocated, and how the risks of dependence on international gas trading will be managed."[22] More broadly and for the purposes of this book, the geopolitics of gas reflects how gas supply, demand, dependence, and transit can determine bilateral relations between importing and exporting states and cause power shifts in specific regions and in the international system as a whole, and at the same time, it reflects how power shifts in the international system, in regions, and in bilateral relations affect gas supply, demand, dependence, and transit.

In analyzing the new geopolitics of gas, this book focuses on the United States and the countries and regions that make up Eurasia, where new gas developments carry the greatest implications for changing worldwide market and political trends. As Victor observes, "Whether gas becomes a truly global commodity and the geopolitical effects of the global gas trade will depend centrally on the United States—the world's largest user of natural gas and the epicenter of most innovation in the industry."[23] Meanwhile, Eurasia—a portmanteau referring to the combined landmass of the European and Asian continents and including over ninety countries such as France, Germany, Russia, China, and India—is both a geographic and geopolitical construct with strategic importance for energy markets. While countries such as Australia, Argentina, and Indonesia, as well as the Middle East states among others, already play an important role in gas markets or will do so, Eurasia will experience the most significant shifts in the new geopolitics of gas.

Eurasia accommodates around three quarters of the world's population and a large portion of its natural resources and wealth. This is where considerable gas reserves are located—specifically, in Russia—and where vast gas demand will continue to grow. According to the EIA, although the Middle East has the world's largest gas reserves at some 2,818 trillion cubic feet (TCF), Russia and the former Soviet republics are a close second at 2,178 TCF; add Europe at 131 TCF and China at 164 TCF, and Eurasia's total comes even closer to that of the Middle East.[24] Eurasia's demand for gas is reflected in Europe's continuing interest in increasing use of the clean-burning fuel, while Asia's growth is driven by China and India, along with LNG demand by Japan and South Korea, two of the most LNG-dependent countries in the world. Moreover, Eurasia is also where the long-standing status quo of gas relations is being transformed between the world's largest importer of gas and America's closest ally—the EU—and the world's largest exporter and revanchist power—Russia. Likewise, the rising powers of China and India are expected to further shift of the balance of power in the twenty-first century.[25]

The geopolitical notion of Eurasia is possibly more important than the geographic one and has for decades, if not centuries, intrigued famous thinkers, even as its definition varied greatly from the vantage points of different countries. Halford Mackinder, an early twentieth-century British professor of geography at Oxford University, was among the first theorists. Focusing on the "Heartland," a geographic area bounded by the Volga River in the west, the Yangtze River in the east, the Arctic in the north, and the Himalaya Mountains in the south, Mackinder argued that those who commanded the Heartland commanded the world.[26] Indeed the regions and countries of Eurasia have competed for centuries for power and territory on their common landmass and the surrounding seas. Today the competition persists for natural resources and their supply routes. It also transcends the landmass and is of paramount importance to the United States.

As Brzezinski argued during the Cold War, "Whoever controls Eurasia dominates the globe. If the Soviet Union captures the peripheries of this landmass . . . it would not only win control of vast human, economic and military resources but also gain access to the geostrategic approaches to the Western Hemisphere—the Atlantic and Pacific."[27] Today this concern remains just as relevant given the changing nature of the global gas sector and the increasing reliance on oceans and seas for gas transport. The growing LNG trade, the emerging trend of building gas pipelines under European waters, and competition for underwater gas reserves in the South China Sea, the Arctic Ocean, the Caspian Sea, and elsewhere all contribute to the likelihood of greater geopolitical jostling in waters of Eurasian states and beyond.

Understanding the competition for power, influence, and energy resources in Eurasia will shed light on Russia's future role in the global balance of power. While Russian thinkers of the turbulent 1920s argued that as a unique civilization, Russia should unify the lands of Eurasia (in Moscow's view narrowly defined as the territories that made up the former USSR), and thus withstand encirclement and efforts to crush it by the maritime (Atlantic) civilization,[28] this school of thought was eventually crushed by Communist ideology. Nevertheless, the Soviets extended their dominance in Eurasia, and, after the collapse of the Soviet Union, the idea of Russia's centrality to the landmass reemerged in the Kremlin's thinking as a civilizational and political counterpoint to the United States and the West.[29] Subsequently, Russia has deployed its own concept of "Eurasianism" since the 1990s in projects such as the creation of the Commonwealth of Independent States (CIS) of 1991, followed by the Eurasian Customs Union and the Eurasian Economic Community, and then, most recently, by the Eurasian Economic Union of 2014.[30] While energy was an important component of these entities and served as the glue for economic relations, in the 2000s gas exports from the Caucasus and Central Asia to Europe and China began serving a counter-integrative role for a number of former Soviet republics, shifting their focus away from Russia. In the medium term, if successful, Russia's own plans to supply gas to China could significantly impact the geopolitics of Eurasia.

Politics, Markets, and Gas

To explain gas politics in light of the transforming global gas markets, this book uses several analytical lenses to view the political and commercial relationships of gas-producing and gas-importing countries, as well as those involved in transit. For gas-producing and gas-exporting countries, the conceptual tool is the *politics of supply.* For gas-importing countries it is either the *politics of demand* or the *politics of dependence.* For gas transit countries it is the *politics of transit.* The importing and exporting states can also share a relationship of *interdependence.*

These relationships are neither one-sided nor static; rather, they are situational and can vary in different contexts and moments in time.

Thus, for instance, while Russia may at times have a relationship of relative interdependence in its gas trade with the EU as a whole, it can simultaneously pursue the politics of supply with Latvia, which in turn experiences the politics of dependence in its relationship with Moscow. Meanwhile, Russia may be unable to pursue the politics of supply with China, given that their relationship is determined by Beijing's politics of demand.

These analytical categories can also be applied to interstate energy relations more generally, though the differences between gas and other resources such as oil or coal should be kept in mind. That is, because oil and coal transport is relatively easier and global market and trade are more robust, they are more fungible as commodities and tend to be less susceptible to politics than gas.

Interdependence

While energy politics implies an inherent element of power and influence, energy relations are complex and not one-sided in favor of exporting states. The exporting states also depend on their markets, or on the importing state(s), for revenue. The resulting relationship can be one of *interdependence*, or of symmetrical degrees of dependence.[31] When gas importers and exporters are interdependent, gas supplies in the long term, if not necessarily in the short term, tend to be generally stable and more resistant to political and security instability.[32] Interdependence is based on a number of conditions, including the size of the importer and exporter's market and the degree to which each side has alternative import or export market opportunities and related infrastructure.[33] Moreover, the degree of symmetry of dependence is not static: it can change with changing gas market conditions, the cycles of glut versus scarcity of supplies, discoveries of new resources or technologies, economic performance, and other domestic, bilateral, and international factors. In the real world of trade and commercial relations, perfect interdependence is a rare occurrence. Thus stable, non-politicized energy relations are elusive.

The Politics of Supply

When the relationship between the exporting state and the importing state is asymmetrical in favor of the former, the supplier can pursue the set of policies or benefit from a position of strength that can be described as the politics of supply. The politics of supply enables large energy-producing countries to pursue their national economic, political, and security interests vis-à-vis energy markets and gas-importing countries. They can include both economic and political policies, especially three highlighted by American diplomat Carlos Pascual: flooding markets, starving markets, and assisting friends.

Flooding markets with increased production can enable the producer to acquire market share or drive out new or existing competitors.[34] Although a number of antidumping regulations exist under the World Trade Organization (WTO), these regulations do not apply to oil and gas trade. Thus, oil dumping, for example, was pursued by Saudi Arabia in 1986 to push Russian and US oil out of the international market and to regain its position as the world's dominant oil supplier. Arguably, Saudi Arabia used this tactic again in 2014 in the global oil market.[35] Conversely, producers can also starve markets or manipulate highly dependent importers by limiting access to supplies. Such was the case in the Arab Oil Embargo of 1973 and in various attempts made by Gazprom in Ukraine, the Baltic states, Bulgaria, Georgia, and elsewhere. In contrast to both flooding and starving markets, the policy of assisting friends can include price discounts, targeted supplies, or financial or technological support to develop energy resources for partner nations.[36] As far as gas exporters are concerned, these three policies depend on gas supply infrastructure, such as long-haul pipelines and LNG terminals, which require sizable upfront investment, years of implementation, and long-term commitments from buyers. Politicking and maneuvering are not in themselves likely to be effective in the commercial reality of gas markets.[37]

At the same time, however, when infrastructure is combined with tactics of flooding or starving markets or assisting friends, political maneuvers can become more effective in enabling gas-exporting countries to gain leverage over importing states. For instance, negotiations of gas supplies or prices can be accompanied by demands that importing nations alter their domestic or foreign policies, join political or economic blocs or alliances with the supplier nations rather than rival ones, or sell off their energy infrastructure or lease their military bases to the suppliers. Similarly, halting gas supply or hiking gas prices can be a form of pressure or type of weapon in times of political tensions or armed conflict. Producers may act globally or bilaterally with international agreements and negotiations. They can also act locally and unilaterally by meddling and supporting lobby groups in the domestic political systems of importing states in order to promote the use of their commodity or to block diversification efforts. Russia, as a dominant piped gas supplier, has in many regards written the playbook on the use of gas for political influence and has employed, over the years, all of these tactics in the politics of supply to various degrees in the EU, Central and Eastern Europe, the Caucasus, and Central Asia. However, the increasingly interconnected and liquid gas markets are reducing some of the efficacy of the tactics gas suppliers use to exert pressure on importing countries.

Not all gas-producing countries can equally pursue the politics of supply. Just as the level of interdependence between the supplier and the importing nation determines whether a producer can pursue the politics of supply, so do characteristics of the producing country similarly constrain or broaden its power and influence in the supply relationship. Smaller suppliers that cannot meet a significant portion of the needs of their customer states lack commercial or political leverage over these importers. Likewise, isolated suppliers that cannot access many markets or importing countries are much weaker players in the politics of supply. The reasons for isolation and constrains in exports, which can be various, include geographical factors such as the lack of coastal access for LNG exports or difficult terrain that makes building pipeline infrastructure expensive or problematic. Diplomatic factors are also relevant and can include international sanctions against the exporting country, as in the case of Iran, while lack of access to reliable transit states when the neighboring region is engulfed in conflict or war can be another problem for exporters. Azerbaijan and Turkmenistan exemplify isolated suppliers that until relatively recently were not able to access the European or Asian gas markets. Nonetheless, they will remain somewhat constrained in their future export capacity due to the lack of coastal access, and in the case of Turkmenistan the shared borders and proximity with the volatile states of Afghanistan, Iran, and Pakistan.

The Politics of Demand and Dependence

When energy-exporting and energy-importing nations are not equally interdependent or their interdependence is asymmetrical, the importing nations can pursue the politics of demand or the politics of dependence, or a combination of both. The politics of demand involves a set of economic and political policy options available to energy-importing countries to pursue their national economic, political, and security interests from a position of strength vis-à-vis energy markets or specific exporting countries or groups of countries. The politics of dependence is a set of economic and political policies available to energy-importing states that are disproportionately dependent on a particular exporting state and thus are pursuing their interests vis-à-vis the exporter from a position of weakness. Three key factors determine whether a country is playing the politics of demand or being played by the politics of dependence: diversification, volumes of imports, and market conditions.

A country is well diversified when it can both produce some of its own gas and import it from a variety of states through a number of pipelines, as well as via sea routes with access to LNG terminals, and is thus in an ideal position in terms of the politics of demand. If, on the other hand, a country or entity is a small or undiversified importer to begin with, it can become highly reliant on a single or a few supplying countries' gas resources, supply routes, and/or infrastructure and thereby become subject to the politics of dependence. In these situations, the policy options of importing states are dictated by the interests of the supplier country or countries, which can also exert further economic and political leverage on the importing states. Examples include many of the Central and Eastern European states that for decades were nearly 100 percent dependent on Russian gas delivered by a single pipeline to their country and faced Moscow's pressure to pursue more cooperative policies toward Russia. The EU has also hovered too close to the politics of dependence in its relationship with Russia. With its twenty-eight member states and all their differing interests, it has been unable to negotiate with one voice with its largest supplier—Gazprom—and its position was thus consistently weaker.

While it may seem that large volumes of energy imports (or high percentages of imports in total consumption) would make a country vulnerable and thus subject to the politics of dependence, it is in fact the diversification levels of these imports that determines the degree of the country's vulnerability in its import relationship. A country with large volumes (or high percentages) of imports from well-diversified sources can be more secure than a country with small volumes (or low percentages) of imports from a single source. Thus, a well-diversified importer such as China can be in a position of strength and play the politics of demand for gas even vis-à-vis a sizable exporting state such as Russia. Moreover, as a significant market, an importer of large and diversified volumes of commodities will benefit from greater bargaining strength in its negotiations with a supplier state. This is especially true if the supplier does not have access to different export markets and / or to the diverse export infrastructure of different pipelines and LNG terminals. In contrast, a small state importing small volumes of gas from a single supplier is often in the most vulnerable position because its market size makes it an insignificant market for the traditional supplier and it often cannot attract new competing suppliers.

Market conditions can also influence a country's position in the spectrum from the politics of demand to the politics of dependence. The cycles of low or high prices and the level of liquidity in the market can strengthen either the buyer's or the seller's position. In this regard, the new global gas markets, including the shale boom, are, in the words of the *Economist,* "turning a seller's market into a buyer's paradise, promising deep and liquid markets with a growing diversity of supplies that improves security for buyers."[38] Such market conditions facilitate the ability of states to pursue the politics of demand rather than the politics of dependence. For instance, as this book will show, due to the transforming global gas sector, the growth of LNG trade, and other important internal and external developments, the EU is increasingly transcending the politics of dependence and moving on to pursue the politics of demand.

The Politics of Transit

The politics of transit is a phenomenon applicable to states whose territories are instrumental in the flow of energy supplies from producing to importing states. Whereas in the past, transportation was conducted mainly by land-based gas pipelines traversing the territories of foreign states, now it is increasingly accomplished through LNG shipments crossing mostly international waters. Unless a producing state is able to directly export energy exclusively through its own territory to the importing state or has coastal access and infrastructure for LNG shipments worldwide, both the producing and importing state must then depend on transit states. This increases the regional strategic importance of transit states such as Ukraine, which has long served as the main route for Russian gas exports to Europe. Such states also gain some negotiating power and revenues from transit fees or operation of energy transit infrastructure, thus making them "rentier states." They can even resort to their own energy weapon by destabilizing the transit of supplies through their territory.[39]

While the politics of transit can be a source of leverage or rents, it often has a detrimental effect on the domestic political and economic conditions of transit countries. Transit countries like Ukraine can get stuck in a "cycle of rents" wherein dependence on significant foreign income, which often allocated to select companies or interest groups, enables corruption and creates a vulnerable political system internally and vis-à-vis the exporting states.[40] Moreover, transit states can be subject to the destabilizing forces of tensions and fluctuations in the balance of power in their regions. This can be further complicated by changes in the status of transit states or in their political alliances, which occurred, for example, following the disintegration of the Soviet Union, when Russian energy transit became reliant on newly independent states. As energy researcher Katja Yafimava has emphasized, the lack of an overarching regulatory body for transit creates a geopolitical, regulatory, and contractual "transit dilemma" between producers and importers and makes transportation of energy resources complicated and crisis-fraught.[41]

The position of transit states is often further complicated by the fact that they themselves can be dependent for gas imports on the same country for which they serve transit functions. Such is the case for Ukraine, Belarus, Turkey, and other countries vis-à-vis Russia. Moreover, gas producers also sometimes serve as transit states, as Russia did in its previous role of aggregating and exporting Central Asian gas to Western markets.

Given the changes in the global gas markets, the strategic importance of transit states will likely decline in the long term with the emergence of new and diverse supply routes for LNG tankers, the construction of additional pipelines, and the implementation of reverse gas flows. Nonetheless, resolving the regulatory dilemma of transit will not be easy even if third-party countries like Ukraine seek to comply with EU energy policies. Regulatory issues of sea transit will also remain an issue; witness Turkey's ban on LNG tankers through the Bosphorus Strait.

Overall, as the new era of gas ushers in a new geopolitics of gas with new winners and losers, an understanding of the changing dynamics of the politics of interdependence, supply, demand, dependence, and transit is important for scholarship, policymaking, and business vested in the key regions and markets of the United States, Europe, Russia, and Asia. Following Chapter 1, which serves as a primer on natural gas and the historic and contemporary developments of gas markets, the country case study chapters in this analysis will demonstrate different types of geopolitically tinged gas relationships in terms of the politics of gas supply, the politics of gas dependence, the politics of gas demand, and the politics of gas transit; they will also show how these relationships are being transformed by the transforming global gas markets. For the politics of supply, Russia as a traditional supplier of piped gas will be contrasted with the United States as an anticipated leader in LNG exports. The EU will be used as a case study of the evolution from the politics of dependence to the politics of demand in light of Brussels's regulatory policies and external market changes. The countries of Ukraine and Belarus will highlight the politics of transit, while Azerbaijan, Turkmenistan, Uzbekistan, Kazakhstan, and their neighbors will shed light on the politics of isolated suppliers. Finally, the strengthened position of importing states under the new market conditions—the politics of demand—will be seen in the discussion of China, which will also extend to other energy-hungry Asian countries such as India, Japan, and South Korea. The political, economic, and gas relations of these supplying, importing, and transiting countries form the new geopolitics of natural gas.

1 The Changing Global Gas Sector

The start of the twenty-first century saw a historic shift in energy markets: new energy powers emerged, new sources of energy were discovered, and technologies for exploring and transporting previously economically untenable gas and oil were improved. Likewise, there were important shifts in energy demand, with slowed growth in consumption among the developed countries while, growth in China, India, and other parts of the developing world increased and look to remain higher than that of developed countries. At the very center of these market changes was the rapid development of unconventional gas and oil, most evident in North America and dubbed the "American shale revolution." The new abundance of natural gas in the markets, resulting from America's boom in gas production and the launch of American liquefied natural gas (LNG) exports in contrast to previous expectations of sizable US gas imports, freed up the available volumes of gas elsewhere to be traded in the gas markets. Even before the upsurge of shale gas, another phenomena was already transforming the gas markets: the growth of LNG trade, which made it possible for gas to be delivered in large volumes in long distances across oceans to any region of the world. Global LNG trade, which still comprises a modest but growing share of the total gas trade, has facilitated increasing interconnection of different regional gas markets, as has the buildup of traditional gas pipeline infrastructure, though progress in different parts of Europe and Asia is still unequal.

Increasingly we appear to be on the verge of a truly global gas market. Certainly, any number of unknowns, environmental concerns, and market forces could create setbacks or new wins, but the long-term prospects for gas markets are being transformed.

The emergence of the global gas market is still debated. As an evolving reality and concept, *the global gas market* is difficult to define and thus it is difficult to ascertain when, or whether, it will fully materialize. Traditionally, the global natural gas market has been understood by the US Energy Information Administration (EIA) in terms of two models. The first considers trade as grouped by patterns of transoceanic shipping in the Atlantic Basin and Pacific Basin trade regions. The second is characterized by trade within subregions that make up the Atlantic or Pacific Basins, including North America, Europe, Asia, and Africa. This distinction between subregional and interbasin trade is determined by how far a cargo ship travels.[1] Gas trade began to change significantly in the mid- to late 2000s when greater gas volumes, specifically in the form of LNG, started moving longer distances and were being traded on shorter-term contracts to take advantage of the price differential between regions. By 2015, the rapid rise in shale production, the increasingly evident global gas glut, collapsed gas prices, and the increasing interconnectivity of the Atlantic and Pacific Basins were slowly allowing natural gas to become more of a global commodity than it ever had been before. This interplay between different regions has continued to transform trade, so that market trends in one region affect markets in the other, and vice versa. It is this type of trade outside of long-term contracts that is slowly globalizing the gas industry, with competing commercial players and increased competition between regions playing a large role in giving global gas trade some of the characteristics of commodity trading in the global oil markets. Such trade is considered liquid, where *liquidity* in gas markets is defined as being able to quickly and easily buy or sell gas without significant price changes or considerable transaction costs.[2]

LNG trade will have a particularly significant impact on the development of a global gas market and the extent to which gas will be considered a truly liquid commodity. Some potential consequences of the extensive volume of natural gas expected to enter the market between 2016 and 2018—and especially the "tsunami" of LNG production coming out of the United States and Australia—include a breakdown of regional pipeline monopolies and increased competition, a move away from pegging gas prices exclusively to the price of oil, more diverse gas trade whereby old and new gas buyers alike will not have to depend on gas supplies from a few nations, reduced price volatility as flexible gas volumes are better able to respond to supply-demand fluctuations, and the establishment of Asian trading hubs that complement those in the United States and Europe.[3]

In 2011, scholar and former US undersecretary of energy John Deutsch put forth the argument that the three main regional gas markets—currently North America, Asia, and Europe—will become more integrated and open as opposed to remaining three distinct regions characterized by different trade patterns and prices, and geopolitical tensions between regional suppliers and importers will decrease. Regarded in this light, a global gas market would imply that demand will be satisfied from supplies from any region of the world and gas pricing will be based purely on supply and demand economics, with transportation costs the only factor creating price differentials between regions. Deutsch foresaw a global market developing for gas just as one developed for oil.[4] As of 2016 some energy experts have been arguing that a global gas market is already underway, given that gas price differentials between different regions reflect only transportation costs. The question remains whether this is a temporary phenomenon due to oversupply or a structural shift.[5]

However, other analysts are more skeptical, believing that a global gas market will not emerge in the foreseeable future because of the nature of gas, which remains more difficult and costly to transport than other commodities and thereby keeps gas from developing the sort of liquidity needed for it to become a true global commodity. Price arbitrage between regions—which can be likely given different supply-demand dynamics between the Americas, Asia, the Middle East, Africa, and Europe—would imply that regional gas markets could not fully converge into one global market. Success in Chinese shale gas exploration; the extent to which nuclear energy returns to Japan and Korea; weak demand in Europe; stricter regional and / or international laws to curb carbon emissions; and the uncertainty as to how much more liquefaction capacity will be built around the world given the global gas oversupply will all determine LNG demand and therefore the extent to which a single global gas market can emerge.[6]

Overall, the concept of a global gas market is still developing as the gas market continuously transforms. Aside from the supply and demand economics of its definition, scholars have yet to study other factors that constitute the new global gas market. According to Jonathan Stern, a gas expert at the Oxford Institute for Energy Studies, these include pipeline-to-pipeline competition; LNG to pipeline competition; the liberalization of domestic markets and the move to market pricing in individual countries; the cost of new pipeline and LNG projects; and individual country developments in relation to the use of other forms of energy, such as nuclear and coal.[7]

This chapter will serve as a primer on the global gas sector and examine the most important developments that mark an age of progress in the global gas markets: the shale boom, the growth of LNG, the build-up of gas pipeline infrastructure, and the implications of natural gas for climate change. We will see how the resulting global gas boom is shifting trade flows from regional to global gas markets, pressuring gas prices, pricing and contract schemes, and raising environmental considerations for natural gas as a bridge fuel.

Natural Gas: A Background

Natural gas, or simply gas, is 70–90 percent methane and 0–20 percent butane, propane, and ethane. Gas has a close relationship with crude oil, or simply oil, which is a mixture of hydrocarbons, and both gas and oil are formed by the process of pressurizing and heating matter decayed under the earth's surface over millions of years. Natural gas can be produced from wells that also contain oil reserves and thus be extracted as a by-product of oil and gas liquid production—called associated gas—or produced in wells unassociated with oil. Particularly at the onset of the global oil industry, this associated gas was often burned off in the oil field, a process called *flaring*. Today natural gas is separated and either processed, vented, or flared. Alternatively, natural gas may also be found in fields isolated from oil reserves, and the extraction process may also include flaring.[8]

Despite their links, oil and gas have not been perceived as equal energy sources. Oil has high energy density (or the amount of energy per unit volume), giving it about 1,000 times as much energy as the same volume of natural gas. Historically, oil has been used in the military, industry, and transportation sectors. Thus, it is widely used in cars, trucks, planes, and ships, and its by-products are used in heating and electricity generation, asphalt, and feedstock, the latter of which is used to produce chemicals, plastics, synthetic materials, and fertilizers. Gas, on the other hand, has tended to be limited to the consumer industry and used for cooking and heating, for example, although it also has applications in fertilizer production, some manufacturing industries (such as in glass, steel, and aluminum production), and industrial products (such as plastics, textiles, and paints). Currently, gas as a cleaner-burning fossil fuel is increasingly displacing coal in electricity production and oil in the transport sector.[9] Due to their higher value as denser energy products, oil exports have been a significantly greater source of revenue generation than gas exports for producer countries. As a result, gas has historically not garnered as much attention from strategists and statesmen even though the peculiarities of the regional gas markets versus the global oil market have of late awarded gas-producing countries such as Russia significant geopolitical influence over the countries they supplied.

The History of Oil and Gas Markets

While the nineteenth century belonged to "King Coal" and the twentieth century was dominated by oil, gas may or may not turn out to be the most important commodity of the twenty-first century. But the fact is that the 2000s have already seen a rapidly changing gas sector, and these changes, along with the potential rise of a global gas market, have tremendous implications, potentially comparable to those associated with the earlier emergence of the global oil market. Indeed, the trajectory of the global gas market development has to some extent followed that of the oil market, albeit with some delay.

Oil was first recognized as an important resource in Mesopotamia in 3000 B.C., when bitumen, a semisolid substance found seeping through the earth's surface, was used in building, some lighting, and fire warfare, but mostly in medicine, although the medicinal power of oil was not identified until the Middle Ages in Europe. Later, Europe imported refining technology from the Arab world. In the New World, while jack-of-all-trades Edwin L. Drake, banker James Townsend, and blacksmith "Uncle Billy" William A. Smith hit oil in Titusville, Pennsylvania, in 1859, it was only during World War I that oil began replacing King Coal and transforming the global energy order. As British Secretary of State for the Colonies Walter Long told the British House of Commons in 1917: "You may have men, munitions, and money, but if you do not have oil, which is today the greatest motive power that you use, all your other advantages would be of comparatively little value."[10] Renowned energy expert Daniel Yergin concluded that Winston Churchill's decision to base British "naval supremacy upon oil" ahead of the First World War best signified how the twentieth century was "completely transformed by the advent of petroleum" and became a century of oil.[11] Britain's decision to import oil for its navy launched the global oil trade. It was in the 1910s when gasoline became "as vital as blood,"[12] fueling postwar modernization in the following decades and creating world powers. Oil also made the United States the leading power of the twentieth century. Once oil was discovered in Texas starting in 1901, the United States soon grew to be world's dominant oil producer up through the 1960s, going from over 200,000 barrels per day of production in 1910 to a peak of over 3.5 million barrels per day in 1970.[13] American oil also gained significant global influence during both world wars, which could not have been won by the Allies without it.[14]

During this era of oil's dominance, natural gas was initially little more than a waste product of oil production—an "orphan of the oil industry."[15] Still the story of natural gas is as old if not older than that of oil. Natural gas seeps were first discovered in Iran between 6000 and 2000 B.C. and in China 2,500 years ago, where it was collected using bamboo pipes and used to boil ocean water for salt. However, in Europe and the United States, the fuel was discovered only centuries later. In England, it was found in 1659, but remained relatively unpopular there and in continental Europe due to the availability of coal (or "town gas") that was used for illumination purposes. In the United States gas was first discovered in 1815 in West Virginia during the digging of a salt well. Subsequently, on that site, gunsmith William Aaron Hart drilled the first US gas well in 1821 and directed the seeping gas using wooden pipes. By 1824 the commercial use of gas emerged in Fredonia, New York, in lighting and cooking, while the first US natural gas company, Fredonia Gas Light Company, was established in 1858. Just as American entrepreneurship and business harnessed early oil development, so too did the United States emerge as a gas innovator. Technological improvements during the post–World War II era as well as the fuel's clean combustion and flexible use made it adaptable to consumer and industrial needs. The United States was consistently the world's largest natural gas producer until 1982, when it was surpassed by the Soviet Union; Canada and the United Kingdom trailed generally behind.[16]

While Russia had emerged as a major player in the international oil market with production in Baku in the 1870s, and while it surpassed the United States in oil output, consolidating its position as the world's largest oil producer from the mid-1970s through the 1980s, the Soviet Union lacked a natural gas industry before the Second World War. During the war, gas was discovered in the Urals and in Western Siberia, and Soviet gas production increased fivefold between 1960 and 1972, with gas exports commencing in the 1960s first to Eastern Europe and then to Western Europe. In 1989, the state-corporate enterprise State Gas Concern Gazprom was established; Gazprom became a state-owned monopoly following the breakup of the Soviet Union.[17]

In contrast to Russia, natural gas production began in Europe as early as the 1920s, but it was the 1956 Suez crisis that incentivized the search for secure oil and gas reserves domestically. Discoveries at the Groningen field in the Netherlands in 1959 were followed by discoveries in the United Kingdom's North Sea field; then, in the 1970s, even larger volumes of gas were found in Norwegian territory, from where gas was piped elsewhere in continental Europe and to the United Kingdom. Coal gradually lost popularity between the 1960s and the 1980s in all of Europe although less so in Eastern Europe and the former Soviet Union, where coal was cheap, hard currency was hard to come by, and environmental concerns were less of an issue than in the West. Domestic gas production in the European Economic Community (EEC) rose to 102 billion cubic meters (BCM) in 1970 and then nearly doubled to 197 BCM in 1980. However, high demand and comparably limited resources meant that Europe was destined to be an importer rather than an exporter, and once domestic gas production started stagnating after 1980, pipeline imports from the Soviet Union and Algeria became increasingly more important.[18]

Like Russia, the Middle East was also a latecomer to the gas markets, although it was central to the development of the global oil market after the First World War, when the British, French, and Americans entered the region to gain access to petroleum. Saudi Arabia's massive oil reserves—discovered in 1938—have been deemed by some historians to be the "greatest geopolitical prize the world has ever known."[19] By contrast, gas was thought to have no economic value until the mid-1970s, when Iran and Qatar emerged as the most important regional gas players. Sharing the single largest gas field in the world, the North / South Pars, Iran and Qatar proved to have the world's second and third largest gas reserves, respectively, after Russia. While Iranian gas production began increasing in the 1980s and grew through the 1990s, gas production in Qatar began in 1949 as a by-product of oil, although most of this gas was flared and vented until 1978, when Qatar began collecting gas. Subsequently, it established its first commercial LNG plants at Ras Laffan in 1997.[20] The growth of LNG shipments in the 2000s ushered in the modern age of gas and launched initial discussions regarding the potential for a global gas market to develop.

Early Oil and Gas Transportation

The rise of LNG was a groundbreaking development because it finally allowed gas to be shipped in tankers rather than by pipeline, as the oil industry had begun doing almost a century earlier. The significance of gas shipping becomes apparent when considering the impact of improvements in oil transportation on the development of the global oil market. In the late nineteenth century, oil transportation began with the advent of pipelines in the United States. More important, the launch of the first oil tanker from Baku, Azerbaijan, enabled the oil industry to grow from a regional to an international enterprise. Previously, when oil was stored in wooden barrels, the barrels were shipped north from Baku via the Caspian Sea to the Volga River, where they were then transferred onto barges and transported to railway terminals, from which the oil was dispatched to its final destination. All this inefficiency began to change, though, after 1878, when the Russian-Swedish engineer and businessman Ludwig Nobel conceived of the bulk tanker for use in the Caspian Sea and even the Atlantic Ocean.[21]

In the United States, oil was originally transported by a system of horse-drawn wagons, barges, and railroads. Land transport was first significantly improved in 1865, when Samuel Van Syckel created the first oil pipeline in Pennsylvania; it was five miles long with a capacity of 2,000 barrels of oil per day; the first larger-diameter, long-distance pipeline, extending 109 miles, was developed in 1879 by Byron D. Benson. On the other side of the Atlantic, Russia's main oil pipeline system, arguably one of the world's first national systems, dates back to the Baku-Batumi pipeline, built between 1896 and 1906, and the Makhachkala-Grozny pipeline, built from 1913 to 1914.[22]

Gas, on the other hand, remained extremely localized during the nineteenth and early twentieth centuries. The first seamless steel gas pipe was introduced in the United States only in the 1920s, after the discovery of huge reserves in Texas and Oklahoma; it was not until 1925 that the first pipeline longer than 200 miles was constructed between Louisiana and Texas. Subsequently, President Franklin D. Roosevelt pushed for natural gas exploration in the early 1940s, but it was only in the post–World War II era that technological improvements made piping gas long distances—that is, more than 150 miles—easier and more reliable. Still, capturing and transporting gas by pipeline was more complicated than for a liquid like oil. Moreover, the first LNG tanker—the *Methane Pioneer*—did not come on the market until 1959; with two storage tanks of 2,200 cubic meters each, it had a much lower loading capacity than its oil counterparts.[23] Transportation difficulties and their associated costs have thwarted and politicized natural gas trade throughout much of its history and still remain factors today.

"Energy Scarcity": The Oil Crisis, the Energy Independence Fallacy, and the Peak Oil Theory

More than any other event, the oil crisis of the 1970s cemented the mind-set of energy scarcity and spawned the ideas of the "energy independence fallacy" and "peak oil" that would prevail until the 2000s. Already by the 1970s, it was clear that a global oil market had emerged, as evidenced by the fact that oil shocks in one part of the world could be felt across the globe. It also became apparent that global demand for cheap oil could no longer be easily sustained. Low oil prices beginning in the 1950s had allowed for tremendous economic growth but simultaneously created an unstable supply-demand balance.[24] As American oil production declined and the Organization of Petroleum Exporting Countries (OPEC) launched an oil embargo against the United States due to its support for Israel in the 1973 Arab-Israeli War, the term "energy crisis" became part of Americans' political vocabulary as demand for all forms of energy—including oil, gas, electricity, and nuclear—rose rapidly. Specific to both oil and gas, artificially low price controls imposed by the Nixon administration stimulated demand and high consumption levels, while supply was tight and financial incentives for domestic exploration and production were low. Gas consumption in the United States, which had risen to 625.84 BCM by 1972, rapidly fell to 553.24 BCM in 1975 during the crisis, and then rebounded to 600 BCM or higher only after 1994.[25]

It was the combination of these events in the 1970s that proved to the United States and other countries of the Organisation of Economic Co-Operation and Development (OECD) that they needed to reduce their dependence on oil imports. The decade was marked by a "scarcity mindset," which made fossil fuels look expensive, unstable, and unreliable. But the oil shock had unintended, positive consequences, as well. It led to the creation in 1974 of the International Energy Agency (IEA), through which countries could coordinate a "collective response" to major disruptions in oil supply. Just a few years later in 1977 the United States founded its Department of Energy (DOE) as an independent, cabinet-level entity. Furthermore, the oil price spike also sparked global initiatives to explore for domestic natural gas reserves, both conventional and unconventional. In the case of the United States, although the fracturing technique in gas production had been used since the 1940s, it was only after this global crisis that the government passed tax breaks and pushed for new innovations in this new pursuit of domestic energy sources. Public investment in new drilling technologies and techniques, including horizontal drilling, drilling at deeper depths, and offshore production, expanded significantly. US government–funded programs and incentives from the 1970s to the 1990s laid the foundations of the US shale boom of the 2000s.[26]

While the United States took the lead in trying to develop its domestic resources, the mind-set of energy scarcity persisted into the 2000s. Indeed, while US shale exploration was beginning to bear fruit, many strategists had already given up hope of achieving energy independence. In a 2006 report, the Council on Foreign Relations (CFR) argued that the very notion of energy independence was a "fallacy" and deemed it a "goal that is unachievable over the foreseeable future."[27] Instead, the report argued, the United States should focus on managing the extent to which it is dependent on other nations for natural resources.[28]

This call to manage dependency was related directly to the theory of peak oil, which, although it most commonly refers to oil alone, claims that demand for all hydrocarbon resources is rising even as global supply has peaked (or will peak soon), thereby bringing about high prices that will further limit access to supplies. More broadly, peak oil theory was one manifestation of the Malthusian school of thought, which sees ever-increasing demand for resources putting pressure on finite supplies and thus leading to a crisis, and which thereby does not give enough consideration to the capacity for markets and technological change to alter behavior.[29] Peak oil has been theorized in a number of ways particularly since the oil crisis of the 1970s and is part of the broader concerns that persisted in the 2000s regarding oil price volatility and its impact on politics, economics, and access to all forms of energy worldwide. On the supply side, the theory of peak oil can refer to the limited amount of oil and gas reserves physically available, the extent to which supplies are available for consumption (determined by the amount of investment, exertion of a monopoly, or inaccessibility due to political instability), and cost. From the demand perspective, peak oil depends on economic and population growth as well as changes in technology, which contribute to changes in consumption patterns and to developments (or lack thereof) of energy efficient measures.[30] These debates on energy scarcity and peak oil lost their potency only with the American shale boom. Meanwhile, the same period from the 1970s to the 2000s saw rising public concern with climate change and its implications for energy consumption.

Climate Change and Natural Gas

The 1970s saw both oil shocks and rising environmental concerns associated with the use of fossil fuels growing so much that by the 1990s and especially the 2000s, climate change became a global political issue. The 1972 Stockholm Conference focusing on pollution, the establishment of the United Nations Environment Programme (UNEP) in the late 1970s, the publication of the Brundtland Commission report entitled *Our Common Future* on the stratospheric "ozone hole" in 1987, the 1990 Second World Climate Conference, and the 1992 UN Conference on Environment and Development in Rio set the stage for the conclusion in 1997 of the Kyoto Protocol, an international agreement that is tied to the UN Framework Convention on Climate Change and that sets mandatory emission reduction targets. In 2007 the fourth Intergovernmental Panel on Climate Change (IPCC) report warned that climate change has serious effects on the planet and the IPCC together with former US vice president Al Gore won the Nobel Peace Prize for "their efforts to build up and disseminate greater knowledge about man-made climate change."[31] At the UN Cancun Climate Change Conference in 2010, countries started making commitments to mitigate climate change and formulating the legal frameworks to achieve it. In the United States under President Barack Obama, the Environmental Protection Agency (EPA) proposed the Clean Power Plan (CPP) in June 2014, aiming to cut carbon pollution from power plants and giving states the flexibility to decide how to meet the CPP's goals. The 2015 Paris Climate Conference was significant in climate negotiations in that 187 countries agreed on emission reductions starting in 2020. Although specific emissions targets were not agreed upon and short-term commitments for climate action were lacking, the Paris conference saw all participants deciding to strive to limit the planet's temperature from increasing by more than 1.5°C, well below the 2°C target voluntarily established earlier. Although this agreement is arguably unrealistic given the international community's past efforts, the broad acceptance of global climate change as a pressing issue is a step in the right direction.

In this time of growing concern over climate change, natural gas has increasingly come to be seen by some as a transition fuel between coal and renewable forms of energy due to its properties as a cleaner fossil fuel. Although some scientists and activists in the environmental community do not regard gas in such positive terms for a number of reasons, including the detrimental effects to the environment deemed to come from natural gas emissions during both conventional and unconventional extraction (called leakage), gas has attracted the attention of both developed and rapidly developing economies such as China.[32]

The Age of Progress in Global Gas

The global gas sector is slowly changing, as natural gas increasingly becomes the fossil fuel of choice. Although oil remains the major global fuel with about a third of the market share, gas consumption from 2005 to 2015 increased by an average of about 2 percent annually, with gas accounting for nearly 24 percent of primary energy consumption in 2015 and LNG supplying about 10 percent of total gas consumption that year. Indeed, gas is no longer just an afterthought of oil. Gas reserves are spread across the globe with nearly 43 percent of the proved reserves found in the Middle East, 30 percent in Europe and Eurasia, 8 percent in Asia Pacific, over 7 percent in Africa, nearly 7 percent in North America, and 4 percent in South and Central America for a global total of nearly 187 trillion cubic meters (TCM) of gas in 2015. LNG demand has grown the most in comparison to gas supplied from pipelines or domestic production, at an average of 6.6 percent per year between 2000 and 2014, with demand continuing to look promising post-2020 and beyond. Importantly, however, is that demand growth will vary from market to market.[33]

In this new age of gas, energy consumption of developing and developed countries has evolved into new patterns. The developed market economies that form the OECD countries outpaced non-OECD natural gas consumption up until 2008 and oil consumption until 2013 and then gave way to the emerging economies. Since the 2000s, the emerging countries have accounted for the majority of primary energy demand due to growth in their economies, industries, manufacturing, and population. From the 2010s to 2030 the non-OECD countries are expected to account for 93 percent of growth in energy consumption, with China,

India, and the Middle East taking the lead.[34] Still the extent to which gas makes inroads into the developing world will depend on a number of factors, including price.

Natural gas production patterns have also changed, especially during the past ten years, with the United States being the most extreme example. Whereas total proven reserves of natural gas in the United States were estimated at 5.8 TCM at the end of 2005, this number increased over 79 percent to reach 10.4 TCM in 2015. At the same time, natural gas production went from 511.1 BCM in 2005 to 767.3 BCM in 2015, an increase of 50 percent. In comparison, the same period saw an overall 27 percent increase in natural gas production worldwide.[35] Much of the new growth in natural gas production—especially in the United States—can be attributed to unconventional gas or the shale gas revolution.

The Shale Revolution

The American shale revolution started to unfold in 2003–2005 and was part of the country's broader search for unconventional resources. Unconventional energy development is essentially anything outside conventional production, which, according to the EIA, occurs via "a well drilled into a geologic formation in which the reservoir and fluid characteristics permit the oil and natural gas to readily flow to the wellbore."[36] However, the definitions of unconventional and conventional resources fluctuate based on factors such as the characteristic of the resource, the exploration and production technologies currently available, the economic environment at the time of exploration and production, and the scale, frequency, and duration of production. Generally, unconventional resources include shale gas and shale oil, extra heavy oil (oil with high viscosity), oil sand (which contains bitumen), oil shale (which contains kerogen), tight gas (gas with low permeability), coal bed methane (gas associated with coal), and natural gas hydrates (gas trapped in the structure of water ice), each of which requires a significant amount of heat to extract.[37]

Shale gas is methane formed from organic matter trapped in 300–400 million-year-old formations of tight layers of sedimentary rock of low permeability. Similarly, shale oil is oil that has been trapped in sedimentary rock that has low porosity and low permeability and is typically found a mile below the earth's surface. Whereas conventional gas sources are located in interconnected pore spaces within shale formations, shale gas must be "mechanically stimulated" to free the gas for extraction through a combined process of hydraulic fracturing and horizontal or directional drilling, more commonly known as *fracking*. The process of hydraulic fracturing involves shooting highly pressurized water and sand (or other fluids based on geological structure) underground to break apart rocks and release trapped oil and gas. Whereas the shallower shale deposits are extracted via wells drilled vertically, shale gas formations are mostly found deeper underground where the shale formation becomes horizontal, hence requiring horizontal drilling technology. Horizontal drilling maximizes the number of natural fractures in the rock and thus the amount of shale gas that can be extracted by allowing wells to be drilled through targeted geological strata, which may be as thin as six meters and as wide as one hundred meters, thousands of meters beneath the surface. Drilling such nonvertical wells requires both longer boreholes and greater volumes of water than conventional gas drilling. As a technological process, horizontal drilling has existed since 1929, but it did not gain prominence until the 1980s, when a French energy company, Elf Aquitaine, began utilizing it in southwest France, with great commercial success. Recent advances in hydraulic fracturing, horizontal drilling, and seismic imaging (which provides more precise information and reduces the number of drilling attempts in search of gas) have been the three most important developments to propel the shale revolution, resulting in a 50 percent increase in reserve finds in the United States, as well as a 34 percent increase in gas production over the past ten years.[38] This shale revolution has been long in the making and its breakthrough came from the combination and improvement of existing technologies rather than new breakthrough inventions.

The History of American Shale

Despite recent breakthroughs, shale gas is not new to American energy history. Shale gas was first found in Fredonia, New York, by William Hart in 1821; hydraulic fracturing techniques were used under R. F. Farris of the Stanolind Oil and Gas Corporation to improve gas production as early as the 1920s; and the first successfully fracked well was the Klepper #1 gas well in the Hugoton field in Grant County, Kansas, in 1947. In 1949, Halliburton Oil Well Cementing Company used Stanolind's technique to perform the first commercial fracturing treatments. The shale method grew over time, with 3,000 wells being drilled per month by 1955 and half a million "jobs" performed by 1968.[39]

Drilling techniques were further explored, researched, and improved in the 1970s and 1980s as conventional US sources began to wane, while the Middle East crises of the 1970s led to shale discoveries in West Virginia, Ohio, and Kentucky. The US government, especially under the Ford and Carter administrations, played a key role in supporting research and development programs and providing tax credits that pushed the private sector to invest in new technologies for expensive and still largely uneconomical unconventional gas development. Government support was also crucial because the nature of the industry made acquiring patents and licenses difficult, thus discouraging investment in unconventional shale gas extraction technologies. At this time, the Morgantown Energy Research Center (now the National Energy Technology Laboratory) and the Bureau of Mines initiated the Eastern Gas Shale Project (EGSP), which, with the support of the government, worked to determine and map the potential for gas in shale formations of the Appalachian, Michigan, and Illinois basins. Between 1975 and 1992, cutting-edge engineering experiments were undertaken, as was testing of techniques such as horizontal drilling and liquid carbon dioxide fracturing. These developments, along with new logging techniques, helped the commercial development of the shale industry in the following decade, as did government funding of such projects to attract investors.[40]

Perhaps one of the more influential figures in the modern American history of shale gas was Texas entrepreneur George P. Mitchell, widely regarded as the "father of fracking." Mitchell's work was funded by the Gas Research Institute, which was itself funded by the Federal Energy Regulatory Committee, and used the techniques studied by the EGSP, which he improved and perfected by 1997. Developing methods for horizontal drilling and hydraulic fracturing, Mitchell was able to commercially exploit the vast stores of gas and oil that geologists had known for decades were trapped in shale formations, prompting widespread shale exploration and jump-starting the shale revolution. By 2007, EGSP research and its practical outcomes were recognized by the energy industry, and given all the new and rapidly improving technologies, shale gas quickly and consistently grew as a percentage of US natural gas production, from 1 percent in 2000 to over 20 percent in 2010 to about 50 percent in 2015.[41]

Shale's Breakthrough

Despite the long history of American shale development, in the 2010s shale production was perceived as revolutionary due in no small part to the fact that it overturned the energy scarcity assumption and made it possible to consider exporting American gas—an idea that seemed implausible before the mid-2000s, especially because LNG import terminals were being built on US shores. After the mid-2000s, however, technological improvements added 47 percent more domestic gas to the resource base, and the EIA raised its estimates for technically recoverable resource supplies by an additional one hundred years' worth. Large-scale shale gas production, which began in 2006 at the Barnett formation in Texas, spread to the Marcellus field of the Appalachian Basin and the Haynesville fields on the border of Louisiana, Arkansas and Texas, and was followed by shale oil production in 2008 at the Bakken formation in Montana and North Dakota. US shale oil production grew by one million barrels a day in 2012, the largest increase in history. Shale gas production went from zero in 2000 to a quarter of US gas production in 2012 and increased to half of total US natural gas production in 2015—a game changer for the US natural gas market. America's shale success is clear and thus far not replicated; while the United States brought 28,354 out of 45,468 globally drilled wells online, the rest of the world (excluding Canada) completed a mere 3,921 wells.[42]

According to the EIA, forty-six countries have technically recoverable shale oil and gas resources, but only four are producing either shale gas or crude oil from tight formations to levels reaching commercial production: the United States (produces both), Canada (produces both), China (produces shale gas), and Argentina (produces tight oil). In the United States, production in the Marcellus and the Utica shale region (which spans across Ohio, West Virginia, Pennsylvania, and New York) has accounted for 85 percent of US natural gas production since 2012, with tight oil production taking place mainly in the Eagle Ford of the Western Gulf Basin in Texas and the Bakken shale. The United States has the fourth largest shale gas reserves in the world, measured at 665 trillion cubic feet (TCF) as of 2013, following China at 1,115 TCF, Argentina at 802 TCF, and Algeria at 707 TCF. After the United States, other large shale reserves are found in Canada (573 TCF), Mexico (545 TCF), Australia (437 TCF), South Africa (390 TCF), Russia (285 TCF), and Brazil (245 TCF).[43] However, it is important to realize that, without drilling, these initial estimates do not necessarily tell us much about recoverable reserves, which could be substantially lower.

Canada, China, and other countries are still far behind the United States' success in shale development. By 2014, tight oil production in Canada grew to about 146 million barrels, while total unconventional gas production grew to approximately 86 BCM, although fracking has been restricted in New Brunswick and prohibited in the Nova Scotia and Quebec provinces, limiting shale production on the country's east coast.[44] Other countries including Algeria, Australia, Brazil, Britain, Colombia, Germany, Mexico, Poland, South Africa, Russia, Ukraine, and a number of North African states are either interested in or planning shale production.

There are a number of reasons why US shale success cannot be easily duplicated abroad: shale gas in the United States is of good quality and is well distributed geologically; it has high concentrations of hydrocarbons and is brittle enough to allow for efficient fracturing; given that subsurface mineral rights are held by landowners, independent smaller operators with both expertise and advanced technology are allowed by local government to invest in shale production; taxation and legal regimes facilitate extraction; drilling techniques are advanced; existing pipeline infrastructure gives producers certainty that produced shale gas can be transported; and water, sand, and most importantly capital are readily available. Furthermore, unlike other countries, the United States does not set price controls at the wellhead for domestic gas, which could serve as a major impediment to gas production and extraction. Arguably, however, the most important factor is the scale or intensity of drilling. Production declines dramatically after the first few weeks (usually around thirty days) of production from a single well—hence the need for the creation of hundreds or thousands of wells to maintain a shale boom. It is this magnitude in the number of wells drilled that allows the United States to adapt quickly to changing circumstances in the global energy market and sets its shale (both oil and gas) production apart from that of the rest of the world.[45]

As successful as the US shale revolution is, there are a number of potential obstacles to what the IEA in 2011 called the beginning of the "golden age of gas."[46] In brief, exploration and production could become less economical due to falling oil prices, which affect the economic feasibility of keeping gas production at high levels; lower than anticipated Asian demand; and an oversupplied global gas market that could imminently turn into a gas glut, whereby the global gas market will be oversupplied by other new LNG suppliers. These conditions have brought into question the long-term sustainability of shale gas production, especially under such economic and financial pressure. In 2013 the EIA predicted that there are 7,299 TCF of *technically* recoverable shale reserves in the world, meaning that although these resources can physically be produced given the technology available today there is no guarantee that resources can be recovered economically, especially in a market such as the aforementioned one.[47] At the same time, research and development initiatives may help in the exploration of alternative means of gas development, such as through methane hydrate, or methane molecules trapped in ice-like structures deep on or under the ocean floor, which may hold up to fifteen times the amount of gas as the world's total shale deposits.[48] Such level of investment will especially depend on favorable market economics.

Global LNG Trade

Though the American shale boom has been important to the transformation of the global gas industry, the first catalyst for global gas markets was the rise of LNG, which is gas held and stored in liquid form at a temperature of –260°F, or –162°C. LNG's key advantage is that, as a highly condensed form of gas, it takes up 1 / 600th the volume of gas in its natural form, making it economical to store in large volumes and transport over long distances via tankers as opposed to being restricted to transportation on land via pipeline.[49] By the end of 2015, the largest year ever for LNG trade according to the International Gas Union, there were seventeen LNG exporting countries—the largest five being Qatar, Australia, Malaysia, Nigeria, and Indonesia—and thirty-three importing countries, with four newcomers (Jordan, Egypt, Pakistan, and Poland) entering the market in 2015. Colombia and Ghana plan on entering the LNG import market in late 2016 and 2017 respectively, while Bangladesh, Benin, and Uruguay are expected to follow suit. In 2015, Asian demand accounted for 57 percent of LNG demand in the global market—down from 75 percent in 2014, but Asia is still the largest market for LNG. Indeed Japan, South Korea, China, India, and Taiwan are the world's largest markets for LNG.[50] The growth of LNG trade, although challenging traditional gas flows between countries and slowly fostering more global gas trade, has raised questions as to whether it can truly compete with piped gas and thus support the rise of a global gas market. Pipeline gas has appeared to be more price competitive than LNG and its corresponding infrastructure in many regions, but whether LNG can compete depends on the gas price environment, governments' political preferences with regard to diversifying their gas sources, and infrastructural and geographical restraints regarding pipeline or LNG access in the regions in question.

A Short History of LNG

While British scientist Michael Faraday first developed LNG technology in the nineteenth century, the first LNG patent for its handling and shipping was awarded in 1914 in the United States. Shortly thereafter, in 1917, the first commercial plant was built in West Virginia, and the first commercial operations began in 1941 in Cleveland. Even though the natural gas industry began looking into liquefaction technologies in the early 1900s, only in the late 1950s did LNG make its way into the global gas industry. The first LNG cargo was shipped out of Lake Charles, Louisiana, to Canvey Island in the United Kingdom in 1959 on the first LNG tanker, the *Methane Pioneer.* This voyage proved that commercial quantities of LNG could be transported safely across the Atlantic. LNG exports from Algeria (and its state-owned energy company, Sonatrach) to the United Kingdom began in 1964. Soon after, France also received its first LNG cargoes from Algeria, while Spain and Italy received theirs from Libya.[51]

Japan became the world's leading LNG importer in the 1970s and 1980s, initially importing LNG from Cook Inlet, Alaska, in 1969. The United States fluctuated in its LNG import and export trends, having imported LNG from Algeria in the 1970s, then reduced imports in the 1980s due to abundant domestic natural gas supplies (a "gas bubble"), then increased imports again in the 1990s, with the primary source of LNG being Trinidad and Tobago starting in 1999. Concurrently with Japan, South Korea and Taiwan became the biggest LNG import markets in the world; between 1975 and 1996, demand by these three countries increased an average of 3.31 BCM per year, whereas that of Europe and the United States increased a combined 0.76 BCM per year over the same time period. Between 1996 and 2004, Asian demand grew 4.22 BCM per year, while Atlantic Basin demand grew 3.97 BCM. The major exporters in the Asian Pacific region during this time of growth were Indonesia (the largest in the world until rapid buildup in Qatar by the late 1990s), Malaysia, Australia, and Brunei; the major exporter from the Middle East was Abu Dhabi starting 1977, which was then overtaken by projects in Qatar and Oman in the late 1990s.[52]

Technology and the LNG Industry

LNG technology consists of three main processes: liquefaction (the process of turning gas into liquid form), transportation (via tanker), and storage and regasification (the process of turning the liquid back into gaseous form). During liquefaction, purified gas is sent to liquefaction "trains" that liquefy the gas, shrinking its volume by a factor of around 600 and to a density around 45 percent of that of water. Liquefied gas is then transferred to specially designed cryogenic storage tanks, which keep the fluid cold and prevent it from regasifying, on LNG carrying vessels, and is then transferred to another set of cryogenic storage tanks once the vessel has reached its destination. Once the LNG is designated for a market, it exits the storage tanks, is regasified and modified based on the needs of the end-user, and is sent out into a domestic pipeline grid. Generally, technological improvements over the years have enhanced value chain efficiency, with increases in the sizes of LNG trains, improving economies of scale, and increasing market competition, coupled with decreased terminal construction and expansion costs, making export projects economically viable. Recent years, especially since the early 2010s, have been an exception to this trend, however, as labor costs, complex environmental approvals, infrastructural challenges, unfavorable exchange rates, and the low oil-price environment make it difficult for new mega-projects to recover construction costs. It is also unlikely that costs throughout the LNG supply chain will soon come down farther than they have since the beginnings of the oil and gas glut, when oil and gas companies already began to aggressively cut costs to remain profitable in the LNG business.[53]

Shipping technology has also come a long way. While the first LNG ships were converted freighters, with aluminum tanks insulated with balsa wood to keep the LNG cool, modern tankers are double hulled and specially designed for the safe and efficient transport of the cryogenic liquid. Vessels today have a cargo capacity five times that of the industry's first ships and can reach speeds of up to 19 knots. By 2015, the global LNG fleet (excluding small vessels) was comprised of 410 active vessels with an average size of over 160,000 cubic meters; most of the 23 new vessels ordered for construction in 2015 were to be 170,000–180,000 cubic meters and are called Panamax class carriers because they are specially made to pass through the newly expanded Panama Canal. The largest vessels in the market are Qatari-made, at 210,000–217,000 cubic meters (called Q-Flex) and 261,700–266,000 cubic meters (Q-Max). There are also vessels that include floating storage and regasification units (FSRUs), floating liquefied natural gas (FLNGs), floating production storage and offloading vessels (FPSOs), floating production systems (FPSs), floating storage and offloading systems (FSOs), and floating storage units (FSUs)—the FSRU designated for the regasification process, FLNGs, FPSOs, FPSs, and FSOs for liquefaction, and the FSUs for storage. All these offshore units are an attractive alternative to land-based LNG projects due to their low costs and shorter construction time, as well as their being capable of being redeployed to new gas fields. As of the end of January 2016, there were 20 operating floating regasification units in 14 countries for a total of 77 million tons per annum (MTPA), almost three times the capacity of 2010 and 10 percent of the total global import market in 2015.[54]

LNG has also become popular in the maritime and transport sectors as emission standards in transportation have become more stringent. As the name implies, small-scale LNG (SSLNG) is particularly useful for regions that want to increase natural gas consumption over coal and oil use but that are located in areas not easily reached via pipeline and where it is not economical to construct large-scale terminals. Twenty

MTPA of SSLNG have been installed as of 2014, mostly in China, with further growth expected. Similarly, compressed natural gas (CNG), or gas compressed to less than 1 percent of its volume at standard atmospheric pressure, is being applied commercially for use in light- to heavy-duty trucks, cars, and buses.[55] These innovations demonstrate opportunities for replacing coal and oil in transportation with natural gas, as well as how additional innovations in gas transportation can spell new changes for how gas is traded across the globe.

Buildup of Pipeline Infrastructure

In addition to the boom in LNG infrastructure innovations, natural gas pipeline construction has seen a steady increase. In the United States, more short- and long-distance pipelines are being built for domestic purposes and to connect the United States with Mexico and Canada. International projects include new Russian initiatives to supply Europe through the Nord Stream II and Turk Stream pipelines. In an effort to diversify from Russia, the European Union (EU) has plans for an alternative infrastructure called the Southern Gas Corridor, which would bring Caspian gas to Europe.[56] Furthermore, interconnectors, or pipelines linking neighboring European countries, have also become increasingly important, especially in transporting gas from major pipelines into smaller grids in individual EU member states.

Both US and European pipelines now increasingly utilize reverse flow capacity, which allows gas to travel bidirectionally through the pipeline. Reverse flow has become particularly useful in Europe, allowing countries like Slovakia, Poland, and Hungary to send natural gas (albeit still Russian) to Ukraine as tensions between Kyiv and Moscow persist. Similarly, US pipelines that normally transport gas from west to east are now being reversed to send gas from the Marcellus and Utica shales west as well as north to the New England region.[57]

Like Europe, Asia is also trying to increase interconnectivity in its gas markets. Plans for major infrastructure projects include the Altai and Power of Siberia pipelines to supply China with Russian gas, as well as the revival of the TAPI (Turkmenistan-Afghanistan-Pakistan-India) pipeline plans in the 2010s.[58] Regardless, Asia still faces a long road ahead to achieve its infrastructure ambitions, especially with countries like South Korea and Japan remaining LNG islands for the foreseeable future. In addition, locking in part of the market in longer-term pipeline contracts in the process of developing gas pipeline infrastructure could hinder the growth of LNG trade and the development of a global gas market. Either way, even as the greater interconnectivity of individual gas markets falls short of being a global gas market, it is moving in that direction, as well as eliminating highly isolated gas markets and their politically charged dependence on a single gas supplier.

Implications of the Global Gas Boom

The shale revolution, the growth of LNG trade, and the buildup of traditional gas pipeline and other gas infrastructure have transformed the global gas sector and created a gas boom and a boom in gas trade. They also point to three notable developments: (1) shifting gas trade flows, (2) changes in global gas pricing and contract schemes, and (3) climate change and the role of natural gas in meeting its challenges.

Shifting Flows: From Regional to Global Gas Trade

While the United States spent forty years worrying that its dependence on traditional oil- and gas-exporting countries would lead to episodes of energy scarcity, the country now has the potential to become a major global energy exporter, especially of LNG, thanks to its shale (both gas and oil) revolution. The United States has one operational terminal (Sabine Pass LNG, which started exports in 2016), four LNG export terminal projects under construction, and more than twenty proposed projects as of late 2016. Similarly, there are twenty-one proposed export projects in Western Canada and six in Eastern Canada.[59] While it is far from certain that these export projects will come to fruition, it is already clear that LNG previously destined for the United States will be redirected to alternative markets. Thus, for instance, gas from Algeria, Trinidad and Tobago, Nigeria, and Qatar will make its way to new and old markets in Asia, Europe, and Latin America. Price signals will determine where the global commodity of LNG will flow, rather than existing routes, which determine the flow of pipeline gas. The abundant volumes of unconventional gas, entering the global market in the form of LNG are, in fact, shifting energy flows globally and impacting energy policies across regions.

Facing this abundance in global gas, China, for one, is trying to diversify its energy sources and to decrease long-term dependence on imports by developing its own resources, while the Middle East is seeking ways to supply its own markets without sacrificing exports and profits. Across Asia, the steady flow of LNG into the world market is allowing buyers to be less concerned with security of supply and enabling them to optimize their portfolios as gas purchases are no longer limited to local or regional trade.[60]

Europe is also working toward acquiring alternatives to Russian pipeline gas and high renewable energy prices. LNG from the United States and other new exporters may help to do just that, as well as to make up for an expected fall in EU domestic gas production. LNG exports in the global market are expected to roughly double between 2014 and 2020.[61] While LNG may not have performed well in Europe in the recent past due to economic stagnation on the Continent, high LNG prices, a relatively limited number of LNG sellers on the market, and a move to renewable energy, it does appear to have a chance to become a fuel of choice as of the mid-2010s. Europe, and particularly the Iberian Peninsula and Central and Eastern states, will increasingly see imports from North America, Africa, and the Middle East, as well as from other regions that offer LNG under flexible contracts and attractive prices.

Gas Pricing and Contract Schemes

The abundance of gas on the international market is beginning to put pressure on traditional contract and pricing structures for pipeline gas and LNG to reflect shorter-term dynamics, at least in more liquid and competitive markets. Traditionally, pipeline gas was sold under contracts with a long-term basis of around twenty to thirty years and an additional "take-or-pay" requirement committing buyers to pay for a minimum annual volume of gas (usually 80–90 percent of annually contracted supplies) regardless of whether they wanted to purchase the entire volume of contracted gas or not. Traditional contracts also included "destination clauses," which forbid wholesalers in an import market from reselling gas outside the countries where they are established. These clauses, which prevented trade across national borders, also perpetuated price differentials between regions, notably between Western and Eastern Europe and arguably have stood in the way of the creation of a common and integrated gas market.[62] Likewise, elsewhere outside of the United States, most LNG contracts have traditionally been both long-term and constrained by destination clauses.

Under those types of contracts, the price of gas was linked to that of oil in a practice commonly called *oil indexation,* which began in the 1960s in northwestern Europe and then spread to Asia. Oil indexation calculates the price of gas as a percentage of the price of oil and oil products averaged over six to nine months, an oil-linked price that is then reflected in the gas price after a three- to six-month lag. The rationale behind oil indexation was that, until the 1990s, oil and gas were largely interchangeable for most end users.[63]

Strict indexation to oil slowly began to erode as trade and liquidity in the gas markets increased and as several important trading hubs emerged in the United States. Here, the rationale for linking gas to oil prices began to weaken. Instead, a pricing scheme based on hub prices—a trade platform where gas can be traded virtually (as in portfolio trading) or physically (where gas transits between trading points via pipeline)—started in the United States and made its way to northwestern Europe via the United Kingdom. A hub is significant in that it facilitates spot trade, whereby, in contrast to long-term contracts, gas is bought and sold for immediate or less than thirty-day delivery. The Henry Hub in Louisiana is America's best-known hub; in Europe, the three major trading hubs are the National Balancing Point (NBP) in the United Kingdom, the Title Transfer Facility (TTF) in the Netherlands, and Zeebrugge in Belgium. Starting in 2008, increased gas volume availability and an economic recession caused European hub prices to fall below the price of oil-indexed gas, intensifying competition between spot-based and oil-indexed pricing. This marked the beginning of a loss of market share from wholesale suppliers of oil-indexed gas. Yet this was still not enough to give European hubs the characteristics of the cheaper, more liquid, and more active US gas hubs.[64]

LNG contracts have similarly been indexed to oil but in addition also use gas-on-gas pricing, or gas priced based on its own supply-demand dynamics rather than indexed to the price of other fuels. This is most prominent in the United States, where the large volumes of gas produced via shale production foster gas-on-gas competition and the main spot price index is the Henry Hub. European LNG prices are traditionally based on a combination of oil indexation and gas-on-gas indexation, reflecting the growing competition between pipeline gas and LNG. As in the United States, market-based prices are linked to regional gas hubs, such as the NBP and the TTF. Asian prices have mostly been indexed to the Japanese Customs-Cleared Crude (JCC, also nicknamed the Japanese Crude Cocktail), or the average price of imported crude oils, due to the lack of pipeline competition. This has meant that Asian prices have struggled to reflect supply-demand dynamics and have therefore followed those of oil, causing gas prices to stay high. Oil-indexed trade is increasingly being supplemented with short-term and spot transactions that reflect gas-on-gas pricing. The most commonly used spot index to assess LNG delivered in East Asia is the Japan Korea Marker (JKM).[65]

As LNG trade grew, the traditional contract model also underwent change. The EU, for one, put regulatory pressure on gas suppliers, including Gazprom, to increase delivery flexibility by annulling contract destination clauses. This was particularly important both because it facilitated the opening of the gas market to greater competition and because it lifted territorial restrictions on the movement of gas. At the same time, commercial buyers, through negotiation or arbitration, received a number of concessions including lowered contract prices by including indexation to hubs for volumes above take-or-pay levels, and adopting a hub index in their price formula. Since 2014, the effect of the gradual LNG oversupply has been even more apparent, with both European and Asian buyers beginning to demand supply flexibility and shorter-term contracts (around half that of long-term contracts) from LNG suppliers. This has also meant increased spot trade, which has already led to the rise of international trading houses and portfolio players—such as Shell, BP, Mitsubishi, Trafigura, and Vitol—which, unlike traditional gas buyers, serve their customers' gas demands without signing contracts with exporters for fixed LNG supply and significantly increase supply flexibility by not linking supply contracts to any particular production field.

Consequently, gas spot prices are increasingly determined competitively, based on the fluctuations of supply and demand of natural gas in the market, as opposed to remaining strictly oil indexed. At a minimum these developments put pressure on traditional suppliers to lower the percentage to which gas is indexed to the price of oil.[66] European buyers have already put pressure on suppliers to lower pipeline gas prices thanks to LNG. In addition, hub prices are beginning to have a stronger impact on pipeline gas prices as LNG continues to compete with coal and oil in power generation.[67] The turn to short-term and spot transactions is significant because, unlike traditional gas purchases, they do not require a continuing long-term arrangement between the buyer and the seller.[68]

Likewise, LNG buyers in Asia, particularly Japan and South Korea, are reexamining current pricing schemes and are beginning to demand lower regional gas prices either through more gas indexation or a lowering of the percentage to which gas is indexed to oil so that LNG prices are comparable to those in the Atlantic Basin. These buyers are also demanding increased flexibility in renegotiating their LNG contracts. In addition, Asian players may consider forming a transparent Asian spot index mirroring the Henry Hub or the NBP, importing LNG at Henry

Hub or European hub prices rather than prices indexed to the JKM, or creating their own regional hybrid price marker. Traders will continue to gain prominence in the global gas market as available volumes of LNG rise. These trends point to the gradual reinforcement of price convergence between Asia and Europe, a lowering of both oil-indexed pipeline gas and LNG prices (both contracted and spot), and potentially even the decline of European hub prices toward US Henry Hub prices.[69]

Even though long-term, oil-indexed contracts ensure that buyers have uninterrupted and steady gas supplies, spot pricing has advantages such as giving buyers the option of purchasing gas at more attractive prices as market conditions change.[70] Nonetheless, oil indexation may remain prevalent in many markets (although at a lower percentage of indexation), especially less liquid ones such as Central and Eastern Europe, particularly because oil indexation serves as a safety net in the face of potentially volatile spot market prices. Still this increase in gas trade globally is pushing gas to increasingly be traded as a global commodity, upsetting more than half a century of traditional pricing models and contract schemes.

Some experts would argue that the US shale revolution has also played a large role in fundamentally changing global oil prices and in turn gas prices. US domestic oil production doubled between approximately 2008 and 2015 and created global oil oversupply. In response, traditional oil exporters, namely Saudi Arabia, pushed even more oil into the market in an effort to keep market share, which then resulted in downward pressure on oil prices. The Brent benchmark oil price—or the primary benchmark for international oil prices—crashed to below $30 per barrel range in January 2016 from the approximately $100 per barrel range in June 2014. Gas, still largely indexed to oil, also felt the downward pricing pressures, and if oil prices remain low, pipeline gas and LNG prices will continue to face downward pressure. This, on top of global gas oversupply, could create additional pressure for both hub- and oil-indexed contracts and potentially make it easier for LNG to compete with pipeline gas and perhaps even coal. However, according to market watchers, if gas prices become too low, LNG could lose its competitive edge vis-à-vis pipeline gas as LNG suppliers find it increasingly difficult to break even on their projects, given also the price of liquefaction and shipping.[71] At high gas prices, hub and LNG spot prices could compete with oil-indexed contract prices although LNG prices would still be more expensive given the additional shipping cost.

Natural Gas: Bridge Fuel or Just Another Fossil Fuel?

The global gas boom and rise of global gas trade make the potential role of natural gas to be a bridge fuel in efforts to move toward cleaner forms of energy more feasible. Undeniably, natural gas is the least carbon-intensive fossil fuel, emitting half as much carbon as coal and oil and emitting no sulfur dioxide and nitrous oxides. The IEA expects gas to overtake coal, which is the world's second most used (and most carbon-intensive) fuel, by 2030. Already in the United States, electricity generated from natural gas surpassed that produced by coal in April 2015, the first time since 1973, although industry experts have noted that in order for this trend to carry on, stronger government regulations will be key. Still, in the expectation that this trend will continue, some supporters of the "green agenda" favor the use of natural gas and complementary carbon capture and storage technology as a "bridge fuel" to replace coal and other carbon-emitting fossil fuels until renewable, nuclear energy, and other zero-carbon sources are ramped up.[72]

However, not all environmentalists or other experts are convinced that natural gas can serve this purpose. For one, increased reliance on natural gas would still produce emissions; nor would emissions be reduced in the long term without policy changes. For instance, a gas boom could lower energy prices and subsequently spur economic growth, which in turn would theoretically increase global gas use. In this scenario, natural gas use could either cut emissions by insignificant numbers or unintentionally increase them due to the significant increase in energy use that results from economic growth.[73] Another unintended consequence is that, while some markets would turn to natural gas, other markets could choose the newly freed-up coal. For instance, in 2012, as the use of coal in electricity production began to decline in the United States, cheap American coal was shipped to energy-hungry markets in Europe—the biggest consumer—and elsewhere in Asia. Thus, some argue that American gas may not reduce coal usage overall, but rather simply shift it and its emissions to other parts of the world.[74]

Other scientists warn that although natural gas is largely recognized as the most cost-efficient energy resource that still reduces carbon emissions, there is rising evidence that it will out-compete renewables and nuclear energy on price, discourage energy efficiency, and take away investment from developing clean energy technologies, including wind, solar, and nuclear energy.[75] Ultimately, others fear that encouraging natural gas infrastructure may "lock us into decades more fossil use that is hard to transition out of," perpetuating a fossil-based energy system for a century.[76]

However, others counter that not investing in natural gas will lead to the advancement of the coal industry, not solar or wind technologies. Indeed, natural gas and zero-carbon nuclear energy, not renewable sources of energy, are the only ample, predictable, affordable, and reliable sources of low-carbon energy for current global consumption (particularly in the industrial sector) and are especially critical given that electricity storage from renewable energy is not yet commercially viable. More broadly, the 2015 Paris Conference underscored that the world is taking seriously the need to shift from carbon fuels to renewables—and especially pushing for a coal-to-gas switch while global gas prices are so low—and that natural gas can play a key role during this transition period, even though gas often lacks policy support because it is a fossil fuel. This lack of support for gas may increasingly raise challenges for countries that have become dependent on oil and gas export revenues as well as for the gas industry as investment priorities shift toward new electricity grids that run on renewable power.[77]

Another possibility is that gas may be able to serve as a balanced complement to the green energy agenda, especially in view of arguments that an unbridled increase in renewable energy may have unintended consequences for the environment. These include the destruction of a river's or lake's habitat from the construction of dams for hydropower; pollution from the mining of rare earth minerals for use in renewable technology; overutilization of land and food resources for the production of bioenergy; over-burning of fossil fuels in the construction of electricity grids compatible with renewable energy; and the need to set aside large stretches of land for thousands of wind farms in order to meet power generation targets.[78]

Shale gas extraction is even more contentious than the rising reliance on gas because there are still perceived environmental risks inherent to the process of fracking. Up to 8 million gallons of water and a total of 40,000 gallons of chemicals are used per fracturing. Fracking fluid consists of 600 chemicals including lead, uranium, mercury, ethylene glycol, radium, methanol, hydrochloric acid, and formaldehyde. Methane gas and toxic chemicals could contaminate nearby water, giving it a concentration of seventeen times more methane than a normal water well. Furthermore, only 30–50 percent of the fracking fluid—which is not biodegradable—is recovered and left to evaporate into the environment, while the remainder stays in the ground. In addition to contaminated groundwater, potential consequences of fracking also include methane leakage in the value chain process, from drilling to production, processing, transportation and distribution; exposure to toxic chemicals; gas explosions; increased waste; and water deficiency. Fracking-induced earthquakes caused by water injection underground and the subsequent shifting of faults (called *induced seismicity*) are another concern, especially in Oklahoma. Moreover, fracking sites have smog levels worse than those in downtown Los Angeles, increasing communities' risks of cancer and birth defects.[79] Indeed many communities have asked if fracking is worth the environmental risks, while some states have overruled public protests to continue shale gas exploration.

In contrast, the industry argues that natural gas emits 30 percent less carbon dioxide than oil and 45 percent less carbon dioxide than coal and has negligible amounts of sulfur, mercury, and particulates. It also maintains that technological improvements have made fracking safe.

Drillers are already boasting the emission reductions they have achieved, though they argue that further regulation would "wipe out marginal drillers." Techniques include using "benign" chemicals like magnesium oxide and ferric sulfate (used in water treatment plants); utilizing recycled frack water, or nonpotable brine, instead of freshwater; replacing diesel-powered drilling with engines powered by natural gas, CNG, or LNG; treating wastewater; and sealing leaks that allow gases like methane to escape. However, scientists, nongovernmental organizations, and some members of the public question the ability of these techniques to eliminate the potential side effects of fracking, and fracking moratoriums, tight regulations, or outright fracking bans have occurred in seven Australian territories, Austria, Bulgaria, four Canadian provinces, the Czech Republic, France, Germany, Ireland, Luxembourg, the Netherlands, New Zealand, Romania, Spain, South Africa, Switzerland, Tunisia, and fifteen US states and the District of Columbia.[80] Indeed, fracking is perceived as such an environmentally risky endeavor that only countries and communities that prioritize economic or geostrategic gains are likely to pursue it.

Environmental considerations mean both opportunities for and risks to the gas boom. Shale-producing countries, especially the United States, will have to maintain an environmentally responsible fracking record and to reassure both their own population and other countries that are skeptical of shale development. Any large-scale fracking or LNG shipping disaster could deliver a huge blow to the industry and thus weaken the prospects of a globalized gas market. To maximize the opportunities for reducing carbon emissions, governments will need to rely on thoughtful policies and regulation if they intend to implement an overall energy strategy where natural gas is complemented by and paves the way for other, low- to zero-carbon emitting energy options.

The Global Gas Market

The rise in gas volumes available on the global market, the shift in trade flows, the increase in LNG trade, and the pressures on gas pricing and contract schemes across the different regions of the globe are altering the dynamics of the gas trade. Also affecting the supply, demand, and pricing of natural gas are political, economic, technological, and natural events taking place in distinct parts of the world, including the US shale boom, the buildup of LNG exporting and importing infrastructure, increased pipeline interconnections between different countries and regions, Europe's economic downturn and slowdown in gas demand, the energy and political crises in the Middle East, and crisis such as the Fukushima nuclear disaster in Japan. Increasing trade between the Atlantic and Pacific Basins as well as between North American, Asian, and European markets is creating a more interconnected gas market, while previously isolated markets are increasingly influenced by developments elsewhere, marking a new phase in global gas trade.

Although some believe gas may never be a true global commodity, it is looking more and more like one. Before spot trade can truly become fluid, LNG will continue to face some serious operational constraints, including a lower level of supply flexibility than that offered by gas pipelines, a vast quantity of LNG volumes allocated to long-term contracts, the need to reserve a "slot" at regasification terminals in advance of a ship's arrival at a port, a terminal's restrictions as to what vessels and what quality of gas it can accommodate at its port, among others.[81] However, the rise of shorter-term contracts, the movement of gas over longer distances, and the ability of sellers/traders to make a profit on gas trade are sharp departures from the historic pipeline trade of gas, and there are numerous other signs that suggest that the global gas market in some form will emerge in the years to come.

Still, there are a number of risks that could derail its development in the short to medium term. For one thing, the effects of the American shale revolution and the rise of American LNG exports are still not foregone conclusions. A lot depends on the demand and the prices of gas. If low gas prices persist, the cost of liquefaction and transportation may make American LNG less competitive on the world market, may make other LNG projects outside the United States less economical, and may dis-incentivize further investment in shale gas development and LNG export. Higher prices and increased gas demand would have an opposite effect. The question remains whether the growth in LNG usage across the world will be large and liquid enough to create a truly global gas market or whether pipeline gas will continue to dominate.[82] The new US presidency of Donald J. Trump, the economic and social difficulties in the EU, and the Syrian civil war are other wild cards that can impact word oil and gas output, energy demand, and implementation of gas infrastructure projects. Nonetheless a global gas market is already underway, and if US LNG exports are coupled with America's leadership in the gas markets, then its future looks bright.

2 The Politics and Commerce of American LNG Exports

The success of the United States as an economic powerhouse with global geopolitical clout is a direct result of its historic leadership in the energy markets and its access to either domestic or imported cheap and abundant fossil fuels. Until very recently, the United States relied on significant imports of oil and growing imports of natural gas to bridge the gap between its strong demand and gradually declining production. By 2010, a combination of factors, including new technologies and innovative practices in the oil and gas industry, facilitated the development of shale resources, which transformed the US energy landscape. An accommodating regulatory environment has aided as well. The shale boom allowed the United States to take Russia's crown as the world's biggest petroleum and natural gas producer. To cite just one example of the transformative effects of the shale boom, the omnibus budget bill that passed both houses of Congress on December 18, 2015, and was signed into law by President Barack Obama the same day included a repeal of the country's decades-long ban on oil exports.

Other impacts of the shale boom are apparent in global gas and oil markets. With the United States no longer drawing on imports of liquefied natural gas (LNG), major global LNG producers are reorienting their export strategies toward Asia and other markets. However, not only is the United States unlikely to import LNG in significant quantities; American companies have been building LNG export infrastructure in anticipation of global demand for American gas. These export projects are widely seen as providing a relief valve for the growing US gas surplus into global markets. Even if not fully utilized, these export terminals have the potential to facilitate a convergence of global gas prices, impacting the growth of the global economy and the energy security of Asian and European energy importers. The long-term success of US LNG export projects, and especially those involving expansions to LNG export capacity beyond what projects currently under construction can accommodate, depends on a myriad of factors, including the sustainability of the US shale boom, global energy prices, competition from other global gas producers, the demand of importers, and overall global economic growth, among many others. By mid-2015, the global oil price decline that started in late 2014 dampened enthusiasm. Fears of gas scarcity of the 2000s had turned into fears of a gas glut. Despite these concerns, in 2016, inaugural US LNG exports made their way to Latin America, Europe, Asia, and the Middle East, and by the end of the same year there was increasing evidence that the US gas industry was successfully weathering the low price environment.

Before examining how the geopolitics of gas could be transformed by the emergence of the United States as an LNG exporter, the most important proposition on which it all rests—namely, that the shale boom can be sustained and that American LNG exports will be able to successfully impact the global market—requires detailed analysis. This chapter will examine US gas production, reserves, infrastructure, and regulation as factors that could drive or hinder the growth of US LNG exports. First we will focus on the supply side issues of the American gas boom and then on the demand side to consider the receiving end issues of LNG imports for European and Asian states. Together, these will give a coherent picture of what the US energy boom means within the changing context of global gas geopolitics.

Sustainability of US Supply Growth

Despite America's historic role in the development of the gas industry, the US shale boom came rather unexpectedly in the mid- to late 2000s. American gas markets of that decade were marked by instability and volatility, with prices skyrocketing and plummeting from year to year, peaking in 2005 in the aftermath of Hurricanes Katrina and Rita but beginning to drop throughout 2009 rather than rebound along with oil prices after the commodities price crash of 2008. Defying most analysts' expectations, natural gas settled between $2 and $3 / one million British thermal units (MMBtu) in 2011 and hardly strayed from that mark through the end of 2016.[1] The continuous rise of US gas production is the main cause of this long-lasting low-price environment. The 2000s shale boom altered gas market dynamics that had prevailed in the US for decades. Moreover, improved methods and technologies in shale development made gas extraction more efficient and economic from gas and oil fields across the continental United States that had previously been difficult to tap.[2] Associated gas from oil drilling also drove the boom in gas production. By 2013, shale accounted for 47 percent of US natural gas production, or 11.34 trillion cubic feet (TCF) per year.[3] Despite low global energy prices since 2014, US natural gas production has continued to increase, reaching an annual record high in 2015 of 79 billion cubic feet (BCF) per day, an increase of 5 percent relative to 2014.[4] Shale extraction continued to rise on the back of fast-growing shale formations, including Marcellus, Utica, Permian, and Eagle Ford, representing over 58 percent of all dry gas production in February 2016.[5] As overall gas production continues to rise and production from traditional wells continues to decline, it is clear that a paradigm shift toward shale has occurred and that it has driven a US gas renaissance.

Producers and Reserves

The American energy industry is unique in terms of its diversity and competitiveness. In 2014, the top forty independent companies in the

United States produced a total of 34.6 BCF of gas per day. In comparison, industry giants ExxonMobil, Chesapeake Energy, and Anadarko produced 3.4 BCF, 2.5 BCF, and 2 BCF per day respectively, collectively accounting for only 23 percent of the total.[6] Overall in 2014, more than half of US production came from producers outside of the top forty, each of which extracted less than 1.6 BCF in a year.[7] This is in great contrast to the limited number of producers in places like Russia, China, or even Canada. With competition in the gas market fostering a culture of innovation among American gas producers, advances in three key extraction technologies—hydraulic fracturing, horizontal drilling, and seismic imaging—have allowed the United States to exploit its reserves of shale gas faster than any other country in the world.[8] This is particularly remarkable considering that with its 665 TCF the United States ranks fourth in the world in total reserves of technically recoverable shale gas, behind Algeria (707 TCF), Argentina (802 TCF), and China (1,115 TCF).[9] Of greater value than technically recoverable reserves, however, are *proved reserves*. Whereas measures of the former take no account of current economics, proved reserves give an indication of how much gas can be extracted in coming years based on the current state of industry technology, infrastructure, and market logic.[10] Seismic imaging improvements have helped US companies chart new proved gas reserves, which have increased from 308 TCF in 2013 to 338 TCF in 2014. Proved reserves were less than half of that figure in 2000 (167 TCF).[11] According to the US Energy Information Administration's (EIA) global estimates, proved reserves make up a whopping 54 percent of all technically recoverable gas in the United States. To put that in perspective, in China they make up 12 percent, and in Argentina they make up only about 1.5 percent.[12] Thus, technologies for extracting gas from shale formations give the United States a critical advantage in gas development of its proved reserves, where shale gas accounted for more than half of total proved reserves.[13] Indeed, not long ago the industry looked to the deep off shore in the Gulf of Mexico and to Alaska as the next frontiers in natural gas development, whereas in the post-2014 low-price environment, production from these regions is seen as too risky and expensive to compete with on-shore development of abundant shale gas resources.[14]

Growing Gas Fields and Hubs

The shale boom is characterized by the exponential growth in output at gas fields across the continental United States. The Marcellus and underlying Utica formations in the Appalachian Basin and the Haynesville formation that straddles eastern Texas and the western Louisiana border have been the most efficient producers of natural gas from shale per rig in the United States since 2014. The Marcellus shale formation lies 4,000 to 8,500 feet below the surface and spans four states: New York, Ohio, West Virginia, and Pennsylvania.[15] It is by far the largest gas field, producing an astounding 18 BCF of gas per day as of October 2016. Its exploration has been rapidly expanded by the aforementioned techniques of hydraulic fracturing and horizontal drilling. The Utica underlies the Marcellus, but is geologically distinct. Like the Marcellus, it has seen its rig count drop precipitously in 2015, but has made up for it with a rapid rise in productivity per rig, from 2.3 thousand cubic feet (tcf) per day at the start of 2014 to 8 tcf per day in October 2016.

Because producers operating in the Utica and Marcellus formations recover natural gas plant liquids in addition to dry gas, they are able to generate more revenue, or uplift, than they would from producing dry gas alone.[16] This drive for liquids has had a negative effect on some fields. The Haynesville formation is the third most efficient producer. It yields almost exclusively dry natural gas and averaged nearly 6 tcf per day per deployed rig in 2016.[17] However, because of the low yields of associated liquids in the Haynesville, production from this formation has been on the decline since late 2011, as producers have shifted to more lucrative plays. Another field of note in Texas is the Barnett shale field. It is one of the oldest to be tapped, starting circa 1990, using shale-exploiting technologies that remain widely used today. This once prolific field has been in decline since 2012 and as of 2016 was only producing about 3.8 BCF a day.[18]

One of the fields that has gained from the shift to liquids-yielding formations is the Permian, which covers large expanses of central Texas. By October 2016, this region produced more gas than the Haynesville field, approaching 7 BCF per day while Haynesville dropped below 5.8 BCF per day. Although the low oil price environment has impacted production growth as producers seek to rationalize their operations, per-rig productivity in this formation continues to grow.

The famous Bakken formation underlying North Dakota and Montana saw a reversal of growth in mid-2015 as a result of the drop in global oil price that started in late 2014. Therefore, in October 2016 it produced barely 1.6 BCF of gas per day, a drop from July 2015 when it produced nearly 1.9 BCF per day.[19] What Bakken does not lack in resource endowment, however, it does lack in the necessary infrastructure for bringing large quantities of gas to market, which has meant that approximately one-third of all produced natural gas has been flared rather than delivered to market.

Ultimately, companies that are producing gas at these fields are interested in garnering the highest price possible for their product. This is where gas hubs come into play. US gas hubs are areas of high gas concentration, with markets where gas is bought and sold according to an area-specific value called *spot pricing*. These hubs, like the Henry Hub in Louisiana, are often established for their proximity to either sources of significant supply or sources of demand, or both. They are the intersectional points of multiple pipelines and as a result pool some of the most affordable gas resources in the United States. Along with city-gates, which are the transfer points between interstate pipelines and local natural gas distribution systems, gas hubs form the backbone of the thriving US gas market.

Located near Erath, Louisiana, Henry Hub has long been considered the "king" of the US gas market and the most important hub both because it is at the intersection of some of the largest interstate gas pipelines in the United States and in the world and because the New York Mercantile Exchange has made this the delivery point of its futures contracts.[20] Increasingly, however, the significant output of natural gas from the Appalachian Basin, along with its proximity to the largest gas consuming markets along the Atlantic coast, has allowed for the emergence of trading hubs in western Pennsylvania, where they sit at the intersection of major pipelines serving the northeastern United States.[21] This shift reflects the changing geographies of US natural gas production and changes within the domestic market. However, it will not affect Henry Hub's relevance in international gas trading. Due to its importance as a benchmark in commodities exchange,[22] Henry Hub will still set the price for future US LNG exports. Most potential exporters have been choosing to site their projects near the Henry Hub because its high volume of gas traded at a standard price makes it an attractive place for foreign importers to buy gas at pricing formulas based on the Henry Hub spot price.

Implementation of Large-Scale American LNG Exports

In late 2013 and early 2014, excitement for American LNG exports almost reached levels of "irrational exuberance," as economists Alan Greenspan and Robert Schiller term investor enthusiasm in a speculative bubble when energy prices are high. Foreign energy companies invested heavily in US export projects, hoping in this way to gain access to American gas.[23] Gas companies in the United States have likewise been eager to generate the highest possible profit for their product with planned exports to Asia and Europe. The domestic economic benefits of gas exports, from job creation to massive current account deficit reductions, have policymakers across the partisan divide willing to back them.[24] Doing so, however, is not as simple as making a set of policy decisions or supply agreements. In order for US gas to reach global markets in large volumes, a number of large investments must be made and hurdles must be cleared, many involving long-term risk. The most significant of these are the investment decisions and regulatory hurdles associated with LNG export facilities, infrastructural bottlenecks, and possible changes to oil and gas regulations.

LNG Import and Export Facilities

While the United States launched LNG exports in 2016, its infrastructure was still better equipped for imports. In the early 2000s, when a domestic gas shortage was expected, gas companies planned to import significant quantities of LNG to the United States and to build coastal regasification terminals. Over $100 billion was sunk into these projects from the early 2000s to the early 2010s.[25] As of the end of 2016, there are eleven such grossly underutilized terminals with a total import capacity of over 5.3 TCF per year.[26] At the same time, American net gas import volumes continue to fall year to year, including those coming by pipeline from Canada.[27] LNG imports witnessed an even more dramatic decline, from 770 BCF in 2007 to 431 BCF in 2010 to only 59 BCF in 2014.[28] This was largely unexpected. For instance, in 2005, the EIA projected that by 2013 imports would account for 29 percent of US gas consumption. However, due to surging US production of natural gas, the United States relied on imports for only 5 percent of its gas consumption in 2013, leaving many import terminals redundant and underutilized. Some regional markets do continue to rely on LNG imports, such as New England, home to the Everett, North East Gateway, and Neptune regasification terminals, which accounted for over 50 percent of all US LNG imports in 2014.[29] However, by then most of the existing regasification terminal operators were seeking to convert their facilities to liquefaction plants, or LNG export facilities, as a means of recovering their investment. Once fully converted, exporting liquefaction plants have the potential to generate ten times the revenue of importing regasification terminals. The cost of converting these facilities, however, can also be ten times greater than the cost of the original regasification terminal.[30]

The United States has two operating LNG export facilities: the ConocoPhillips' Kenai liquefaction plant at Nikiski, Alaska, and the Cheniere's Sabine Pass facility on the border of Louisiana and Texas. The Kenai plant represents the minor role the United States played in the LNG markets of the past. The Sabine Pass represents its future role.

Commissioned in 1969, Kenai has been the only operational American LNG export facility for over forty years, even though it has never exported more than 70 BCF of LNG in a single year, which it did to Japan back in 1996. Over the past ten years, it has exported a total of 328.4 BCF.[31] Geographically, Nikiski poses an interesting case study for the future of American LNG exports. In addition to the Kenai plant, it is the site of a new proposed LNG export facility—Alaska LNG—to export 929 BCF per year of gas over a thirty-year period.[32] The project involves the construction of an 800-mile pipeline[33] connecting the Kenai Peninsula to Alaska's North Slope, where over 200 TCF of conventional gas reserves have been discovered. Former US Energy Secretary Ernest Moniz had stated that he would treat this gas differently from gas reserves in the lower forty-eight states and not subject it to the lengthy regulatory review because, given its immense geographic distance from US gas hubs, Alaskan exports would not affect the gas market of the continental United States in the same way.[34]

The Sabine Pass liquefaction plant in Cameron Parish, Louisiana, is the largest LNG project in the United States, and it has the potential to become one of the largest LNG liquefaction projects in the world, with planned total annual capacity of about 1.2 TCF per year.[35] Initially built as an importing regasification terminal from 2005 to 2008, Sabine would cost an estimated $10–$18 billion to convert. The facility is operated by Cheniere Energy, which led the field during the early shale boom in preparing to export LNG. Between its two LNG facilities of Sabine Pass and Corpus Christi in Texas, Cheniere has a total of seven LNG liquefaction trains either in operation or under construction, with plans for additional expansions in the future. Sabine Pass LNG is the most developed US LNG facility, with train 1 operational (put into long-term service by the Federal Energy Regulatory Commission [FERC] on May 3, 2016).[36] Train 2 completed the commissioning (or testing) phase on September 15, 2016, and has been issued its permit to commence LNG exports by the FERC, and train 3 began its commissioning phase in October 2016.[37] Corpus Christi has two liquefaction trains under construction. Both plants had secured all permits from the FERC and the Department of Energy (DOE), including for the planned sixth liquefaction train at Sabine and the planned third train at Corpus Christi.[38]

Three other liquefaction plants in the United States have FERC approval and are under construction with a total export capacity of over 3.5 TCF of gas per year.[39] In order of export capacity, these are: the Freeport LNG facility in Freeport, Texas, the Cameron LNG facility in Hackberry, Louisiana, and the Dominion Cove Point facility in Cove Point, Maryland.[40] Meanwhile the sixth, Lake Charles in Louisiana, has cleared FERC approval though Shell has not taken the final investment decision on this project due to the low oil price environment since 2014. All but the smallest of these facilities (Dominion Cove Point) are located on the Gulf of Mexico coast in Texas and Louisiana, near abundant natural gas resources and several major gas hubs, including the ubiquitous Henry Hub.[41] All of these facilities are also brownfield projects, meaning they are located at sites where established LNG infrastructure already exists. Three of these—Freeport, Cameron, and Dominion Cove Point—are former functioning import terminals. Sabine Pass and Corpus Christi were planned as import terminals, and while Sabine Pass did receive a few LNG cargos from overseas, Corpus Christi did not even reach completion before the market turned. By repurposing the existing LNG infrastructure at these sites, terminal operators can save time and money in what is shaping up to be a costly and time-sensitive global competition to export LNG. Brownfield terminal projects are cheaper to build than some of their greenfield competitors, which require all infrastructure to be put in place along with the liquefaction trains. Hence, brownfields can take about five years to build and cost between $15 billion and $35 billion. The relatively lower cost of construction for US exporting liquefaction plants due to conversion potential could make them more competitive vis-à-vis their international peers, which may have a geographical advantage in exports. For instance, the almost 740 BCF per year Gorgon LNG terminal in Australia shipped its first cargo for Asia in early 2016. At about 3,700 miles from Shanghai, Gorgon is just a ten-day sail away. This distance is almost three times closer than that between Sabine, Louisiana, and

Shanghai, and even with transit through the newly expanded Panama Canal, the trip would still take about twenty-five days to complete. Nonetheless, the Gorgon terminal experienced technical difficulties soon after starting operations despite the fact that some $54 billion were sunk into the project, or roughly three times more than the conversion of the Sabine Pass.[42]

According to the FERC, there were twenty-three proposed projects in the United States as of early 2016, eight of which have pending applications, one of which has been rejected, thirteen of which have not yet filed their FERC applications, and one that is an offshore project outside FERC jurisdiction. Of these projects, two are on the West Coast (Oregon, including the initially rejected Jordan Cove plant), one in the Northeast (Maine), two in the South (Georgia and Florida), one in Alaska, and the remainder on the Gulf Coast, mostly in Texas and Louisiana.[43] Similarly, there are eighteen proposed export facilities in western Canada and five in eastern Canada, none of which is under construction and only one of which (East Coast) is brownfield.[44] In the boom period, developers submitted tentative projects for regulatory approval even though the financing and much of the preparatory work for construction had not been completed. Due to the low oil price environment, low demand globally, and the plethora of LNG capacity under construction in other markets, analysts predicted that the majority of these planned and proposed projects in the United States and Canada would not come to fruition.[45]

To see how established LNG export facilities operate, we can take the Freeport LNG facility as an example. Access to expansive gas pipeline networks makes it easier for facilities like Freeport to receive regular supplies of feed gas. This is critical in order for a liquefaction plant to reach its export capacity. Once all three liquefaction trains come online, Freeport LNG will require over 2 BCF of feed gas each day for it to reach its planned annual export capacity of about 635.6 BCF per year. At brownfield facilities such as Freeport LNG, which was built as an importing regasification terminal on the Gulf of Mexico in Texas in 2008, metered lateral connections and pipelines capable of receiving gas are already in place. From the gas meter, the feed gas moves to a pretreatment facility, where impurities such as carbon dioxide are removed before the liquefaction process. After treatment, the gas passes to a liquefaction train, where it is super-cooled until it condenses into a liquid state. Finally, the LNG is depressurized and moved into storage tanks from which it can be pumped for loading onto marine tankers to be shipped anywhere in the world.[46]

Infrastructural Bottlenecks

Outside of LNG exporting facilities, there are other domestic infrastructure needs to overcome if the United States is to capitalize on its gas renaissance. Major bottlenecks still exist, preventing gas from new, large shale fields from reaching consumers and potential export sites. One such bottleneck is in New England, where as in the rest of the United States, the gas and electricity markets are closely interdependent. Even before the gas boom began to take off in the United States, New England began switching more of its electric generation to gas-fired power plants in response to environmental regulations (all New England states are part of the Regional Greenhouse Gas Initiative) and their complementarity with renewable sources of power generation. While this would not be a cause for concern in most of the country, because pipeline capacity into New England is primarily committed to long-term industrial users and public utilities, power plants have access only to service that can be interrupted if customers with firm capacity call on that same capacity. Consequently, during periods when temperatures are low and heating demand is high, little gas remains in the pipelines for power generators. So, in order to get fuel for power generation, and sometimes just to supplement the natural gas supplies available in order to meet heating demand needs, New England utilities have had to compete with other consumers on the global market, importing LNG at times of peak demand in the winter months. These infrastructural bottlenecks have the residual effect of sending electricity prices upward.

New pipeline projects such as Access Northeast have been proposed to increase natural gas delivery capacity in these regions to during peak demand. Still, it is unlikely that public funds secured toward projects in areas where seasonal supply disruptions occur will be extended to export projects.[47] Furthermore, in April 2016, Texas-based pipeline company Kinder Morgan cancelled the construction of the Northeast Energy Direct–Tennessee Gas pipeline due to a lack of commitment from New England utility companies (which cannot sign binding commitments for pipeline capacity) to buy piped Marcellus gas and the lack of creditworthiness of some Marcellus producers (which puts in question pipeline supply) due to the low oil price environment. At the same time, New York State denied the 124-mile-long Constitution Pipeline a construction permit for not meeting the state's water quality standards.[48] In effect, then, seasonality of demand issues may prevent the conversion of import infrastructure to serve the export market in regions like the Northeast.

Another challenge universally faced by the natural gas industry is the practice of flaring gas, which in the United States takes place in large shale fields like Eagle Ford and Bakken. *Flaring* is the practice of burning off gas that is difficult to use, capture, or transport at the wellhead; the process burns excess hydrocarbons, producing water vapor and carbon dioxide.[49] This tends to occur during the exploration and development phases of well drilling, when the wellhead is not yet connected to gas-gathering infrastructure. It is also widely used to dispose of natural gas produced in association with shale oil that cannot be delivered to a gathering system for processing due to pipeline or gas plant capacity constraints—or lack of a sufficiently sizeable market. This problem is particularly acute for producers in the Bakken formation, where only two major (24-inch or wider) gas pipelines run through the region, as opposed to the dozens running through the regions overlying the Marcellus, Haynesville, Permian, and Eagle Ford formations.[50] In 2014, 129.4 BCF of natural gas was flared in North Dakota, where the vast majority of Bakken production occurs.[51] To put that in context, that is slightly more gas than Gazprom sold to Finland in 2013.[52]

The extension of the interstate natural gas pipeline infrastructure, or in some cases, reconfiguring gas flows in different directions to reflect the changing geography of natural gas production and consumption, is therefore vital for the long-term viability of the shale oil and gas industry.

To be sure, the United States has more natural gas pipelines than any other geographic region, but the country's transformation from a natural gas importer to exporter and of the Midwest and Northeast into natural gas production powerhouses have resulted in infrastructure bottlenecks that will need to be resolved alongside the development of new LNG export terminals to ensure the efficient flow of natural gas from newly emerging production regions to the export outlets. If US president Donald J. Trump succeeds with his plans to invest in the country's infrastructure, the gas pipeline systems could also see an upgrade.

Politics and the Regulatory Process

Before looking at the regulatory process surrounding natural gas exports in the United States, it is important to understand its broader context. Regulation cannot be viewed in isolation from politics, and the decision on whether or not to mobilize government support for the export of significant quantities of gas from the United States has created a serious policy debate, which is closely tied to the DOE and FERC approval process. On one hand, supporters of exports argue for easing regulation, claiming that gas exports will bring jobs to the economy, improve the balance of trade, and displace coal in the global energy mix, ultimately reducing carbon dioxide emissions worldwide.[53] On the other hand, critics believe that exports will raise the domestic price of natural gas to consumers and thereby reduce the shale boom's benefits for American families, gas-intensive industries, and domestic producers in general.[54] Beginning in 2012, the DOE commissioned a series of studies to gauge the potential impacts of LNG exports on domestic gas prices and the US economy, the latest of which, made in 2014, projected an increase in domestic gas prices between 4 percent and 12 percent depending on the scale of exports. The report found that increases in production will balance the market effects of exports and that real gross domestic product (GDP) growth will more than offset the detrimental effects of gas price increases for the US economy.[55] Yet another perspective in this debate comes from those who argue that the United

States should focus on expanding renewable energy options rather than continuing to extract fossil fuels such as natural gas, especially considering the perceived environmental implications of fracking. And, in addition, a myriad of secondary issues also comes into play, such as the role of LNG exports as an incentive to pass international free-trade agreements with Asian and European Countries[56] and the security benefits of LNG exports to energy-insecure North Atlantic Treaty Organization (NATO) allies in Eastern and Central Europe.[57]

In the context of this political debate, the steps in the regulatory process have been clear: in brief, in order to export gas, a company must first gain FERC approval before the DOE will consider an export license. The first step, however, is lengthy and cumbersome, taking between sixteen and twenty-four months and costing from about $100 million per project.[58] It covers the construction of LNG import and export terminals and can become complicated when a variety of state, local, and federal agencies become involved. The first step is the prefiling phase in which the FERC gathers documentation from the applicant for the terminal project. This phase's completion depends upon the applicant's ability to file the necessary documents and in the cases of the Freeport and Lake Charles terminals, took over nineteen months.[59] Next comes the application review phase, which brings with it an environmental review and a host of new bureaucratic actors, including the US Coast Guard, the Army Corps of Engineers, the Environmental Protection Agency, the Department of Fish and Wildlife, the DOE, and the Department of Transportation's Pipeline and Hazardous Materials Safety Administration.[60] In addition, because applications are reviewed in the order they have been submitted rather than on the basis of their viability or merit, the review process has slowed further for many on the list.[61]

By comparison with the FERC's regulatory process, the DOE's is fairly straightforward. In order to approve gas exports from a facility, the DOE has to deem them within the public interest. Generally, this hinges on the exporters' ability to show that once exports from their facility begin, the US market will remain sufficiently supplied with reasonably priced natural gas and the economic benefit to the US economy will more than offset the impact of higher prices on US natural gas consumers. This process includes a consideration of the domestic need for gas, implications for domestic energy security, and a project's potential to promote market competition without raising serious economic or environmental concerns.[62] In the case of deep-water ports, a potential exporter may also need Maritime Administration approval to construct and operate its export facilities and to use existing pipelines and other infrastructure to source the gas.

As far as the destination of exports is concerned, current practice is that exports to countries with which the United States has a free trade agreement (FTA) are routinely deemed within the public interest and are granted the export/import licenses without modification or delay. There are twenty such FTA countries, though few of them aside from South Korea are sizable LNG importers.[63] However, not all FTA countries receive this so-called National Treatment for Trade in Natural Gas regulation, which was passed in 2012. The two exceptions to this rule—FTA countries Costa Rica and Israel—do not receive an expedited process. Applications for exports to FTA countries generally cost around $100,000 and essentially require an economic impact study from a consulting firm and fees to lawyers. DOE approval for exports to non-FTA countries is more complicated. The US Natural Gas Act of 1938 requires the FERC and the DOE to authorize gas export to non-FTA countries, although since 2014, the DOE considers applications only after they have cleared an environmental impact assessment from the FERC under the National Environmental Policy Act (NEPA).[64] Because the law deems exports to non-FTA countries not consistent with public interest by default, the DOE must also issue a Federal Register Notice seeking comments, protests, and motions to intervene to make a public interest finding. Importers in non-FTA countries that are interested in buying US LNG must wait for the terminals from which they look to import gas to be granted authorization "with or without terms and conditions."[65]

Overall, however, the US government has been largely supportive of LNG permits and American exports. Before the Jordan Cove project in Oregon was rejected by the FERC in April 2016, no LNG export projects had been rejected by either the DOE or by the FERC. In 2015, legislation sped up the regulatory process, limiting the DOE's and the FERC's review time for export projects.[66] Also in that year, then US energy secretary Ernest Moniz expressed high expectations for US LNG exports: "Certainly in this decade, there's a good chance that we will be LNG exporters on the scale of Qatar, which is today's largest LNG exporter."[67] However, the FERC review process may potentially become slightly more complicated in the future because in August 2016 the White House Council on Environmental Quality included more stringent environmental conditions to be met by projects being reviewed under NEPA, though whether this is upheld by President Trump's administration remains to be seen.[68] Overall, the export approval process can be cumbersome, redundant and lengthy to navigate, but it does provide a degree of certainty for investors and communities alike.

Current State of Exports

The year 2016 saw the first export of LNG leave the US lower forty-eight, with Cheniere's Sabine Pass terminal sending its inaugural cargo to Brazil in February and subsequently to India, the United Arab Emirates, Argentina, Portugal, Kuwait, Chile, Spain, China, Jordan, the Dominican Republic, and Mexico.[69] All other facilities are in the planning and construction phases, with the exception of the long-running but not highly active Kenai facility in Alaska. However, much of the groundwork for growing LNG exports has already been completed. Besides Sabine Pass, the DOE has given final authorization for LNG exports from four other terminals: Corpus Christi, Freeport, Cameron, and Cove Point. Together, by 2014, all five terminals had secured commitments from international off-takers for 38.5 million tons per annum (MTPA) of LNG per year, through twenty-year liquefaction tolling agreements (LTAs).[70] These agreements allow foreign companies to purchase gas directly from the US domestic market, where it is priced off the Henry Hub rather than indexed to crude oil, as it is in Asian and European markets. By 2015, committed volumes had reached 58 MTPA to countries such as Japan,

India, South Korea, France, Spain, and the United Kingdom.[71] A considerable American export volume—in the range of 25 to 35 MTPA—had been bought by traders rather than actual end users, suggesting that it is still far from certain where American LNG exports will go.[72]

The Demand Side of American LNG Exports

Even if all the conditions enabling American LNG exports are met, their success and scale will depend on the demand side of the equation, which is multidimensional and extends beyond the political pronouncements in Europe or Asia of a desire to import American gas as part of a diversification drive. Among policy analysts and political scientists, analysis of demand often goes no deeper than a consideration of political aspirations, such as a move away from Russian piped gas. In reality, however, infrastructural, commercial, and market factors play an equally (if not more) important role. Here the analysis of demand-side factors affecting American LNG exports will begin with an examination of infrastructure and regulatory needs, followed by an overview of commercial and market factors such as global gas price differentials, currency exchange rates, global oil prices, global demand for gas, and availability of alternative gas suppliers and supplies.

LNG Import Infrastructure: Terminals, Pipelines, and Storage

The supply side section above has briefly touched on LNG importing plants but here the infrastructure needs of LNG importing countries will be examined in depth. It is crucial to realize that without this costly but necessary infrastructure, states are unable to take advantage of the shale boom and import LNG. LNG import terminals—also known as regasification terminals—are the final destination points of LNG carriers. These facilities return LNG to a gaseous state so that it can be fed into transmission and distribution systems. The regasification cycle begins with liquefied natural gas being offloaded from LNG carriers and placed into cryogenic storage tanks, which ensure that the gas remains in a liquid state at a temperature of –162°C. Subsequently, under immensely high pressure (60 to 100 bar), the liquefied gas is gradually reheated, usually via seawater trickle-type heat exchangers, until it again reaches a gaseous state.[73] The cycle ends with the gas being blended to conform to regulatory and consumer specifications by adjusting the concentration levels of various gases such as nitrogen, ethane, and propane.[74]

In principle there are three main types of LNG regasification and import facilities—onshore facilities, offshore gravity base structures (GBSs), and offshore floating storage and regasification units (FSRUs).[75] As the name implies, onshore LNG facilities are land-based structures that receive LNG for storage and regasification. These have a large footprint and have huge cryogenic storage tanks—the size of fifty Olympic-sized swimming pools—that hold more LNG in liquid form than the capacity of a common LNG carrier. Offshore GBSs are fixed, self-contained concrete structures that rest on the sea floor and are equipped with LNG storage tanks and regasification equipment.[76] Finally, FSRUs, or so-called floating terminals, are ships as long as three football fields that anchor semi-permanently at a jetty in a harbor. LNG carriers align side by side with the FSRU for the unloading from the LNG carrier to the FSRU, then the FSRU regasifies the liquid back into gas to be pumped ashore.[77]

The price of all such LNG facilities varies greatly, depending on the local construction costs, the cost of land, the regasification technology used, and the total amount of storage installed.[78] Nonetheless, FSRUs are on the lower price spectrum of LNG import terminals, costing about $200–$300 million and taking roughly two to three years to build from concept to completion. This time frame can be shortened if, instead of ordering a newly built vessel, project sponsors are able to find a FSRU ship available for lease. For example, operators of the new Lithuanian LNG floating terminal lease the FSRU from a private operator that charges $189,000 per day, or $68.9 million per year.[79] Overall, as of 2015, there were sixteen operating floating regasification units in eleven countries, for a total regasification capacity of around 54 MTPA, twice the capacity of just five years before.[80]

Compared to FSRUs, onshore import terminals are pricier—they can cost around $1 billion for terminals with 1 BCF per day throughput capacity—and they may take around ten years from inception to completion.[81] Project sponsors must find partners for their projects and sign memorandums of understanding (MoUs); identify markets from which to buy or to which to sell; create legal and regulatory parameters to attract investors and to show stability in the project; develop create appropriate tax regimes; and set up supply-chain management.[82]

Pipelines and Coastal Access

For a country to be able to benefit from LNG imports, it has to have access to open water. Landlocked countries need to build cross-border pipelines to LNG importing countries or implement reverse flow on existing pipelines in order to take advantage of LNG imports. Here, an example of such a transit mechanism is the discussed, but still largely unconfirmed, Adriatic Gas Corridor project, through which gas imported through the planned LNG import terminal on the Croatian island of Krk could reach Hungary and potentially move on to other landlocked Central European states such as Slovakia.[83] Still, the cost of pipeline infrastructure crossing several states remains a salient obstacle. While the investment required for natural gas pipeline construction can vary widely due to terrain, capacity, population density, and other factors that greatly impact the overall capital outlay, the average cost of a gas pipe in 2013 lingered around $100,000 per inch-mile (the cost per pipeline diameter inch per mile of length), and due to rising labor and material costs is expected to reach around $170,000 per inch-mile by 2035.[84] Given the high cost of gas pipelines, some combination of political will and strong commercial interests will be required for LNG imports to reach landlocked countries and regions.

Gas Storage

Furthermore, gas storage facilities are also needed to take advantage of LNG imports. Underground storage of natural gas requires certain geological characteristics such as a layer of porous and permeable rock to contain the gas, along with a layer of impermeable rock overhead in order keep the gas within the porous formation of the storage area.[85] Three types of underground formations are commonly used for underground storage of natural gas: depleted natural gas or oil reservoirs; aquifers—porous and permeable rock formations that are topped by an impermeable cap rock; and salt caverns that are formed by drilling a well down into an existing salt deposit and pumping water through the completed well in order to wash out an empty void for storage.[86] It has been estimated that as a general rule of thumb, the development costs in the United States of developing natural gas storage can be as low as $8.6 million per BCF for depleted reservoirs, around $10.9 million per BCF for salt caverns, and approximately $17.2 million per BCF for aquifers.[87] These costs can be significantly higher in countries where population density is higher, where the geology is not as favorable, or where the regulatory regime is not as accommodating.

Nowadays it is also possible to store natural gas in liquid state above ground in LNG facilities, albeit at a greater financial cost, by liquefying the natural gas in a cryogenic facility and thereby reducing its volume approximately 600 times compared to uncompressed gas. Both liquefaction plants and regasification terminals have at least one insulated cryogenic storage tank; in the former, gas is converted into liquid form and then stored in such a tank, before it is loaded onto an LNG carrier to be shipped to an import market, while in the latter the liquid gas is first stored in a cryogenic storage tank after unloading from the LNG carrier and later taken out of storage to be converted into gaseous form to be sent out into the natural gas transmission grid.[88] Finally, a new gas storage reservoir requires an initial fill of a large quantity of gas to build it up to a workable pressure, which is likewise a considerable expense.

Current Global LNG Import Capacity

Because LNG imports often occur where pipeline infrastructure is unavailable or uneconomic, and because regasification facilities require significant capital outlay for infrastructure and technology, LNG import terminals are unevenly distributed across the globe, with most of them concentrated in the Eastern Hemisphere. As of late 2016, there were more than one hundred operating large-scale LNG receiving terminals around the world, with more than half of them in the Far East. In Europe, of the twenty-five large-scale operating LNG import terminals only two were located in countries formerly behind the Iron Curtain in the more energy vulnerable states of Poland and Lithuania, with the rest located in Western and southern Europe.[89] In Europe there are currently another six new LNG import terminals under construction, with two large-scale terminals meant for the Canary Islands, and four small-scale facilities meant for break-bulk receipts; another twenty-one new regasification terminals are planned across Europe within the broad timeframe of the next eight years.[90] However, the LNG outlook for the states of the former Soviet Union (outside the Baltic states and Russia) contrasts starkly with the rapidly developing LNG capabilities of the European Union. Currently in this region there are no functional LNG import or export terminals, primarily because many of these states, including the five Central Asian states as well as Armenia and Azerbaijan, are landlocked.[91] Meanwhile, Ukraine and Georgia have expressed interest in building regasification terminals.[92] Russia, currently the leading exporter of natural gas via pipelines, has ambitions to expand its LNG export capacity with four more facilities. Currently Russia's only LNG liquefaction plant is located on Sakhalin Island in the Far East.

Asia is the largest and most vibrant LNG market in the world. In total, there are fifty-eight operating LNG import terminals, most of which are located in Japan, China, South Korea, and India. Additionally, twelve LNG import terminals in Asia are either being newly constructed or are having their capacities expanded. Compared to Asia or Europe, Latin America and the Caribbean play a relatively minor role in the global LNG market. The region has only nine operating LNG terminals with nearly all capacity located in the Southern Cone countries of Chile, Argentina, and Brazil.[93] Only three new LNG import terminals are planned for construction or expansion in the whole region, in Chile, Colombia, and Uruguay, respectively.[94]

Regulation and LNG Imports

For a region to maximize the benefits of LNG imports, the regulatory frameworks for free trade of energy across international borders must be taken into account, in addition to infrastructure. While successful energy trade between countries has been taking place for decades, issues affecting free trade across regions do emerge and need to be managed according to the rules of such supranational bodies as the World Trade Organization (WTO) and the General Agreement on Trade in Services (GATS). The most common of these include various forms of most-favored-nation treatment, transparency, domestic regulation, monopolies, and exclusive suppliers.[95] Even though WTO agreements have general provisions for energy trade, and the United States and the EU have already signed other treaties, such as free trade agreements that include provisions for energy trade with other countries, additional free trade treaties such as the Transatlantic Trade and Investment Partnership with the EU or the Trans-Pacific Partnership with Pacific Rim countries could bolster energy trade between countries, though both are unlikely to find much support from the Trump administration in the near to medium term.

Across the Atlantic, the EU has made it clear that it seeks to liberalize the disparate gas industries of its member states by reducing internal trade barriers, integrating isolated markets, and developing international pipeline infrastructure, but much work remains. In Central, Eastern, and southern Europe, countries remain disconnected from broader European energy networks.[96] In Europe the so-called Energy Union (which seeks to integrate the internal energy market on the Continent, as well as to ensure energy security and diversification, improve energy efficiency, reduce emissions, and support research and innovation across member states) and the Third Energy Package (which aims to increase transparency and promote liberalization of the energy market) are important elements in the regulation and integration of gas markets.[97]

Market Drivers: Supply Meets Demand

Global Pricing Differentials

The odds of large volumes of American LNG exports circumnavigating the globe and reaching the distant shores of such markets as Asia and Europe depend largely on the profitability of exports as commercial ventures rather than on political interests. One of the factors will be whether or not the price differential between US pricing hubs and importing markets is large enough to cover the costs of capital, liquefaction, and transportation and still be profitable. From the 2000s to the early 2010s, when demand for natural gas in Japan and other Asian countries was growing faster than regional supply, prices were relatively high and stable. In Europe at that time, prices generally reflected the EU's relationship with its primary suppliers—Norway, Russia, Algeria, and Qatar. In the mid-2010s this paradigm shifted. LNG prices in the two major LNG gas markets, namely Asia and Europe, plummeted starting in late 2014, thus weakening the argument for making shipments from the United States across the Pacific and the Atlantic. At the same time, the EU has made efforts to strengthen its energy security by creating a cohesive internal gas market across member states and by seeking out a variety of gas suppliers. This inward and outward integration has reduced the premium for natural gas in Europe, which in mid-2015 witnessed a fall from around $12 / MMBtu in 2013 to slightly over $6 / MMBtu.[98] Similarly the price of LNG in Northeast Asia fell from a record high of $20.125 / MMBtu back in early 2014, to just $8.10 / MMBtu by mid-2015.[99]

To the dismay of players involved in the US LNG market, it has been estimated that US LNG producers require prices higher than the "new normal" to profitably export their production abroad, because they have to take into account the costs of capital, terminal operations and maintenance, and shipping.[100] A Henry Hub price between $2–$3 / MMBtu translates into a delivered cost for LNG sourced from the United States right around the $6–10 / MMBtu range. Between the end of 2015 to mid-2016, European and Asian buyers were willing to pay prices in that range, which may remain competitive in some markets unless European and Asian hub prices fall further. In any case, this break-even range varies, with calculations depending on factors such as the price of oil, Henry Hub and other hub prices, whether shipping costs are included, and terminal costs that are considered sunk or unrecoverable (such as the price of liquefaction).[101] However, an increase in the Henry Hub price, a fall in global LNG prices due to oversupply, or buyers' unwillingness to pay premiums above domestic spot prices for US LNG could negatively impact the competitiveness of American LNG exports abroad. At the same time, the preoccupation with the immediate levels of spot prices in the United States, Asia, or Europe is a little misleading. LNG exports are often long-term projects with twenty- to thirty-year off-take or tolling contracts being signed; project financers, bankers, and project partners providing long-term credit or other forms of funding; and thirty- to fifty-year facilities being built with time to recover project costs. The commercial viability of US LNG exports will thus also depend on what long-term buyers and long-term sellers perceive is a fair long-term price, as well as the competitiveness of the tolling model, whereby sellers continuously pay a terminal service reservation charge (which is a form of finance for the terminal that is unique to US LNG projects) but can at any time choose not to actually off take gas.[102]

Determining where Henry Hub or National Balancing Point hub prices will go is a near-impossible task, but it is possible to see how pricing differentials determine exports by examining the process. A customer first purchases an LTA, which is a fixed fee that a buyer will pay for a fixed period of around twenty years. As the name "tolling model" implies, this part of the pricing model is the cost of access to a US liquefaction plant. The liquefaction is provided as a service, with the off-taker procuring gas on the free market and paying for the feed gas, transportation to the liquefaction plant, and all transportation costs from the terminal dock to the regasification facility overseas; the facility operator collects a tolling fee for its liquefaction services.[103] This flat liquefaction fee of about \$2.50–\$3.50 / MMBtu—which includes the plant's fee for the actual liquefaction process and a fee for tolling—has to be paid whether or not the buyer exercises the option to utilize this call on US gas. This model gives the buyer the flexibility to decide when to import gas, according to when it is needed, for example, or when the price is right. This puts the risk of gas price volatility on the buyer, but in return the buyer gains off-take and destination flexibility. If the buyer does choose to purchase US gas, it will then pay 115 percent of the Henry Hub prices in addition to the aforementioned flat fee. While most US projects use this tolling structure, Cheniere differs in that it also procures the natural gas used as feedstock, making it more of a "full-service" company. However, this adaptation of the US tolling model means mandatory minimum LNG off-take by the buyer. The rationale for required off-take is that Cheniere signed agreements with utilities for gas purchases while the various international energy market players (such as Japanese utilities) that own interest in other US export projects sell tolling agreements for terminal access.[104]

In an example, using the Henry Hub price of $2.89 as a baseline, a company would pay a procurement cost of about $3.32 / MMBtu for natural gas. After that, shipping, insurance, and other transportation costs run between $0.50 / MMBtu for Europe and $1.30 / MMBtu for Japan as of early 2016, plus a regasification fee of about $0.50 / MMBtu once the LNG is delivered at its final destination. This leaves the final price at the destination somewhere between $6 and $9 / MMBtu.[105] Thus, for exports to be commercially viable, either prices at European or Asian gas hubs would have to be above the outlined costs, or the price of US LNG would have to be lower than oil-indexed gas contracts, which would require the wholesale Henry Hub price, in this example, to be at or below the $2.89 baseline. However, some experts calculate that US LNG can sell at an even lower price because the flat liquefaction fee can be considered a sunk cost, as explained above. Considering the flat fee—which can be thought of in terms of a fixed take-or-pay charge—a sunk cost lowers the price of US LNG by the fixed $2.50–$3.50 / MMBtu amount, making it more commercially viable to sell cargoes in alternative LNG markets.[106]

The various methods, calculations, and variables used to estimate costs make it difficult to ascertain whether or not US LNG exports can break even in the low price environment. Currency fluctuations could also impact the price and thus the economics of American LNG exports, making purchases in dollars of American LNG cheaper or more expensive for international consumers. Cost-wise, US LNG may not be competitive for countries in Europe outside of Spain, Portugal and the United Kingdom, but then countries in Central and Eastern Europe may be willing to pay a premium for American LNG if they can hedge against supply disruptions from Russia.[107]

Global Oil Prices

Global oil prices are possibly the single greatest variable that will influence the demand for American LNG exports because they impact the gas markets in several ways. For one, gas pricing outside North America is still largely linked to oil pricing, so declining oil prices depress international gas prices. Low oil prices make the costly practice of shale exploration a more risky and less profitable venture. Because of the high cost of drilling, drillers will normally consider the price of drilling a well as a sunk cost. Because that cost is already unrecoverable, the driller may as well produce gas. However, low oil prices may mean losses on incremental costs and therefore losses in profit, which may in turn discourage drillers from producing natural gas at all. Thus, perhaps the biggest threat to the future of US shale production and LNG exports is a period of sustained low oil prices.

Nonetheless, despite such a period from 2014 to 2016, the US shale industry proved resilient enough to weather the storm, with US natural gas production continuing to reach new highs. Against the expectations of the Organization of Petroleum Exporting Companies and many market analysts, shale producers in the United States have turned to innovation as a way to reduce costs, improve output, and make production more economical. Although the US rig count continued to fall, production rose in 2015.[108] From 2014 to mid-2015, drillers cut drilling costs from $4.5 million to $3.5 million per well and reduced drilling time from twenty-one to seventeen days through better planning, more targeted drilling, deployment of new technology, and workforce reductions.

Squeezing further cost savings out of an already efficient industry was the result of the full application of traditional manufacturing techniques to oil and gas fields to reach economies of scale. These techniques include the use of "walking" rigs with tracks or pneumatic lifters that can drill multiple wells from one pad; drilling longer laterals and sometimes drilling multiple laterals at various depth intervals in the same well; increasing the number of hydraulic fracturing stages in each lateral; using innovative proppants, such as resin-coated sand and ceramics; and reducing water consumption through water recycling or the use of subsurface brine. Producers have also been redrilling or recompleting existing wells, thereby maximizing these "super-size" wells rather than focusing on total well count in the country.[109] This has allowed drillers to lower the breakeven price of production from $75 per barrel to around $60 per barrel, with some wells reaching breakeven prices at as low as $22 per barrel in 2016.[110] General breakeven costs for US independent producers have been suggested at $42 per barrel, with wide variability among producers and plays.[111] Patrick Pouyanne, CEO of French energy company Total, summarizes this trend: "You are more efficient because you are forced to be more innovative."[112]

Near-term shale production may be declining, mostly because output declines at legacy wells gains are currently outrunning new production coming online, as far fewer new wells are being drilled. Nonetheless, the shale industry in the United States is still much healthier than experts could have foreseen.[113] Analyses such as the EIA's *Annual Energy Outlook 2016* support the feasibility of US LNG exports even under a low oil price scenario and project that LNG could account for most US net gas exports, although US companies would be less incentivized to export gas under such a scenario because their LNG would be less competitive on the global market.[114]

Global Demand for Gas

Global demand for gas is the broadest variable that can impact US LNG exports and is further complicated by differences between long-term and short-term demand, the latter being much more volatile. Looking at decades rather than years, most assume that the global demand for natural gas will continue to rise absent an abrupt global policy shift on energy issues.

Over the last few decades, demand for natural gas has followed a gradual upward trend. Only such historic events as the collapse of the Soviet Union in 1991 or the financial crisis of 2008 have caused temporary declines in demand. According to the EIA, between the years 1980 and 2010, global consumption of natural gas more than doubled, from 53 TCF to 113 TCF.[115] This increase in natural gas demand, which took place mostly in the developing states, was underpinned by continuing consumption growth in the developed word. The Middle East had the highest growth rate, increasing more than tenfold, from 1.3 TCF in 1980 to 13.2 TCF in 2010, followed by Asia, where consumption increased more than eightfold, from 2.2 TCF in 1980 to 19.2 TCF in 2010.[116] Although many factors could explain the growth of global natural gas demand, Andrea Gilardoni, Associate Professor at the Bocconi University in Italy, succinctly argues that this trend was chiefly influenced by "rising GDP levels of developed economies, the decline of oil in power generation and its substitution mainly with natural gas, and strong economic expansion of emerging countries like China, India and Brazil."[117] In the years to come gas demand for power generation is expected to rise even further. Based on the long-term projections carried out by British Petroleum (BP) in 2015, global natural gas demand is expected to grow by 1.8 percent per year, reaching around 174 TCF by 2035.[118] Growth in global natural gas demand will also be driven by the perception that it is a cleaner form of energy than oil or coal, which were at the foundation of economic development over the last centuries.

While global natural gas demand is indeed poised to witness significant growth in the decades to come, the medium-term outlook is far from rosy or predictable, especially in Asia and Europe. In a 2015 report, the International Energy Agency (IEA) significantly slashed natural gas demand growth in developing Asia, excluding China, from a 2013 growth forecast of 3.9 percent annual growth rate for the six years to

2018, to 2.9 percent per annum.[119] A significant share of this slowdown can be attributed to restarts at Japanese nuclear reactors, which were shut down after the Fukushima disaster of 2011. The medium-term outlook for natural gas demand in China is also slightly bearish as it is projected to grow 10 percent annually, down from an average annual growth rate of 14 percent from 2010 to 2015.[120] High gas and LNG prices have made local coal and renewables more competitive in Asia. In the EU, by mid-2015, gas demand was falling to 23 percent below its 2010 peak.[121] This drop was largely attributed to structural shifts in the EU economy as manufacturing was becoming displaced in some states by knowledge-intensive sectors, changing consumption patterns, progress on renewables and energy efficiency, and the relatively mild winters of 2013 and 2014.[122]

Competing Suppliers and Supplies

A plethora of other uncertainties related to competing suppliers of gas and competing energy supplies may also have a profound effect on potential US LNG exports to distant corners of the world. Arguably the greatest questions concern Russia—specifically, its planned pipeline exports to China, its potential to strategically cut prices of gas exports to Europe, and its own LNG export capabilities that are currently under construction. Both in Europe and in Asia, US LNG exports will have to compete in the short term with potentially cheaper piped Russian gas; in the long term, they may even have to compete with Russian LNG capabilities. In the future, US LNG exports will also face competition from the development of shale gas and other forms of unconventional energy resources in other countries, as well as new and existing LNG exporters. In this regard, the LNG market is bound to be dominated by a handful of major suppliers. Qatar, which sits on the third largest gas reserves in the world, totaling around 882.8 TCF, has become a foundational supplier and can arguably offer more competitive prices to hold onto its market share.[123] Other potential foundational suppliers are Australia, which by the end of this decade might reach an export capacity of 3.764 TCF per year and become the largest LNG exporter in the world, or Malaysia, which sits on vast gas reserves and by 2025 could export as much as 3.125 TCF of gas per year.[124] Mozambique and Tanzania could potentially also become game changers. Meanwhile, China is starting to develop its own shale gas reserves, which are estimated (though mostly unproved) to total nearly as much as the United States' and Canada's combined.[125]

Iran could be the greatest LNG production wildcard: the country that holds the second-largest natural gas reserves in the world (about 18 percent of the world total) has, for most of the last decade, been isolated from international trade.[126] With the lifting of Western sanctions in 2016, foreign direct investment in the Iranian energy sector will start to flow, given that companies in both Europe and Asia have already expressed interest. LNG and possibly piped gas exports could follow, albeit with a significant time lag for financing and construction of the necessary infrastructure.[127]

Finally, future US LNG exports could face uncertainty due to competition from the multitude of other energy sources such as nuclear, coal, and renewables. If the Japanese nuclear energy industry slowly recovers from the 2011 Fukushima disaster with the government's encouragement to restart mothballed nuclear reactors since 2016, the world's top LNG importer may witness a drop in LNG demand in the years to come. The nascent comeback of coal could also pose a challenge for potential LNG trade as it has regained its status as a source of cheap energy. Out of the Group of Seven (G7) countries Britain, Germany, Italy, Japan, and France together burned 16 percent more coal in 2013 than 2009 and are planning to further increase construction of coal-fired power stations.[128] Finally, a massive surge in popularity for renewable energy sources in the EU and China could weaken global LNG demand.

US LNG Exports and Global Gas Markets

Since the US shale boom took hold and American LNG exports became a more concrete possibility, a plethora of think tanks, NGOs, government organizations, universities, and large energy and financial firms have run numerous studies and simulations to gauge their potential global impact.[129] Back in the early 2010s, energy expert Daniel Yergin recognized the significance of the US shale revolution and argued that it may upset the emergence of an oil-indexed global gas market, as US shale lowered gas prices in North America significantly below oil-indexed prices in other parts of the world. In favor of unconventional natural gas, he argued that "the emergence of this new resource in North America is certainly having a worldwide impact—demonstrating that the gas market is global after all."[130] By 2015, many early supporters of American LNG exports such as the Atlantic Council pointed to their global energy security implications, which were expected to provide "enhanced diversification, competition, and energy security for global and regional markets through the expansion of the overall volume of global LNG supply."[131]

Certainly, open access to the US gas and oil markets will go a long way to depoliticize energy trade and to increase the energy security of US allies around the world, many of which rely heavily on imports for their energy needs. However, the Brookings Institution, among others, has been more cautiously optimistic. They argue that because US government support for exports is limited to essentially not imposing any major restrictions through the regulatory process, the success of future exports is entirely up to the global market. In their view, while "the United States is poised to become a major global supplier of LNG," it will face both LNG and pipeline competition from countries with geographic advantages, as well as from domestic production and alternative energy sources that are suppressing the demand for LNG in major markets like Asia.[132]

On the other side of the argument are experts like Gal Luft, who believe the oversupplied global market and falling prices since 2014 are no longer favorable to US LNG exports: "The answer to the North American gas glut isn't building multi-billion-dollar LNG terminals along U.S. coasts with the hope of exporting gas to distant markets where it is no longer wanted." Instead he and others argue for promoting the use of domestically produced gas and LNG in America's transportation sector, where as of 2015 only 1 percent of domestically produced gas was used as automotive fuel.[133] China's growth slowdown, the European recession, the potential eventual restart of Japanese nuclear plants, and the rise of LNG exports from, for example, Australia, are other considerations for the US LNG industry. Despite these short-term difficulties, in the longer term, the buildup of export infrastructure will better position the United States in the market when demand starts to overtake supply.

Downward Pressure on Prices

One area of agreement among most scholars and analysts is that increases in US gas production and the shift by the United States toward gas exports and away from imports are placing noticeable downward pressure on global energy prices. Gas that is no longer being tagged for export to the United States is finding its way into other markets, and in the process increases the volumes of available gas and diversity of supply. This not only decreases the cost of gas for major LNG consumers, but also erodes the market validity of long-term, oil-indexed contracts or contracts with a high percentage indexation to oil', and moves the world closer to a global price convergence for LNG.[134] Even without substantial LNG exports from the United States, the LNG that US domestic production has displaced in the global markets will have a substantial impact.

Still, US LNG exports could expedite this price convergence process and improve global energy security. In 2014, CitiGroup projected that by 2020, gross US exports of LNG would reach around 2.9 TCF per year. This may not seem like all that much gas, but it will have an outsized impact on the larger global consumers and producers of LNG, because while 2.9 TCF makes up only about one-tenth of US gas consumption, it equals 20 percent of the current global LNG trade.[135] In purely economic terms, a global gas price convergence may not be within the US national interest. In 2015, the International Monetary Fund ran a set of simulations to discern the macroeconomic impacts of the US shale boom over roughly the next twelve years.[136] Their findings, simply put, are that domestic energy prices will certainly be much lower if the United States does not export its gas and prices around the world do not come down. Lower prices will increase industrial productivity, making employers want to increase hiring. The increased demand for labor will, in turn, raise the cost of labor (that is, wages), and as employment and wages rise, a boon for individual workers and consumers will be created across the United States. Still, the study concluded that keeping gas within the United States gives a smaller additional benefit to the US economy relative to the *large* global economic benefit that would be realized by exporting the gas and smoothing international price disparities.

Boosting Global LNG Markets

Even before the American shale boom, the rapid growth of LNG markets in the 2000s already signaled that a global gas market was in the making.[137] In the post-shale boom, the United States may be in the best position of any country in the world to further boost the global LNG market for three reasons: (1) its contracts do not have destination clauses, so they allow reexports and cargo diversions; (2) it has the capability for rapid surge of wellhead production; and (3) it traditionally projected the international personality of a promoter of free trade. The freedom to reexport gas across international borders or redirect cargos while in transit increases liquidity in the gas market and leads both to increases in supply security and acceleration of price convergence between regional markets.[138] Increased liquidity in the gas market, while not eliminating long-term oil-indexed contracts entirely, allows buyers the leverage to push for prices that more closely resemble spot pricing and even to renegotiate contracts when major changes in the gas market occur.

Finally, the United States has a reputation as a leader in promoting free trade, though its track record with natural resources has been uneven at best; it was only in 2015 that the US government lifted a four-decades-long ban on crude oil exports.[139] It would be counterproductive for the United States to restrict gas exports under conditions of abundance. Promoting the growth of a free and robust global market for gas would go hand in hand with other US trade policy goals and would go a long way toward incentivizing nations like China and Russia to partner with free market institutions like the WTO. Greater trade, interconnectedness, competition, and interdependence would support the global energy market and in the end would improve US energy security more than a restriction on exports or any effort to achieve "energy independence."[140] While the Republican administration led by President Trump as well as the Republican-controlled Congress will likely prioritize domestic natural resource production, especially of gas and coal, and likely seek to reduce environmental regulation, their final position vis-à-vis exports and free trade remains to be seen.

Conclusion

Developments in the US natural gas market, as well as international gas market dynamics, will both contribute to a growing role for US LNG producers going forward. While expert opinion is not settled on whether American-sourced natural gas will be able to successfully win sizable market share from the likes of Qatar, Australia, or even Russia, it is clear that through its sheer presence and the introduction of a different way of doing business, the global LNG market will be indelibly altered. Unlike most other natural gas exporters, the United States boasts a transparent domestic market, a significant resource base, a competitive industry, and a robust natural gas infrastructure. Combined with some of the lowest-cost liquefaction technologies in the world and ready access to capital, US LNG exporters find themselves in a leading position from the word "go."

Market uncertainty remains, though, from variables such as the trajectory of international crude oil prices (and by proxy, international natural gas prices); the success of rival projects in Qatar, Australia,

Canada, and East Africa; and unexpected technological innovations in shale production or LNG shipments. More recently, world oil and gas output, and particularly that of the United States, has become vulnerable vis-à-vis Syria's civil war. Some have suggested that the war, which has been going on since 2011, may cause a proxy war in the oil markets between supporters of Syria's rebel groups, Saudi Arabia and the Gulf coast countries, and Syrian President Bashar al Assad's allies of Russia and Iran, which would involve overproducing oil and keeping global oil prices and subsequently US production low.[141]

Nevertheless, since 2014 and 2015, the mere prospect of US LNG entering the international market has already changed the way existing exporters and importers behave. Japanese LNG importers have been more willing to allow for a larger portion of their volumes to be priced off of an emerging LNG spot market,[142] Singapore has aspirations to become an LNG trading and pricing hub,[143] and Gazprom has proven more flexible than its past behavior would suggest when it comes to offering deeper discounts and more lenient contract terms.[144] None of these developments would be possible without US LNG exporters virtually changing the rules of the game by offering contracts that provide flexibility, market pricing, and transparent terms. Thus, while it may not be possible to project just how much of a market share US LNG exporters can capture, given the volatility of global energy markets, it is a safe bet that their entry into the market in 2016 will exert downward pressure on gas prices across the globe and spearhead a transformation of the global LNG market. In turn, the United States will be the crucial player in transforming natural gas into a more integrated, flexible, and depoliticized global commodity and in moving the global market toward its new reality.

3 The Politics of Supply: Russia and Gazprom

Before the American shale revolution of the 2010s and before the growth of the liquefied natural gas (LNG) trade in the 2000s, the story of the geopolitics of gas in Eurasia had been largely centered on Russia, which through its state-owned gas company, Gazprom, dominated both gas exports and the pipelines for these exports. Regardless of how the global gas sector changes in the future, Russia will remain a crucial actor because it has the largest natural gas reserves in the world, it is the world's second largest gas producer, and, at least for now, it is still the leading gas-exporting country. Russia mastered the politics of supply in its past. Supplying oil and gas to Europe gained Russia and its predecessor, the Soviet Union, much needed cash and political clout during the Cold War. After the Cold War, its energy exports helped keep newly independent Russia and its fledgling democracy under Boris Yeltsin afloat. Under Vladimir Putin's leadership and the coinciding rise in global energy prices in the 2000s, Russian gas exports were not only important sources of revenue but were also used to strengthen Moscow's relationships with old European Union (EU) member states such as Germany, France, Italy, and the United Kingdom, as well as to maintain influence in the energy-vulnerable states of Central and Eastern Europe and the former Soviet republics. Russian gas and infrastructure were king, and importing and transit countries were eager to do business with Gazprom.

During the 2000s, until the global financial downturn in 2008, Gazprom rode the wave of rising oil and natural gas prices and growing global demand for energy resources. Russia was persistently the largest supplier of gas to Europe, while Gazprom acquired European downstream energy assets and launched new gas pipelines such as the Nord Stream and Blue Stream. At the same time, as more countries of the former Soviet bloc joined the EU or the North Atlantic Treaty Organization (NATO) or simply gravitated farther out of the Russian sphere of influence, as in the case of the Central and Eastern European states in the 1990s, the Baltic states in the 2000s, and some of the Commonwealth of Independent States (CIS) subsequently, they lost their former privileges of cheaper Gazprom prices, which resulted in sharp gas price hikes and aggravated political tensions. Moreover, Moscow and Gazprom were increasingly perceived as freehanded in their exploitation of energy dependency of some vulnerable European states both for political leverage and revenue gains. Moscow's energy diplomacy was increasingly likened to a political tool or even a weapon, prompting increasing wariness in the EU.

By the mid-2010s, there were increasing signs that the era of Russian gas dominance could be coming to an end. Already, the role of Gazprom in Europe has been more constrained due to the transforming global energy markets and EU regulatory efforts. The commercial environment has resulted in more competition and more pressure on contract schemes, prices, and margins. However, Russian gas is not going down without a fight. Plans to launch exports to China, boost LNG production, and ramp up resource exploration of the Arctic all show Russian efforts to adjust to the new realities. Success has been slow, which points to the struggle that Russia has in maintaining its preeminent position in global gas markets.

Before telling the story of Russia's weakening politics of supply, we will first examine how Russia emerged to be a gas power. We will look at its gas reserves and production, the historical expansion of its exports westward, Moscow's use of gas diplomacy and the gas weapon, and the heyday of Gazprom's power in post–Cold War European gas markets.

Subsequently the future of Gazprom and Russian gas will be assessed by considering recent key developments: Western sanctions, Gazprom's domestic woes, pressure on gas prices, China's gas demand, the potential of Russian LNG exports, and the unfolding competition for resources in the Arctic. The case of Russia and Gazprom best illustrates how the politics of supply has operated in the past and how the modus operandi is shifting in the transforming global gas markets.

Russian Gas Reserves, Production, and Exports

Russia boasts the largest natural gas reserves in the world, totaling some 48 trillion cubic meters (TCM), or around a quarter of the world's proved reserves, which are concentrated mainly in the fields of Yamburg, Urengoy, and Medvezhye in West Siberia.[1] The modern history of Russia's gas production boom began in the early 1980s, when, due to the lower extraction and transportation costs of gas, it replaced oil as the Soviet "growth fuel."[2] Although oil remained the dominant source of Russian hydrocarbon revenues, natural gas production increased by over 50 percent in the 1980s. Additionally, unlike the temporary, but dramatic, fall off in Russian oil production immediately following the collapse of the Soviet Union, natural gas production and exports actually increased throughout the early 1990s. Much of this is attributable to the relative stability in the gas sector once the former Ministry of Gas was converted into a single state-owned joint-stock company, Gazprom. In contrast, the oil industry exhibited a much more tumultuous transition marked by vertical disintegration, privatization in the hands of oligarchs, and the entrance of multinational energy companies.[3] In fact, Russian natural gas production never dipped below 500 billion cubic meters (BCM) per year throughout this period. A further surge in the 2000s saw production soar to over 600 BCM per year.[4] The global recession in 2008, accompanied by a slow economic recovery, has dampened gas demand and subsequently plateaued Russian natural gas production levels ever since, leaving them fluctuating from 668 BCM per year in 2013 to 620 BCM in 2014, while Gazprom's production levels have actually decreased from 550 BCM in 2008 to 419 BCM in 2015, with independent Russian gas producers making up much of the gap.[5] Russia is not only a large producer but also the largest consumer of its own gas. Throughout the 1990s until 2007, Russia's domestic consumption also increased consistently (by 2–2.5 percent annually) until the 2008 economic crisis, which resulted in eventually flattening consumption. This returned to 2007 levels only in the recent peak of 2011, with around 473 BCM, or about 70 percent of total Russian gas production.[6]

Since the 1980s, Russia (and its predecessor, the Soviet Union) has been the largest gas exporter in the world.[7] In 2013, Russian exports totaled around 226 BCM, slightly less than the combined volumes of the next two largest exporters, Qatar (around 125 BCM) and Norway (106 BCM).[8] Russia first began exporting gas via its pipelines to the European Soviet satellite states in the 1960s. In 1970, Soviet exports started with a meager 3 BCM of natural gas to Poland, Czechoslovakia, and Austria; however, by 1980, exports grew to 56 BCM and extended to Bulgaria, East Germany, Hungary, Romania, and Yugoslavia, as well as to Western European countries such as West Germany, Italy, France, and Finland.[9] By the end of the 1980s, Russian gas exports to Europe had doubled, reaching 110 BCM per year in 1990.[10] Total Russian gas exports continued to increase (although at a slower pace), reaching 248 BCM (159 BCM to Europe) in 2008.[11] But from the mid-2000s, Russian gas exports to Europe began to plateau largely due to declining energy demand (mostly due to increased energy efficiency), increased use of renewables, cheap coal, and high prices.[12] As a result of reduced European gas demand and the conflict in Ukraine, Russian gas export volumes fell to 197 BCM by 2014.[13] In many ways, the numbers of Russia's exports tell the story of Russia's fortunes and woes and the rise and fall of its position on the international stage. As John Lough, Russia energy sector expert, has insightfully noted, "Without its ability to produce and supply energy, Russia would not have the status it has today. It would not have the same influence as a G8 member, it would not command the attention of the United States as it does, and it would not have the same privileged relations with several leading European Union countries, notably Germany. It would also be of less interest to China."[14]

Meanwhile, oil and gas exports, including refined petroleum products, have come to play an increasingly important role in the Russian economy, accounting for about half of the country's total budget revenue and about 70 percent of its total export revenue in 2013.[15] Russia's rising dependence on natural resources did not go unnoticed. As US senator John McCain glibly noted in 2014, "Russia is a gas station masquerading as a country."[16] Financially, oil remains much more significant for Russia than gas; in 2011–2012 it contributed about five times more than gas to the country's gross domestic production (GDP), seven times more to budget revenues, and almost three times more to export revenues.[17] Likewise, in 2013, natural gas accounted for $73 billion while crude oil together with petroleum products totaled $283 billion in Russia's total export revenues. However, due to the fixed and long-term nature of the gas infrastructure and trade, it has been gas more than oil that has allowed Russia to gain political influence in Europe and Eurasia. Oil can be transported more easily and cheaply over long distances to a variety of markets, while the close proximity and long-term pipeline infrastructure projects made many EU and CIS countries highly dependent on Russian gas.

Historic Expansion of Russian Gas Exports

Russia has historically been an important global energy player, first as an oil power and only much later as a gas exporter. In 1872, when tsarist Russia opened oil production and exploration to private investment, freeing it from state control, Russia quickly became one of the world's largest oil producers.[18] Among the important figures in Russian oil history are Robert Nobel, who invested in and modernized small local refineries, and his brother Ludwig Nobel, who was considered "the Oil King of Baku" and a founder of the Russian oil industry in the second

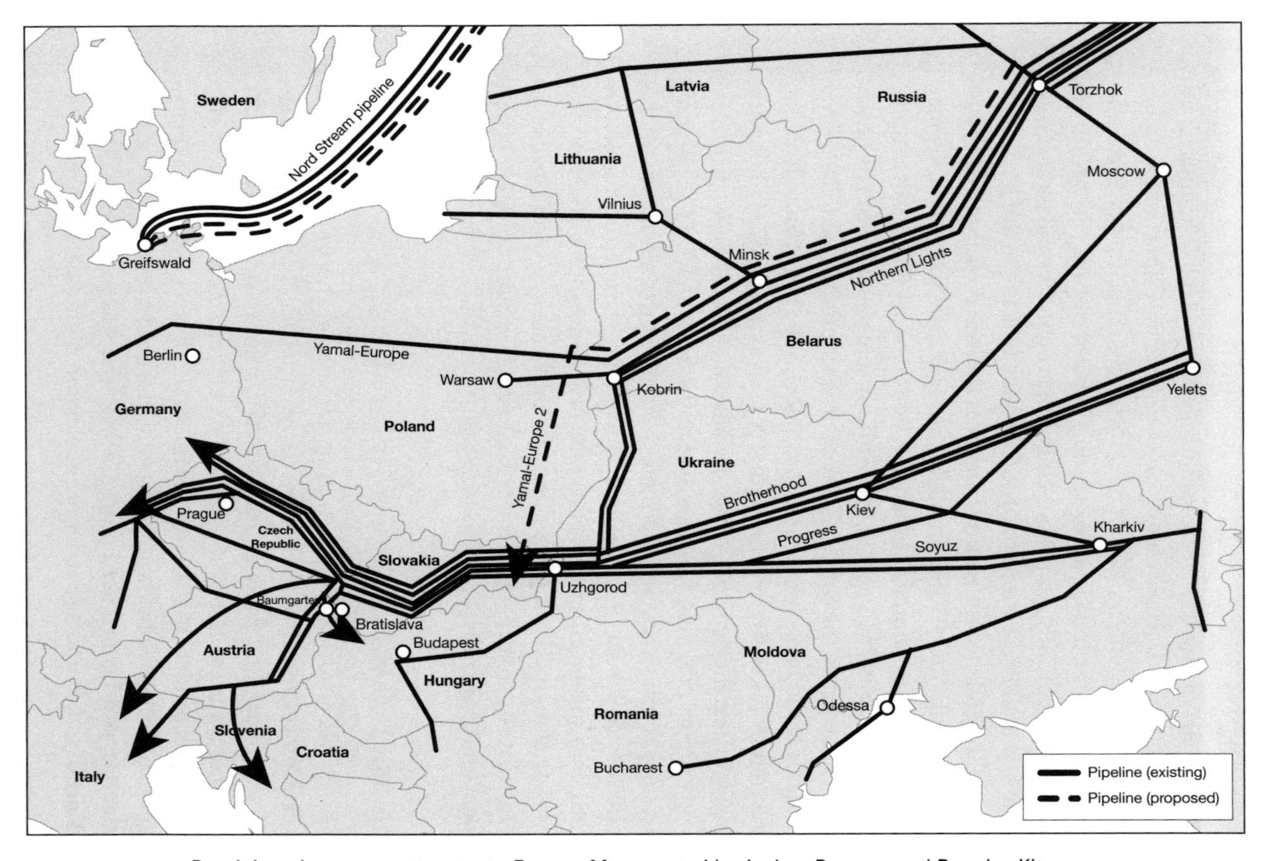

Russia's main gas export routes to Europe. Map created by Joshua Posaner and Douglas Kitson.

half of the nineteenth century. It was during this time that Russia produced 600,000 barrels of oil per day and was surpassed only at the beginning of the twentieth century by the United States, which then dominated the global oil market.[19] Russian oil (and natural gas) regained prominence in the global market after World War II, with oil output doubling between 1950 and 1960 and surpassing that of the United States. Subsequently, Russia began using oil as an instrument of soft power, particularly in Eastern Europe. Specifically, by initially supplying oil at very low prices, Russia was able to shift the dependence of Eastern Europe's heavy industry sector from cheap coal to cheaper oil. This switch backfired, however, when oil prices peaked in the mid-1970s; by then, relations of dependency were established.[20]

In comparison, the Russian gas sector developed later, largely during the Second World War, when gas was discovered in the Urals and in West Siberia and its production proved to be cheaper than that of coal and oil for the cash-strapped country. The growth of the Soviet and Russian gas sector, and particularly the boom in its exports, has been inextricably linked to Europe's gas demand in the postwar period. The Soviet gas sector developed and grew largely to meet this demand starting in the 1960s, while Russia continued expanding the Soviet pipeline network in the 1990s. The most important pipeline systems supplying Europe from the East are Brotherhood, Northern Lights, Yamal-Europe, and Nord Stream.

The Brotherhood System: The First Soviet Gas Exports to Europe

The Soviet Union started an expansion of gas pipelines to Europe in the 1950s and 1960s, with the first ones being built to supply the Soviet zone of influence in Central and Eastern Europe and then extended farther west.[21] The first states that received Soviet natural gas from the Dashava gas fields in western Ukraine via Belarus were at the time members of the Soviet Union: Lithuania in 1961 and Latvia in 1962.[22] When gas production in Ukraine's Kharkiv region emerged, the Soviet Union chose

Czechoslovakia as the Soviet "window" for gas exports to Western Europe via Kyiv and the Ukrainian-Czechoslovakian border town of Uzhgorod, and in 1967, the first segment of the Brotherhood pipeline ("Bratstvo" in colloquial Russian) to Czechoslovakia was built there.[23] Subsequently, the Brotherhood system became the Soviet Union's most important gas transportation system to Soviet satellite states. In the 1970s, separate lines started transmitting gas northwest to Poland, and south to Hungary.[24] Next was Romania, and by 1974 Bulgaria started receiving Ukrainian gas via a separate southwest route from the Kharkiv region.[25]

Soon the Western European countries were clamoring for Russian gas, and Moscow was happy to oblige. Initial USSR-Western contracts consisted of "pipes-for-gas," whereby the technology poor Soviet Union sought to receive pipes and other energy industry equipment in exchange for gas. Just a year after the Brotherhood line to Czechoslovakia was completed, in 1968, the first Western European company, Austria's Österreichische Mineralölverwaltung (ÖMV, now known simply by the acronym), signed a three-year contract with Soviet authorities for supplies to Austria.[26] An extension of the Brotherhood system was completed in the same year, and Austria became the first European Communities (now the EU) country to import Soviet gas. According to energy historian Per Högselius, Austria was a "test-case for Europe," and as he further notes "It is far from certain that Western Europe would have come to import Soviet gas at all, had Austria not opted to do so." Doing its best to fulfill its ÖMV contract and establish a good track record of supply, the Soviet Union accomplished this at a high cost to itself. During the winter of 1971–1972, it left its own Ukrainian, Belarusian, Lithuanian, and Latvian citizens to freeze, closing schools and other municipal institutions, in order to deliver the agreed-upon amount of gas to Austria.[27]

Soon after Austria's deal, a flurry of interest from the other Western European countries followed. In November 1969, West German gas company Ruhrgas signed a twenty-year contract for Soviet gas, forging a close relationship with Sojuzgazexport (later part of Gazprom) and laying the foundation for Ruhrgas and Gazprom's collaboration, which still exists. West German Chancellor Willy Brandt viewed gas relations with the Soviets as part of a sound geopolitical strategy, the so-called *Ostpolitik,* which could reduce Cold War tensions and also help reestablish contact with communist East Germany.[28] "Economics, therefore, remains for the foreseeable future an especially important element of our policy in Eastern Europe," Brandt wrote in 1969.[29] Beginning in 1973, West Germany started receiving natural gas from the Soviet Union via the extended Brotherhood system.[30]

A few years earlier, in 1964, the Soviets discovered the Vuktyl gas field in Komi Autonomous Soviet Socialist Republic in northwestern Russia. This discovery prompted the construction of another major Soviet transmission system, called Northern Lights ("Sijanie severa" in Russian). The Northern Lights system reached Minsk, Belarus, in 1974, and there it linked with the older (1960) pipeline from Ukraine, connected with the Brotherhood system, and thus brought Komi gas to the Czechoslovakian border. Supplemented in 1976 with additional volumes from West Siberia, Northern Lights became the first pipeline to bring Russian, rather than Ukrainian, gas to the West. Three decades later, the future Yamal-Europe pipeline would run alongside this system.

The successor to Brandt, Chancellor Helmut Schmidt (1974–1982), continued the same gas import policy from the Soviet Union. Under his leadership, the Soviets were able to further expand the Brotherhood gas system via Uzhgorod to Czechoslovakia with the Orenburg-Uzhgorod (also the so-called Sojuz, or Union) pipeline, originating from Russia's Orenburg field close to the border with Kazakhstan.[31] This pipeline expanded the Brotherhood capacity even more, and other Western European countries and companies no longer wanted to be left out. Back in December 1969, Italian Ente Nazionale Idrocarburi (ENI), another of contemporary Gazprom's closest Western partners, had signed a twenty-year agreement for Soviet gas to be delivered via Ukraine to the Austrian-Czechoslovak border but now the infrastructure made it possible.[32] Deliveries of Soviet gas to Italy started in 1974 after Moscow had expanded the Brotherhood system in 1973.[33]

In September 1969, shortly before West Germany's deal, France also signed an agreement for the "delivery to France of [a] large quantity of Soviet natural gas and reciprocal deliveries of pipe and equipment for gas industry to the USSR," which started in 1976.[34] Before Brotherhood was extended to West Germany in 1980, France used swap arrangements to receive Soviet gas: Italy received the Soviet supplies contracted by France, while France received the Dutch supplies from the newly discovered Groningen gas field in 1959 destined for Italy.[35] These swap deals were the early signs of functional European gas markets wherein the increasingly interconnected Western European countries could receive gas from each other.

Thus, step by step, and yet without "supergiant gas fields" in Siberia in play, Western European countries became heavy Soviet gas importers.[36] In the 1980s, political scientist Thane Gustafson extensively studied Soviet negotiating strategy for expansions of the Brotherhood system capacities and counted eleven gas contracts signed by the Soviets with the Western European countries between 1968 and 1977. Not all were pleased with these developments. In 1975, foreshadowing the debates on energy security that would take hold in Europe several decades later, American economist Arthur W. Wright was concerned that, despite the fact that the "Soviet natural gas industry scarcely existed when war broke out in 1941," a couple of decades later gas exports via pipelines to Eastern and Western Europe were "rising noticeably."[37]

Brotherhood's Urengoy-Uzhgorod Pipeline: The Rise of Siberian Gas in Europe

In the early years, the Brotherhood system delivered gas from Soviet territories adjacent to Europe, such as Ukraine and the eastern-central region of Russia. However, Ukrainian gas fields started to be depleted at the same time that Western European demand continued to grow.[38] In the 1960s and 1970s, when the potential of Russia's West Siberian gas fields, like Urengoy and Yamburg, became evident,[39] the main Soviet gas producing regions shifted north to Siberia, while the Soviets continued to use already existing export pipeline networks to send Siberian gas west via Ukraine. This, in turn, solidified Ukraine's status as the crucial transit territory for Soviet and then Russian gas exports to Western Europe that lasts until the present day.

The first West Siberian gas reached the West in 1976 via the Northern Lights system, which was connected to the Brotherhood system at Uzhgorod. By late 1970s, the Soviets proposed to its Western partners to deliver Siberian gas to the border of Czechoslovakia. At that time the project was called the East-West, or Yamal, pipeline, but it was really an addition to the Brotherhood export system via Czechoslovakia and eventually became known as the Urengoy-Uzhgorod pipeline. With West Germany taking the lead, European companies and banks accepted the deal to construct the "world's largest natural gas pipeline project" at the time.[40]

The most vocal criticism of the plan came from the other side of the Atlantic Ocean. In 1981, while attempting to prevent "the possibility of over-reliance" on the Soviet Union, the United States urged Western Europe to consider alternative energy sources, such as American coal.[41] In 1982, Milton R. Copulos, a Cabinet-level advisor in both the George H. W. Bush and Ronald Reagan administrations, called the planned pipeline a "Soviet natural resources noose" that "Western Europe is about to put its neck in."[42] The words were prescient but did not convince the Europeans.

The newly inaugurated President Reagan first raised the Urengoy-Uzhgorod pipeline issue in July 1981 at the Group of Seven (G7) economic summit in Ottawa, but failed to get traction.[43] After the Soviets imposed martial law in Poland on December 13, 1981, Reagan imposed sanctions on the Soviet Union—sanctions thought to have objectives beyond mere solidarity with the Poles and to include blocking the construction of the new pipeline. In 1981, in the midst of Washington's efforts, the Commission of the European Communities (future European Commission [EC]) issued a "Communication Concerning Natural Gas," in which it expressed worries about increasing gas import dependency on "third countries" (nonmembers of the Communities such as the Soviet Union and Algeria) from 11 percent of consumption four years earlier to the prospect of 46 percent in 1990, which eventually proved to be an underestimation.[44] By mid-1982, Reagan's administration extended the sanctions to certain American-made or American-licensed energy equipment, specifically so that it would not be used for the Urengoy-Uzhgorod pipeline to Western Europe. However, with the global recession of the 1980s, which was tough on the economies of West Germany and other Western countries that wanted both natural gas and contracts for its manufacturers of energy-related equipment, the Reagan administration's efforts against the pipeline failed, and in November 1982, it lifted its sanctions against the Soviets.[45]

The new Urengoy-Uzhgorod line of the Brotherhood system became operational in 1985.[46] After its completion, and until the inauguration of the Nord Stream pipeline three decades later, the Brotherhood system was the primary means for supplying Western European markets, and Ukraine was a major export route for Russian natural gas to the West. The Brotherhood system exemplified the geopolitics of Russian gas on the European continent for the next half century. Making Europe and Russia mutually dependent on one another for gas imports and export revenues, respectively, it both led Europeans to seek a cooperative relationship with Russia and sowed fears about energy security and Europe's vulnerability.

Post–Cold War: The Era of Gazprom and Elimination of Transit States

The end of the Cold War created a new era for Russian gas—the era of Gazprom. The Russian national gas company has been the dominant player in Russian gas politics for more than a quarter of a century. The company originated in the Soviet Ministry of Gas Industry, established in 1965, which in 1989 was restructured into the Gazprom State Gas Concern—the first state-corporate enterprise.[47] But that is only a small part of the long and intertwined history of relations between the Russian state and Gazprom. The company's first chairman, Viktor Chernomyrdin, who was known better among the Russians for his numerous sayings such as the infamous "We wanted the best, but it turned out like always,"[48] did not remain at Gazprom long. In 1992, he was appointed prime minister by President Boris Yeltsin and became the longest tenured prime minister (six years) in Russian history. Under Yeltsin's leadership, Gazprom was privatized and made new gains in capturing Western markets, while the company's links with the Kremlin persisted. In November 1992, by Yeltsin's presidential decree, Gazprom was transformed into a joint-stock company, with the state holding more than 51 percent of the shares.[49]

Gazprom remained the largest Russian joint-stock company until 2016, when it lost the position to oil giant Rosneft, whose market capitalization of $51.7 billion slightly exceeded that of Gazprom's $51.5 billion.[50] Currently, the Russian state together with other Russian state-controlled energy companies still owns more than 50 percent of the company's shares (50.2 percent), with the rest held by American Depositary Receipt holders (28.4 percent) and other legal entities and individuals (21.4 percent).[51]

The Kremlin's involvement in the management of the company also continued from the Yeltsin to the Putin eras. At the start of his leadership in the 2000s, Vladimir Putin focused his attention on Gazprom in order to break the oligarchs' power in the energy industry and to usher in his trusted allies, Dmitry Medvedev and Alexei Miller, to the management of the company.[52] Before he was elected Russian president (2008–2012), Medvedev had served as Gazprom's chairman of the board in 2000 and then on the board of directors between 2000 and 2001 and between 2002 and 2008.[53] Meanwhile, Miller not only worked for Putin in the St. Petersburg Mayor's Office in the 1990s, but was also a loyal friend and in 2001 was appointed to be chairman of the management committee of Gazprom—a post he still holds. Miller is also the deputy chairman of the board of directors, while Gazprom's chairman of the board is another political figure, Viktor Zubkov.[54] In 2008, when Medvedev took the presidential office, then Prime Minister Zubkov went on to succeed him at Gazprom.

Under Putin's leadership, Gazprom has been operating as the so-called national champion, not only seeking to maximize revenues, but also to advance Russia's national interests.[55] However, these political pursuits sometimes conflicted with Gazprom's commercial operations. Already in 2005, leading Gazprom expert Jonathan Stern clearly saw that Gazprom's "narrow considerations of the 'bottom line' will always be tempered, and sometimes overturned, by the political requirements—both domestic and international—of the Russian government." Major structural reforms, including corporatization and legal unbundling, improvement of pipeline reliability, replacing gas production in declining fields, and managing an expanding EU export portfolio, all of which posed complex and significant challenges to Gazprom, were forced to compete with the strategic priorities of the Kremlin.[56] In 2015, energy scholar Andreas Goldthau observed that Gazprom is increasingly "squeezed between a rock and a hard place"—a challenging situation stemming from changes in both the international and domestic Russian gas markets.[57]

Eliminating Transit States: Yamal-Europe and Nord Stream

After the dissolution of the Soviet Union, Russia altered its strategy for new export pipelines to bypass the newly independent states it had used for transit. Thus, instead of adding new sources and routes to the Uzhgorod cross-border point, as before, Russia built two new pipeline systems, Yamal-Europe and Nord Stream, circumventing Ukraine. Yamal-Europe, with its 33 BCM per year capacity, was the first major pipeline to bypass Ukraine to bring Russian gas to Germany. The pipeline, which was completed in 2006, runs along the Northern Light system but goes west through Belarus and Poland, instead of linking with Ukraine's Uzhgorod.[58]

While Russia's efforts to bypass Ukraine stemmed largely from the desire not to be dependent on foreign territory for the flows of its energy exports, another factor was Ukraine's tendency to build up large debts for gas in the 1990s, as well as allegations that it siphoned off substantial volumes of gas in transit to Europe. Of course, Moscow and Gazprom found it increasingly intolerable to overlook these liabilities as Ukraine became more politically aligned with the West. Moreover, in 2009 Russia started to raise gas prices for Ukraine and other CIS states to more closely reflect market prices or European levels, which further exacerbated tensions over transit and debts between Kyiv and Moscow. Ultimately, "transit avoidance" by Gazprom reflected a hedging strategy, and, as scholar of gas markets Simon Pirani observes, it was "a direct consequence of the two decades-old failure of Russia and western CIS countries to find a lasting solution to the transit problem."[59]

The mid- to late 2000s saw the second stage of Russian pipeline expansion to European markets in the post–Cold War era with the Nord Stream mega-pipeline. Unlike the other two major export routes, the Urengoy-Uzhgorod pipeline via Ukraine and the Yamal-Europe pipeline via Belarus, Nord Stream was the first one to directly connect Russia to Western Europe without any transit countries. The pipeline runs 1,224 kilometers from the Russian city of Vyborg on the northeast banks of the Gulf of Finland to the German northeastern city of Greifswald on the bend of the Baltic Sea. Nord Stream's two pipelines (in operation since 2011 and 2012) have the capacity of 55 BCM, which is equivalent to 37 percent of Gazprom's exports to Europe and 10 percent of annual gas demand in the EU.[60]

The Nord Stream consortium was formed in 2005, with Gazprom (51 percent) and two German energy companies, E.ON (24.5 percent) and Wintershall (24.5 percent), emerging as the main shareholders, though the German stakes were soon lowered to 15.5 percent after the French natural gas company Gaz de France and the Dutch natural gas infrastructure and transportation company Nederlandse Gasunie joined the project in 2008 and 2010, respectively.[61] What was unique about the Nord Stream project from the start was not only that Gazprom went into direct partnership with Western European companies for the first time but also that for the first time Western European companies became shareholders of the Russian gas export pipeline outside of their countries' territory.

As was the case in the 1980s, when the Urengoy-Uzhgorod pipeline was built, another global economic crisis (2007–2009) was underway when Nord Stream was commissioned. At that time, while the project enjoyed support from large and old EU member-states such as Germany, it met political resistance from newer ones, such as the Baltic states and Poland,[62] which were furious about the plan for the new pipeline to bypass their territories. Back in 2003 and a year before EU enlargement, the EC had already expressed support for a "gas pipeline connecting the United Kingdom–northern continental Europe–Russia either across the Baltic Sea or overland alongside the existing Yamal–Europe pipeline."[63] By contrast, in 2006, Poland's minister of defense, Radosław Sikorski, compared it to a 1939 deal between the Soviet Union and Nazi Germany to divide Europe: "We are especially sensitive when it comes to corridors and when Eastern Europe is treated differently than the West . . . that's reminiscent of Locarno and the Molotov-Ribbentrop Pact. That is 20th century." Later, in 2010, when he was foreign minister, Sikorski raised the further objection that the Nord Stream was a "waste of European consumers' money."[64] Others criticized Nord Stream as a political project chosen at the expense of cheaper alternatives like modernizing existing pipelines or as a way for Moscow to gain further bargaining leverage over former transit countries.[65] Nevertheless, the Nord Stream project was implemented, and the EC went further to grant it the privileged status of a Trans-European Network Project, which is a designation for projects that are key to increasing competitiveness in the energy market and security of energy supply, and which is also considered essential for attracting investors in early stages of construction.[66]

The special relationship between Vladimir Putin and Gerhard Schröder, the German chancellor (1998–2005) and leader of Germany's Social Democrats, significantly affected the decision to proceed with the Nord Stream project.[67] Just weeks before the German parliamentary election in September 2005, Chancellor Schröder signed an agreement to build the gas pipeline, and then just weeks before leaving office in October, he extended a guarantee that Germany would cover $1.4 billion of Nord Stream costs in case Gazprom defaulted. After finishing his term as chancellor, Schröder made another highly controversial move by agreeing to become chairman of the shareholder's committee of Nord Stream.[68] Since then, Schröder has been one of Putin's strongest allies in the West, even during the Ukrainian crisis of 2014.[69] According to Michael Thumann, Moscow bureau chief (2006–2011) of the German newspaper *Die Zeit*, Putin and Schröder's close relationship reflects the fact that they are "brothers in biography": both had servicemen fathers, grew up poor, and made their careers starting from the bottom. Communication between the two is direct, given Putin's knowledge of German language and culture, and they share similar views on global politics, giving priority to political stability over human rights and democracy, as seen from Schröder's praise of Putin's domestic policies. In addition to repeatedly calling Putin a "flawless democrat," Schröder responded to Chechnya's controversial 2004 elections by stating he saw "no major flaws."[70] Furthermore, both Putin and Schröder have an affinity for the energy trade, which is treated more than simply as a business, but also as a way to exhibit state power. So it was that Schröder became a key figure linking German energy security interests to Russian supplies, which resulted in Germany's being dependent on Russia for as much as 40 percent of its gas after Nord Stream was built.

Over time, gas deliveries via Nord Stream rose rapidly, replacing gas supplies to Europe via the Brotherhood export system through Ukraine and Slovakia. By 2014, Nord Stream was transporting 32.8 BCM of gas, which is slightly more than the 31.4 BCM Europe received via the Brotherhood routes, though Nord Stream's rated capacities of 55 BCM per year are even higher.[71] In September 2015, Gazprom and the Western energy companies of Badische Anilin und Soda Fabrik (BASF), E.ON, ENGIE, ÖMV, and Shell signed a shareholders agreement on the Nord Stream II project to build two more pipelines along the original Nord Stream. Scheduled for completion in 2019, the additional new strings would have an aggregate annual capacity of 55 BCM and would thus double the existing Nord Stream capacity to 110 BCM per year.[72] While the Western European countries seem enthused about the project, Eastern European countries such as Slovakia and Ukraine, which will be eliminated from the gas transit business by it, have been critical. Indeed, there is real uncertainty whether the pipeline will be built. In November 2015, ten Eastern European countries wrote an official letter to the EC stating that the Nord Stream II project runs counter to EU interests.[73] Furthermore, in June 2016, the president of the European Commission, Jean-Claude Juncker, weighed in and warned that "Nord Stream 2 could alter the landscape of the EU's gas market while not giving access to a new source of supply or a new supplier."[74] However, there has been little action from the EC to back up Juncker's words. As a result, countries like Poland have been spearheading the opposition against Nord Stream II and claiming that if need be, they will go to court to maintain European energy security.[75] Ultimately and possibly worse than deepening Europe's dependence on Russian gas, this project may well put a wedge between the new and old EU member states, challenging the unity that Brussels has struggled to formulate vis-à-vis EU foreign and energy policy.

Southern Routes and Turkey: Blue Stream, South Stream, and Turk Stream

At the same time as Russia explored new routes to supply northern Europe, it also considered southern routes to European markets that would likewise avoid Ukraine. In 2006, when Gazprom and the Italian company ENI established a joint venture to build a pipeline across the Black Sea, the proposed South Stream line was to be the largest European gas infrastructure project aimed at bypassing Ukraine as a transit state. In May 2009, when the gas companies of Russia, Italy, Bulgaria, Serbia, and Greece signed an agreement on the construction of South Stream, Putin said that this €8.6 billion pipeline project would contribute to Europe's energy security.[76] However, pipeline construction never actually launched, and almost six years later the EC threatened to start legal actions against Bulgaria, where South Stream would have accessed Eastern Europe from the Black Sea, due to the project's noncompliance with EU law.[77] After Bulgaria stopped the construction in its territory in August 2014, Russia canceled the project and turned to Turkey in yet another attempt to deliver gas to Europe directly.

The substitution of Turkey for Bulgaria in a new pipeline project to deliver Russian gas to Southeast Europe—the Turk Stream—was no surprise. For one, Putin and Turkey's president, Recep Erdogan, seemed to share common interests and a penchant for authoritarianism, while Moscow and Ankara were already engaged in energy trade.[78] In 2003, Turkey had started receiving Russian gas via the Blue Steam, the world's deepest gas pipeline,[79] running 750 miles under the Black Sea, with an annual capacity of 16 BCM. In addition to fulfilling promises given to Ankara by Gazprom back in 1997, gas exchange through the Blue Stream also demonstrated the growing strategic cooperation of both countries. Indeed, by 2013, Turkey had become the second largest European regional market for Gazprom.[80] Thus, when the Russia–Bulgaria South Stream pipeline was withdrawn, Moscow and Ankara had the wherewithal to rapidly unveil a new project.

The proposed plan for a Turk Stream pipeline seemed to mark a new era in Turkish-Russian energy and political relations, as Turkey was also a key country in Moscow's plan for a Eurasian energy union, which, within the framework of the Eurasian Economic Union, is designed to coordinate energy policies over much of Eurasia. Still, due to Russia's financial woes in mid-2015 and disagreements over gas prices, the Turk Stream increasingly appeared as much "a pipe dream" as "a pipeline."[81] Relations between Russia and Turkey then became more divided in 2015, owing to their divergent policies regarding Syria and the Islamic State, highlighted by Turkey's downing of a Russian fighter jet on November 24, 2015. Only nine days later, Russian energy minister Alexander Novak said the negotiations over Turk Stream "have been suspended."[82] However, by 2016, a rapprochement and a revival of the project was in the making. The failed July 15 coup attempt against President Erdogan resulted in ambivalence of Turkey toward the United States, which was perceived as somehow complicit, and signaled an opening to normalize relations with Russia.[83] In October 2016, Putin and Erdogan signed the agreement to build Turk Stream, though the on-again, off-again project still depends on future political relations between Russia and Turkey.

The Politics of Supply: Gas Diplomacy and Gas Weaponry

The connection between gas and politics is anything but simple, especially in the case of Russia. Before the changes in the global gas markets started to be felt in the late 2000s and before they started to constrain Gazprom, the Russian politics of gas supply was characterized by Russia's commercial power and political influence over gas-importing states. Certainly Russia was likewise dependent on the European gas markets for revenue, but the nature of gas markets and pipeline infrastructure had made Gazprom a monopolist in a number of EU and Eastern European states. These conditions spurred a debate among scholars and analysts as to whether and to what extent Russia abused its monopolist position and used energy as a weapon against importing states. Even the term "energy weapon" has become a topic for debate, prompting energy expert James Henderson to suggest that "rather than 'weapon' (a means of destruction or punishment) a more appropriate term would be that gas is an instrument of pressure, of attempts to keep other countries in the Russian sphere of influence."[84] Of course, there is a fine line between "pressure" and "weapon," but "weapon" appears to be the preferred term for those critical of Russia's use of gas as an instrument of its foreign policy.

This viewpoint was particularly prevalent in the late 2000s. Thus, in 2008, former US ambassador to Lithuania Keith C. Smith argued that "Moscow's deployment of the 'energy weapon' dates from 1990" and was manifested in the cases of the Baltic states, Ukraine, Belarus, and the Czech Republic.[85] In the same year, Anita Orban, Hungarian ambassador-at-large for energy security, wrote in her book on the new Russian imperialism that since 1991, the "energy weapon" was systematically used in Russian foreign policy toward former Soviet republics and satellites.[86] Likewise, in 2008, Marshall I. Goldman, a professor of economics and Russia specialist, analyzed Putin's use of gas as a secret economic and political weapon in a "pipeline poker game."[87]

A number of other scholars and analysts, such as Janusz Bugajski, have argued that in Russian hands, gas is "soft power," used to manipulate European energy supplies and thus to influence policies of selected governments.[88] Russia scholar James Sherr goes further, seeing Moscow's using a mix of "hard diplomacy" and "soft coercion" to create co-opted interest groups in energy dependent states.[89] Professor Margarita Balmaceda similarly argues in her work on energy dependence in the former Soviet republics that countries like Ukraine cannot be seen as "simply victim[s] of Russia's energy blackmail" because there Russia takes advantage of weaknesses, such as corruption, in domestic politics.[90] Stressing the important limitations to the energy-as-weapon approach, Balmaceda notes that in the multipurpose nature of energy leverage, political pressure is only one among other tools. However, she also admits that a purely economic approach is not sufficient in explaining Russian energy policies: "throughout the post-Soviet world, energy issues remain highly politicized as foreign policy issues."[91] Another proponent of the view that Russia "ruthlessly" uses energy as a weapon is Edward Lucas, senior editor at the *Economist,* who states that, together with its use of money to divide the weak countries of the West, Russia's deployment of the energy weapon is characteristic of the new Cold War between Moscow and the West.[92]

Indeed, Russian politicians themselves have declared the connection between Gazprom and Russian foreign policy on numerous occasions. Back in 2003, which was Gazprom's tenth anniversary, Putin stated that "Gazprom is a powerful political and economic lever of influence over the rest of the world."[93] In the same year, a memo written by a team of Russian foreign policy and energy experts pointed out that if "the leaders of this or that country decide to show good will toward the Russian Federation, then the situation with gas deliveries, pricing policy and former debts changes on a far more favorable note to the buyer."[94]

Nonetheless, some analysts have not been convinced that Russia uses the "energy weapon," even in the case of Gazprom's gas halts in Ukraine in 2006 and 2009—possibly the most infamous of Russia's interruptions of energy supplies to a target state. In 2006, Stern argued that the Russian giant was primarily motivated by commercial considerations (gas prices) rather than political ones. Later, together with his colleagues Simon Pirani and Katja Yafimova, Stern took the position that in the 2009 Russo-Ukrainian gas crisis "the often-cited desire of the Russian government to use energy as an economic or political 'weapon' *against European countries* [emphasis in the original] did not even play a part."[95] Another energy analyst, Andreas Goldthau, also shared this view, arguing in 2008 that Moscow could not use gas as a weapon against Europe because it would risk losing its main customer.[96] In practice, Gazprom never acted against the aggregate European market, but rather on individual, weaker states with smaller undiversified markets; it also turned to price hikes rather than permanent cut offs from supplies. Thus, in 2015, energy expert Tim Boersma suggested that a Russian energy weapon "barely exists" and that Gazprom is a gas market player searching for profits and taking advantage of its monopoly position in some countries to charge higher prices.[97] While this is certainly true, Gazprom has a tendency, rather unique for a commercial player, to offer lower gas prices in its monopoly markets in exchange for political concessions or invitations to join Moscow-led political unions, as seen in the case of Ukraine, Belarus, Armenia, and others.

In reality, though, the best way to understand Gazprom's behavior in European gas markets is to recognize that it is shaped simultaneously by its commercial considerations and by Moscow's political objectives. In my first book on the Baltic-Russian energy relations, I demonstrated that Moscow did not shy away from enacting energy cuts, mostly when its political objectives coincided with commercial preferences (or such policies were made possible by new commercial and infrastructural conditions). Russia typically engaged in prolonged energy cutoffs in countries after it had secured alternative transit routes for its energy resources. The timing of the cutoffs coincided with Moscow's interests in punishing uncooperative political or commercial behavior of the target states. This can be illustrated by a number of examples. For instance, Moscow pursued temporary gas cutoffs to Ukraine following the Orange Revolution of 2005, before the Yamal-Europe pipeline was completed in 2006. Subsequently, Moscow started voicing threats to permanently reroute gas flows from Ukraine after Maidan revolution and the 2014 conflict. Although Russia has not yet been able to successfully divert *all* gas flows away from Ukraine, this objective has been made more feasible by the completion of the Nord Stream pipeline in 2011. Likewise, Russia has permanently halted supply to oil pipelines feeding Latvian Ventspils Nafta since 2003 and Lithuanian Mažeikių Nafta since 2006, after Moscow had built up alternative oil export infrastructure and no longer needed to rely on Latvian or Lithuanian territory. Similar commercially and politically motivated energy halts include suspicious pipeline explosions that disrupted gas flows to Georgia in 2006 and gas flows from Turkmenistan in 2009.

Essentially, Russian energy companies, including Gazprom, have flexed their monopolist muscles when it was in Moscow's political interest to do so. They have rewarded vulnerable countries like Belarus, Armenia, Kyrgyzstan, and previously Ukraine with cheap gas, but they have also allowed them to run up massive gas debts, which were later used to extort the countries either to sell their energy infrastructure or to make political concessions, or both. This has been part of Gazprom's acquisition strategy, which serves the company's commercial interests while awarding Moscow potential influence over the target countries via Gazprom's ownership of strategic assets. Likewise, in the 2000s, when prices for Russian gas were peaking and at times exceeding average European levels in countries already outside of Russia's zone of political influence, such as Macedonia, Poland, the Czech Republic, and Lithuania,[98] Gazprom was flexing its commercial muscles in those monopoly markets while also retaining for Moscow the ability to promise lower gas prices if these countries made political or commercial concessions. But then, when Gazprom lost its monopolist position in some markets, it reacted to changing market conditions and sought to maximize revenues by making commercially minded deals. As Stern has pointed out, since 2014, despite the political tensions between Kyiv and Moscow,

Gazprom has had to progressively adopt to market pricing in Ukraine, "where reverse flows from Europe have forced a reduction in prices in order to restore the competitiveness of direct sales to one of its major markets rather than see its gas sold 'second hand' to that market by its European customer."[99] In other words, with Ukraine now able to import cheaper Russian gas via reverse flow from other European countries, Gazprom opted to lower prices rather than risk losing direct sales to the Ukrainian gas market.

Russian Gas Interests in EU Downstream Markets

For Russia, gas exports were not only a source of revenue but also a tremendous source of direct economic and indirect political influence on the European continent. By the 2000s, pipeline projects were not enough for Gazprom's growing ambitions, and it sought other ways of expanding its presence in Europe. Consequently, this became the most important decade in Gazprom's efforts to build an "energy empire."[100] Gazprom and a variety of other Russian companies and interests moved aggressively to gain a foothold in downstream EU oil and gas markets. This was achieved through investments in all kinds of activities related to the oil and gas industry in Europe, including the aforementioned pipelines, as well as involving storage and trading companies and striking alliances with European energy companies for projects within the EU and beyond.[101] By and large Gazprom's operations in major European markets like Germany, France, and Italy remained in the commercial domain, but even though they at times resulted in commercial tensions, they also served to strengthen political ties between Moscow and the governments of the target states.

Leveraging its position as a piped gas supplier to Europe, Gazprom has sought to acquire ownership of pipeline segments on the territories of European countries. Gazprom has invested in the Nord Stream pipeline in Germany (51 percent share) and the Blue Stream pipeline in Turkey (50 percent share), and it possesses a 10 percent share in British Interconnector Limited, which links the United Kingdom and continental Europe. Gazprom also owns a 48 percent share of Polish company EuRoPol Gaz, which operates a nearly 700-kilometer-long Polish section of the strategic Yamal-Europe gas pipeline, and a 100 percent share in the German section of the same pipeline.[102]

Gazprom has met some resistance in these expansion efforts, including from the Polish political establishment, and this resistance highlights many of the inherent dilemmas of Russian investment for target states. The other shareholders of EuRoPol Gaz are the state-owned Polskie Górnictwo Naftowe i Gazownictwo (PGNiG), with a 48 percent share, and Gas Trading, with 4 percent. Gas Trading is also a joint venture in which ownership was considered Russian-influenced because more than a third of company was owned by the trading company Bartimpex, which in turn was partially owned by Gazprom Export (Gazexport) with 16 percent of shares. This ownership arrangement led to speculations that Poland did not have control over the key gas pipeline on its territory.[103] In July 2015, this issue was resolved when PGNiG purchased Bartimpex's shares, thus acquiring a total of nearly 80 percent of Gas Trading and gaining the majority share of 52 percent of EuRoPol Gaz.[104] Another complicated moment in Russian-Polish pipeline relations came in 2013, when Gazprom and PGNiG signed a memorandum of understanding for the construction of the new Yamal-Europe 2 gas pipeline, planned to run through Poland to Slovakia and Hungary.[105] This unexpected announcement was met with bewilderment by Prime Minister Donald Tusk, who said, "Strategically speaking, we do not want to expand the pool of Russian gas."[106] Because the Ministry of Treasury supervised the preparation and signing of this agreement by PGNiG, Tusk later dismissed Treasury Minister Mikołaj Budzanowski, explaining that he has lost confidence in top management of PGNiG to pursue public interests effectively.[107] The episode was illustrative of the influence and leverage Gazprom had gained in a number of European markets where national and local energy companies sought collaboration with the Russian major, even at the expense of their home countries' interests or political considerations.

Gazprom also invested in European underground storage facilities (UGSs). Gazprom owns shares in some of the largest UGSs in Europe, such as the 4.2-BCM-capacity Rehden UGS in Germany and the only functioning UGS in the Baltic states, the 4.4-BCM-capacity Inčukalns in Latvia.[108] Gazprom also owns shares in the 2.6-BCM-capacity Haidach UGS in Austria, which is the largest in the country and second largest in Central Europe, as well as in the 0.45-BCM-capacity Banatski Dvor UGS facility in Serbia.[109] In addition, Gazprom owns a lease of 40 percent of the massive 4.1 BCM Bergermeer UGS facility located in the Netherlands, and in 2016 it built the 0.448 BCM Damborice UGS facility in the Czech Republic together with the Czech oil and gas company Moravské Naftové Doly (MND) Group.[110]

Furthermore, in the 2000s, Gazprom also sought to acquire shares in the national gas companies of new EU members. It came to own more than a 30 percent share in the three national gas distribution companies in the Baltic countries, although in Lithuania and Estonia Gazprom was forced to divest these shares in the 2010s. In Latvia, although Gazprom had agreed in 2016 to comply with EU regulations and sell its shares of gas company Latvijas Gaze, there is still uncertainty if it will actually do so.[111] Gazprom still owns 50 percent of the shares in Bulgarian Overgas, a private monopoly company that imports Russian gas, but it 2015 the Finnish state bought out Gazprom's 25 percent of the shares of the national gas company, Gasum Oy.

Gazprom's role in Western Europe has been mostly limited to participating in gas-trading companies like WINGAS and Wintershall Erdgas Handelshaus (WIEH) in Germany and Wintershall Erdgas Handelshaus Zug (WIEE) in Switzerland, as well as establishing numerous subsidiaries, such as Gazprom Germania GmbH and Gazprom Energy France. Wintershall and WINGAS, in particular, are exemplary of the complex ownership structures Gazprom created with companies that sought easier access to Russian gas. Wintershall is Germany's largest oil and gas producer and a subsidiary of German chemical giant BASF. In the early 1990s, BASF sought to secure Russian gas companies as suppliers in order to ease pressure from the monopolistic, supraregional gas supply company Ruhrgas AG.[112] This cooperation produced a number of joint ventures and subsidiaries such as W&G—a joint venture between Wintershall and Gazprom—which in turn came to own other subsidiaries in Eastern Europe. In 2012, the two companies announced plans for an asset swap that would give Gazprom Wintershall's shares and full ownership of the gas-trading company and of Germany's second-largest gas importer, WINGAS. Gazprom would also receive another Wintershall subsidiary and gas-trading company, WIEE, plus 50 percent shares in Wintershall Noordzee B.V. (WINZ), which conducts exploration and production operations in the North Sea.[113] In exchange, Gazprom had to grant Wintershall access to the Urengoi gas field in West Siberia.[114] However, following the Ukrainian crisis and political tensions between Russia and the EU in 2014, BASF announced that, due to the "difficult political environment," it would not go through with the asset swap.[115]

Some of Russia's and Gazprom's relationships with European energy companies go back half a century. Gazprom has formed particularly close partnerships with Germany's E.ON (previously Ruhrgas), Italy's ENI, and France's Gaz de France (GdF), using these relationships to enter downstream European gas markets. Germany is Gazprom's largest European customer, with E.ON, through its predecessor Ruhrgas, having been a historic importer of Russian gas since 1969. E.ON Global Commodities SE (known as E.ON Ruhrgas AG until 2012) imports around 60 percent of Germany's total gas consumption.[116] E.ON had been the largest foreign shareholder of Gazprom starting with its first stock acquisitions in 1998 and continuing up to 2008 when it owned 6.4 percent of Gazprom.[117] In 2008, however, E.ON and Gazprom agreed to an asset swap that resulted in E.ON's stake in Gazprom dropping to 3.5 percent.[118] Two years later, E.ON sold the remaining 3.5 percent stake in Gazprom seemingly due to a disagreement over Gazprom's oil-linked gas pricing, which has likewise been a source of tensions with Germany, France, Italy, and Turkey.[119]

Despite this episode, E.ON and Gazprom's close relations were affirmed when a member of E.ON's management board, Burckhard

Bergmann, who had also served on Gazprom's board of directors since 2000, was appointed advisor for international affairs for Gazprom's CEO, Alexei Miller, in 2011.[120] The relationship between Gazprom and E.ON reminded some scholars of a "bilateral monopoly between a producer and a buyer / reseller" wherein both companies sought to dominate German markets.[121] E.ON has also invested, together with Gazprom and other Russian gas companies such as Itera, in the national gas companies of the Baltic states during the 1990s and 2000s.[122] Meanwhile, Gazprom's relationship with German companies has also been influenced by Russia's need for foreign technology and investments, as can be seen from the fact that both E.ON and Wintershall have stakes in the Yuzhno Russkoye gas field in Siberia. With respect to these two partners, E.ON has become for Gazprom its major foreign stakeholder, while Wintershall has emerged as a major partner in exploration and production.

After Germany, Italy is Gazprom's second largest customer in Europe, and ENI has been importing Soviet and Russian gas since the late 1960s.[123] After the Cold War, Gazprom continued to expand into the Italian market and deepen its relationship with ENI. In the 2000s during the premiership of Silvio Berlusconi, who developed a close relationship with Putin somewhat similar to that of Germany's Schröder, closer energy ties between Italy and Russia were established. In 2006, in one of the defining moments of the relationship between Gazprom and ENI, a strategic partnership agreement enabled Gazprom, beginning in 2007, to directly supply the Italian consumer market, including by way of power plants (among others).[124] The partnership also seemingly facilitated ENI's 2007 purchase of shares of former Russian energy company, Yukos, which had license to develop gas fields in West Siberia. Yukos had earlier belonged to Putin's political opponent, the jailed oligarch Mikhail Khodorkovsky, and the bankruptcy sale of the company's assets was tinged with controversy. When ENI acquired these assets, the resulting company was renamed SeverEnergia, whose 51 percent shares ENI immediately offered to sell to Gazprom.[125] Two years later, in 2009,

Gazprom finally made the acquisition, which drove speculation that ENI's initial acquisition was only made to facilitate Gazprom's takeover.[126] In that vein, others have argued that the ENI-Gazprom partnership and dealings were all part of Moscow's effort to gain Italian support for the Russian planned South Stream project.[127] In May 2014, ENI and Gazprom signed a new agreement that indicated a major concession for the Italian company as Gazprom agreed to use spot-market gas prices in long-term contracts.[128] Gazprom had long resisted this change in the gas pricing mechanism in Europe, so awarding it to ENI both represented a major prize for the Italians and the good relations between ENI and Gazprom, and also signaled new pricing pressures in the changing global gas market.

France's second largest gas supplier, Gaz de France (GdF) (after the 2008 merger with Suez, renamed GDF Suez), has been the main importer of Soviet, and later Russian, gas to France. The hallmark of the modern relationship between GdF and Gazprom was their signing of a memorandum of further cooperation in St. Petersburg in 2003, the initial results of which came in a 2005 agreement on the swap of Algerian LNG for Russian pipeline gas. Gazprom had to supply GdF extra gas in exchange for a shipment of LNG from the Med LNG, a GdF joint venture with Algerian Sonatrach.[129] The deputy chairman of Gazprom's management committee, Alexander Medvedev, described the deal as "strengthening Gazprom's position in the traditional natural gas market in Europe."[130] Further agreements in 2006 granted Gazprom the right to directly deliver gas to French consumers in the amount of up to 1.5 BCM annually. Additionally, GDF Suez agreed to the procurement of extra gas volumes by 2010 once the Nord Stream pipeline became operational. Besides granting Gazprom access to French downstream markets, these deals contributed to the importance of the Nord Stream project beyond Germany and increased the dependence of France on Russian supply.[131] Miller described these agreements as "vivid example(s) of Gazprom's successful implementation of its strategy aimed at entering the final consumer market in European countries."[132] The CEO of GDF Suez, Jean-François Cirelli, referred to the deal as improving "gas supply security" and stated that "deeper business cooperation with Gazprom stimulates strengthening the gas supply security for our consumers in France and Europe in a long-term perspective."[133] He largely maintained this view as late as 2014.[134] Unlike in Germany and Italy, though, the strengthening gas connections between Moscow and Paris did not result in closer relations between the leadership in Élysée Palace and Putin.

Meanwhile, although UK-Russia energy relations are less seeped in history than those of Russia and Germany, Russia and Italy, or Russia and France, they have geared up in the last decades. In 2003, British oil and gas giant British Petroleum (BP), together with Russian businessmen, created the joint venture Tyumenskaya Neftyanaya Kompaniya–British Petroleum (TNK-BP); then, after its sale at the end of 2012, BP acquired 19.75 percent of Rosneft as part of the deal.[135] Moreover, a Gazprom subsidiary in the United Kingdom supplies around 14 percent of country's annual gas demand, and this dependence will increase by at least 5 percent more, given Britain household supplier Centrica's deal with Gazprom in 2015.[136] Britain's "Brexit" vote in July 2016 to leave the EU creates an opening for Putin to exploit a weaker Europe and to potentially reset London-Moscow relations without the previous constraints of Brussels.

Despite the geopolitical undertones of Gazprom's ventures into some of the EU's largest economies, commercial and economic pressures as well as a slow but steady EU gas market restructuring have fundamentally altered the relationship between Gazprom and its major European markets. Echoing Stern's work, energy scholars James Henderson and Tatiana Mitrova explain that the widespread number of arbitration cases in the years 2014–2015 taken by many of Gazprom's most important partners, including E.ON, had profound effects on Gazprom's pricing strategy and, therefore, its relationships with EU energy providers. Renegotiations were highly confidential but marked a departure from the oil-linked prices that underpinned Gazprom's business model.[137]

The Future of Russian Gas and Gazprom

The future for Russian gas and Gazprom is filled with challenges both external and domestic, but also with new opportunities. "Russian Energy Strategy up to 2035," released in September 2015, immediately acknowledges the "main external challenge for Russia's energy sector is increased competition at external energy markets." In light of these constraints, the Energy Strategy lays out a clear set of priorities: (1) "stable relationships with traditional consumers of Russian energy" (such as the European markets), (2) faster "access to the Asia-Pacific market," (3) "export diversification" (ostensibly through larger volumes of LNG exports), (4) "integration of Russian companies into the international energy business," and (5) "formation of a common energy market for Eurasian Economic Union members."[138] While Russia's position in the European markets will be detailed in Chapter 4, this one considers obstacles to the long-term success of Russian gas and Gazprom in their agenda—namely, Western sanctions, Gazprom's domestic difficulties, and pressure on gas prices—and then goes on to look at new opportunities in Chinese markets, LNG exports, and unconventional and Arctic resources.

Western Sanctions

Russia's annexation of Crimea in March 2014, its subsequent military support of secessionist rebels in eastern Ukraine, and other war-related incidents were met with multiple rounds of US and EU sanctions against Russian government officials and businesses. Many of these sanctions directly and indirectly targeted the Russian energy sector and its participating companies and businessmen, and their fallout will continue to constrain Russian gas interest in the medium to longer term. However, the limits of sanctions also demonstrated the leverage Russian energy continues to have in Europe. Sanctions did not include restrictions on the purchase of Russian gas or oil, as this would have been difficult to implement and enforce within the EU. Similarly, while the blacklist included Igor Sechin, chairman of the world's largest traded oil and gas company, Rosneft, they did not include the head of Gazprom, Alexei Miller, possibly because the EU feared gas supply disruptions.[139]

The sanctions primarily restricted financing of, or providing financial services to, Russian companies (including energy companies), thus impacting their ability to raise capital, as well as blocking the export of items and technology that could advance energy exploration.[140] Among the impacted companies were Gazprom, Gazprombank (financing Gazprom), and Russia's second and third largest gas producers, Novatek and Rosneft, respectively,[141] as well as other Russian oil players such as Gazprom Neft and Surgutneftegas.[142] In addition, many of the EU and US energy sector sanctions against Russia were targeted at deep-water drilling capabilities, specifically Arctic energy source explorations, and shale projects,[143] while the December 2014 round of EU sanctions also outlawed providing technical assistance to Crimean companies engaged in energy exploration or production in and around the occupied peninsula.[144] Overall, however, the EU showed less inclination for tough energy sanctions. In 2015 and 2016, Gazprom successfully continued to make deals with European companies and in September 2015 agreed to terms on the construction of Nord Stream II.[145]

Nonetheless, Western sanctions have impacted the Russian economy and, indirectly, the Russian energy sector. Since 2014, Russia's economic and fiscal numbers have been dismal while inflation has soared. In 2014, official capital flight from Russia totaled more than $130 billion, but estimates of unofficial sums were much higher.[146] In 2015, capital outflow continued at the same rate. Prime Minister Medvedev stated that Western sanctions had cost Russia $26.7 billion in 2014.[147] A US State Department official estimated that sanctions were responsible for 1.0 to 1.5 percent of the 3.8 percent that Russian's economy shrunk in 2015.[148] Moreover, in 2015, Russia faced its lowest trade surplus since 2010 and the lowest export level since 2009.[149] Finally, in 2016, Russia's economy has contracted by another 0.8 percent, and only in 2017 is it expected to demonstrate some modest growth.[150]

The impact of the Western sanctions had more bite, considering that they coincided with a period of low global energy prices, creating the "perfect storm" for the Russian economy. The low global oil prices of 2014, 2015, and 2016 brought more woes because oil and gas are the backbone of Russia's economy, contributing more than half of all government revenues. Crude oil prices continued to fall to record lows over these years, down to just under $30 per barrel in early 2016. Inasmuch as sanctions and low oil prices have had a constraining effect on the Russian gas sector, they are not permanent conditions. Moreover, the sustained lower oil and gas prices have also hurt Russian gas competitors and could likely impact the strategies of US and European energy companies as well as others, possibly slowing down investment in diversification projects and the rate and volumes of American LNG exports.

Gazprom's Domestic Woes

In this context of Western sanctions and low prices, Gazprom has been facing its own difficulties as well. While the 2000s were a boom time for the company, the 2010s have been less favorable. In 2014, Gazprom's net profit declined by 86 percent, and in 2015, its revenues from Europe sunk to the lows of the decade.[151] The shortfall since 2014 has been attributed largely to the collapse of the ruble's value, which increased Gazprom's net debt, but other factors have been critical as well, including lower global energy prices, declining production and declining exports in light of the conflict in Ukraine, warmer winters in Europe, the EU's reduced consumption, and its energy efficiency and renewables policies.[152] Gazprom's expansionist strategies and overspending, such as the $40 billion spent since the 2000s on what commentators have regarded as unnecessary politically driven infrastructure projects, have also compounded its problems.[153] Moreover, during the 2000s, Gazprom was forced to commit to high-cost gas fields in the Yamal peninsula to make up for both declining production in its core fields and the fact that accessing Central Asian gas became more expensive.[154]

Another problem is that Gazprom's and Russia's gas interests have been starting to show some signs of divergence, especially as the company is facing new competition from other Russian gas producers. The greatest challenges come from the largest independent natural gas producer, publicly traded Novatek, and state-owned Rosneft, both of which are extremely close to the Kremlin. Novatek is owned by Putin's close friend Gennadiy Timchenko, while Rosneft is led by an insider of the Kremlin's security services faction, Igor Sechin—a man many consider to be the second most powerful person in Russia. Comparing Gazprom's, Novatek's, and Rosneft's gas output best illustrates the changes underway. Between 2008 and 2014, Gazprom's gas production for domestic sales went from 290 to 238 BCM while Novatek and Rosneft together nearly doubled their production, from 63 to 111 BCM.[155] As a result, Gazprom's share in domestic gas production has been squeezed from 95 percent in 2000 to 76 percent in 2014.[156] This trend continues, with independent Russian producers (including, in addition to Rosneft and Novatek, Lukoil, Gazprom Neft, and international stakeholders) planning to double their output by 2020 to 330 BCM per year. This increased domestic competition has forced down Gazprom's output estimates. As recently as 2012, Gazprom was forecasting production levels of 650 BCM by 2020, but that is becoming increasingly unlikely without major changes in export forecasts. Even if markets for independents are limited to domestic consumers and foreign purchasers of LNG, Gazprom's future production strategy could still be significantly impacted.[157]

These rivals would also like to gain access to Gazprom-owned export pipelines, but it is unlikely that Gazprom's monopoly on exports will be broken any time soon. Thus, in 2015, when Minister of Energy Novak was asked about such prospects, he responded definitively that there was no basis for such an idea, although he supported the general trend of liberalization in energy policy and cited "liberalization of export of liquefied natural gas" as an example of liberal achievements.[158] Already in 2013, the government approved amendments to gas export laws, which would free other companies, such as Rosneft and Novatek, to pursue

LNG export facilities.[159] But liberalization of piped gas exports are unlikely because competition among gas suppliers would inevitably push prices down and result in lower revenues for the Russian state budget.[160] At the same time, even if Gazprom were somehow to maintain its monopoly on exports, the company has questionable potential for growth due to its uncompetitive practices, which are common among nontransparent state monopolies.

Gazprom is also bearing the financial brunt of heavily subsidized Russian domestic residential gas consumption, a carryover from Soviet days. Gazprom has continuously charged below market gas prices domestically—for instance, in 2014, Gazprom's price on the domestic market was a third of the export price.[161] Viewed by most Russians as a public good, low natural gas prices within Russia are a political necessity, but they are hurting Gazprom's bottom line and are thus making it harder for the company to compete in other spheres with independent gas producers that do not have the same revenue drains.[162] While Gazprom and the Russian government had planned to raise domestic prices to market or export levels by 2015 to achieve "equal profitability in the foreign and domestic markets," Prime Minister Medvedev signed a decree that postponed this "equal profitability of gas prices" until 2017, most likely due to the deteriorating Russian economy since 2014 and its effects on public sentiment.[163] While Gazprom has been subsidizing the domestic market, it is also increasingly facing pricing pressures and lower margins in its main export markets. However, even though the future looks less favorable to Gazprom, the company's woes are at least to some degree separate from the pressures facing other Russian gas producers.

The Future of Russian Gas Prices and Contracts

As the main current exporter of Russian gas, Gazprom is the primary company impacted by increasing downward pressure on Russian gas prices. The declining oil and gas prices since 2014, EU's diversification efforts, the broader transformation of global gas markets, and the boom in gas trade are hurting gas prices as well as challenging the traditional gas pricing models and contract schemes. Gazprom has long tried to resist delinking the price of gas from oil, shifting from long-term contracts to hub-based pricing, and removing destination clauses for exports. At the same time, rising gas production due to shale gas, abundant LNG volumes in the global markets, new LNG import / export terminals, and the ability for countries to reverse gas flows make it increasingly difficult for Gazprom to exact high gas prices from its European customers. For instance, Lithuania's new LNG terminal has strengthened the bargaining position of the country vis-à-vis Gazprom, and in mid-2014, for the first time ever, it negotiated a discount for its long-term gas contract with Gazprom—some six months before its Klaipėda LNG import terminal was even completed.[164] Overall, Russian gas has already witnessed a significant drop in the 2010s as the price for its European consumers fell from $403 per thousand cubic meters (tcm) in 2012 to $238 per tcm in 2015, the lowest level in eleven years. Moreover, according to some estimates, in 2016 average gas price in Europe could drop to just below $200.[165]

However, Gazprom's position in Europe should not be underestimated. For the near and mid-term, Gazprom benefits from long-term "take or pay" contracts that commit Organisation of Economic Co-operation and Development (OECD) countries to no less than 110 BCM per year of Russian gas imports as far out as 2035.[166] With existing pipeline infrastructure and under many market conditions, Russian piped gas would still likely be cheaper than the price of LNG imports to many European countries. Indeed, analysts and the media started speculating that, as a low cost producer with spare production capacity, Gazprom could launch a "price war" by lowering prices to defend its market share in Europe from competing supplies such as American LNG exports.[167] The Russian gas giant has already proven that it is willing to make concessions on oil-linked prices and pricing models with its EU customers, which would impact its profits. In January 2015, because of the changing market conditions, Gazprom amended its supply contract with Austria's ÖMV and, reportedly, changed the pricing model to one based on spot market prices.[168] By September 2015, Moscow even offered gas price discounts to Ukraine, which will receive gas at European prices instead of at the elevated prices that resulted from political tensions, and in 2016, Russia has agreed with Armenia to reduce the cost of gas by 9 percent.[169] Furthermore, in 2016, Gazprom made an agreement with France's ENGIE to decouple gas indexation from oil prices and link the pricing to natural gas hubs.[170] Russia's future pricing strategy may include flooding European markets to thwart investment in new LNG import projects.[171] Nonetheless, lower prices will hurt Gazprom's and Moscow's bottom line.

Thus, Gazprom's ability to increase prices or exert monopolistic power over its markets looks increasingly curtailed in view of competition and abundant supplies in the gas markets. Looking forward, it is clear that Gazprom will be more constrained in its politics of supply and will likely have to make long-term changes in its gas prices and pricing models. Either way, the domestic and international pressures together are signaling that the era of Gazprom's dominance is coming to an end and that the era of Russian gas dominance in the European gas market is likewise being challenged. Will Gazprom and / or Russian gas fare better in new markets, in LNG exports, or in accessing additional sources of gas?

New Markets: China

In view of the many global and domestic pressures on Gazprom and Russian gas, China has emerged on Moscow's agenda as the main alternative market to Europe. Back in 2006, Putin and Chinese president Hu Jingtao signed an agreement to build two strategic pipelines from Russia to China: eastern (Power of Siberia) and western (Altai), both expected to deliver 68 BCM of gas annually.[172] After eight years of negotiations, in 2014 Russia and China agreed on a $400 billion deal for Gazprom to supply China National Petroleum Corporation with 38 BCM of gas annually for thirty years beginning in 2018 via the Power of Siberia pipeline.[173] According to Miller, this deal was "the biggest contract in the entire history of the USSR and Gazprom—over 1 trillion cubic meters of gas will be supplied during the whole contractual period."[174] However, as Chapter 7 will elaborate, Russian gas exports to China still face numerous hurdles, including delays in construction. The completion of the Power of Siberia pipeline is planned for 2019, while the Altai project was postponed seemingly indefinitely. In the end, it is highly debatable whether Russian gas will find its new main market in China; instead, it may serve merely to close the gap between China's supply and demand. As of now, neither China nor Asia more broadly appears to be the panacea for Gazprom's troubles or a guarantee of the dominance of Russian gas in the new era of changing global gas markets.

New Frontiers: Russian LNG Exports

While access to new markets in Asia or maintaining a monopolist role in Europe via piped gas looks increasingly problematic, Russia has also tried (with little success) to become a significant LNG exporter, even though its LNG industry has lagged significantly behind others. One of the main reasons for this is that historically Russian gas transportation infrastructure was designed to serve the European market, where territorial proximity made pipeline exports feasible.[175] Only in the mid-2000s did the combination of falling transportation and processing costs per unit of LNG and of increasing demand for LNG incentivize Russian leadership to speak more openly about possible LNG projects.[176] In comparison to other major LNG exporters, such as Malaysia and Qatar, which launched their operations in 1983 and 1997, respectively, Russia did not officially launch its first LNG export terminal—the Sakhalin II plant—until 2009.[177] This LNG plant, which is located offshore of Sakhalin Island in the Okhotsk Sea and receives natural gas from the Lunskoye gas field, possesses a modest annual liquefication capacity of just over 13 BCM and, since 2009, has been supplying mainly Japan and South Korea.[178]

In the future, Russia looks to boost its LNG export capacities, including to China and beyond. For instance, in 2017–2018, it plans to launch the Yamal LNG export terminal in the port of Sabetta, supplying it with gas from the Yuzhno-Tambeyskoye gas field located northeast of the Yamal Peninsula; this will bolster Russian LNG exporting capacity by about 23 BCM a year.[179] Novatek is the primary shareholder and is looking internationally, including to India, for financial partnerships.[180] Although Russia has plans for two more LNG export terminals, namely the Vladivostok terminal (with gas to be provided from the Sakhalin, Yakutia, and Irkutsk gas production centers), with a planned capacity of nearly 21 BCM, and the discussed Baltic LNG terminal, with a planned annual capacity of nearly 14 BCM, their future is far from certain.[181] In 2014–2015, amid Western sanctions and low global energy prices, Russian energy companies found it more difficult to develop their LNG exporting capabilities—as exemplified by the indefinite postponement of the Vladivostok LNG terminal, initially scheduled for completion in 2018.[182] Even the Yamal LNG terminal, which should be online in the end of 2017, is experiencing financing difficulties due to the sanctions. Thus, the French energy company Total SA, which acquired a 20 percent stake in the Yamal LNG terminal, is borrowing as much as $15 billion worth of funding for the project in yuan and euros through Chinese banks in order to avoid Western sanctions that affect dollar-based fundraising.[183] The shortage of technological know-how and of advanced equipment, which is manufactured mainly by Western companies (especially for liquefaction), are other hindrances. For the near to medium term, there are too many hurdles for Russia to cross for it to become a significant LNG exporter, meaning it will remain largely dependent on its piped gas exports to Europe.

Unconventional and Arctic Resources

Despite the setbacks in LNG exports and in gaining new markets in Asia, Russia will continue to play a significant role as a gas producer in the global context especially in light of its determination to develop Artic resources and its unconventional gas reserves. Russia probably has larger shale gas reserves (8 TCM), which are concentrated in the

Bazhenov reservoir of West Siberia, than any European country.[184] In Eurasia, only China's reserves are larger—by four times. However, Gazprom has not developed its shale reserves, stating in 2013 that at least for the next decade "its vast conventional reserves are much cheaper to produce."[185] In reality, the technology, know-how, access to capital markets, and land mineral rights, which, among other factors, have made the US shale boom a success, are still lacking in Russia.

Instead, Russia has fared better in another energy frontier of the Arctic, which boasts 30 and 13 percent of the world's gas and oil reserves, respectively.[186] The countries bordering the Arctic—the United States, Canada, Russia, Denmark, Norway, and China—are increasingly competing over the Arctic territories and their resources. Russia seems to be leading this geopolitical and energy race. In 2015, the Russian deputy prime minister in charge of the defense industry, Dmitry Rogozin, called the Arctic "the Russian Mecca," and Moscow resubmitted its claim to the United Nations to broaden its continental shelf in the Arctic Circle.[187] Indeed, for Russia, its Arctic territories not only make up around a quarter of its continental land, but also hold 90 percent of the nation's estimated gas reserves and 60 percent of its oil reserves and could generate about 30 percent of Russia's GDP in ten to fifteen years.[188] As Konstantin Simonov, director of the Moscow-based National Energy Security Fund, stated, "Gas fields discovered in the 1960s laid the foundation for Russia's decades-long dominance of the global natural gas market. Now these Soviet era giants are in decline. Moving farther north into the Arctic is the next logical step."[189]

In December 2013, Gazprom started to produce oil in the Prirazlomnoye field located in the Arctic offshore and, despite Western sanctions, brought its second well into production in October 2015.[190] In 2014, Gazprom also started the flows of Arctic gas from Bovanenkovo's natural gas deposit in the Yamal Peninsula, which by 2030 is expected to supply a third of Russia's gas production.[191] Overall, Gazprom owns some thirty field licenses in the Russian Arctic shelf.[192] However, Western sanctions as well as low crude oil prices have somewhat slowed down Russia's efforts in the Arctic. Despite talks of enlisting Chinese companies, Rosneft has postponed drilling its second well in the region until at least 2018.[193] At the same time, however, Moscow's energy and other interests in the Arctic are already protected by some 6,000 Russian soldiers, with additional plans to create a united fleet of combat ships and submarines permanently deployed in the Arctic.[194]

The largest barrier to Russian prospects for Arctic oil and gas development is cost. The extraction cost of Arctic oil is, on average, $75 per barrel, while break-even points are around $100 per barrel. Even under more amenable conditions for extraction, such as when oil traded for over $100 per barrel and natural gas for over $13 per MMBtu in the early 2010s, the margins for oil and gas extraction proved risky.[195] Well before the current drop in energy prices, many analysts were certain that Russia's Arctic development could succeed only with high oil-price expectations and that gas discoveries would be worthless due to the high costs of development.[196] Still, although market conditions are certainly slowing down the rate of new Russian Arctic hydrocarbon projects, they are by no means halting Russia's desire to pursue new development areas to compensate for dwindling Siberian natural gas and oil fields. For instance, Moscow offers special tax rates for offshore Arctic development, and Rosneft still plans to invest around $400 billion in Arctic offshore programs through 2036.[197] As Russia's Natural Resources Ministry explained, "The Arctic shelf, despite some project delays related to oil prices, remains a strategic direction for development."[198]

In contrast to Russia, other Arctic states have been more ambivalent. In October 2015, the United States canceled its two lease sales for offshore oil and natural gas exploration in the Arctic for 2016 and 2017. China is only investing in Arctic research and building icebreakers.[199] Without a doubt, developing the Arctic is crucial for Russia's ambitions to remain an energy superpower, especially as it has not succeeded in unconventional gas production. If it fails, over time it will increasingly lose its influence in the global gas markets and the geopolitics of gas to the United States.

Conclusion

The dominance of Russian gas in Europe and the post-Soviet space in the second half of the twentieth century demonstrates how in import-dependent gas markets, being a gas supplier (and a monopolist supplier in some markets) awarded Moscow both sizable commercial power and political influence. Using its control over other countries' access to Russian gas, Moscow was able to assist its friends and punish its enemies. At the same time, Gazprom's behavior in European markets was shaped both by the company's commercial considerations and by Moscow's political objectives. This behavior oscillated between two (at times seemingly contradictory) goals: on the one hand, of maximizing revenues, and on the other, of maximizing political and economic influence over importing states. However, it was when their commercial and political interests coincided, that Gazprom and Moscow made some of their most important and strategic decisions and gains.

Going forward, the Russian gas sector will continue to reflect the goals of the Kremlin. However, while revenues and political clout will remain important to policy, conditions in the global gas markets have changed. Gas is abundant, competition thrives, and Gazprom's traditional captive markets are now successfully diversifying while also reducing their overall demand. Although seeking new markets in Asia, increasing LNG exports, and harnessing unconventional and Arctic resources are priorities for Moscow, all are proving difficult or costly to achieve. Thus, as the global gas markets change and become more integrated, the future of Gazprom and Russian gas will depend largely on how well they can adapt to and compete in these markets, where the politics of supply will hold less sway and the politics of demand will determine gas trade relations.

4 The Politics of Dependence Transformed: Europe

Europe—the European Union (EU) together with its neighboring states—is the largest importer of gas in the world. In 2015, Europe imported almost 461 billion cubic meters (BCM) of natural gas while the EU imported around 408 BCM, which was three times more than the world's second largest importer, Japan, and seven times more than China.[1] The EU's neighbor, Russia, is the world's largest exporter of natural gas and has both quenched its energy thirst and provoked energy security concerns. While the Russian gas sector demonstrates the dynamics of the politics of gas supply, the EU has long represented the politics of dependence in its relationship with Russia. However, over the last decade the EU has started to leverage its buying power to start slowly transforming its vulnerable state of dependence into a power of demand. This chapter will focus on the twenty-eight EU member states, but also discuss other relevant countries like Norway, which is a European Economic Area member (though not an EU member), and neighboring EU aspirants like Ukraine and Turkey. The lessons learned from EU states can also be applied to other countries that are dependent for gas on potentially unreliable or even hostile countries. The EU demonstrates not only the vulnerabilities of gas importer states, but also how both supplier and consumer nations experience and manipulate the commercial import-export relationship and the political strategies that accompany it.

The EU is unique relative to the other case studies selected because it is neither a nation-state nor an international organization. Rather, it is primarily an economic union of states, aggregating functions formerly reserved for national governments, with internal politics, foreign policy, and energy relationships continuously evolving. Because of those features, EU integration theorists call the union sui generis—one of a kind.[2] Each of the twenty-eight member states has its own national interests that often do not align with the others, especially with regard to energy policies, the gas sector, and their relationship with Russia. The EU is marked by numerous divides—Western and Eastern, northern and southern, old and new members, gas producers and gas importers, gas-secure states and gas-vulnerable ones, those favoring green energy and those favoring shale gas development. Despite these differences, the EU has forged a new agenda for its energy sector, including plans for an energy union and an increasingly visible political and regulatory strategy to stand up to Russia's gas dominance. First, this chapter will provide an overview of the historical and contemporary developments of the European and the EU's gas markets and their relationship with Russia. Then we will focus on the EU's multidimensional agenda of regulation to achieve integration and diversification of energy and gas markets, followed by a consideration of the future driving forces or new wild cards of the EU gas sector.

The Historical Development of European Gas Markets

Historically, Europe's early urbanization and industrialization, along with its climate, have dictated the Continent's energy thirst. Gas emerged as an important source of energy both through imports that largely ramped up in the 1950s and subsequently through domestic production. This coincided with the beginnings of the economic and political project of the EU itself, which was founded in 1957 as the European Economic Community. In Western Europe, Italy was the first to discover natural gas in the Po Valley during World War II and subsequently became the largest gas market in Western Europe by the mid-1960s.[3]

A discovery of natural gas in the Groningen field in the Netherlands followed in 1959, which marked the beginning of "the natural gas era in Europe."[4] It inspired Great Britain to begin oil and gas explorations in the North Sea in the mid-1960s. By the end of the same decade, Norway's large oil and gas reserves in the North Sea were discovered.

Despite these discoveries, European countries also sought out gas imports and were in fact the earliest liquefied natural gas (LNG) importers. In the 1960s, Western European countries started importing North African LNG. The United Kingdom housed the first import LNG terminal in the world—the now defunct Canvey Island LNG terminal, which pioneered commercial LNG trade by importing LNG in 1959 in a one-off-case from Louisiana on the US Gulf Coast and, in 1964, from the Arzew LNG export terminal in Algeria.[5] Just a year later, in 1965, France was the first continental European country to import Algerian LNG.[6] Algeria-Spain LNG trade followed in 1969. Libya became an LNG supplier in the 1970s, first to Spain in 1971 and later to Italy in 1975.[7] Europe's piped gas imports from North Africa began only in 1983, when Italy and Algeria inaugurated the Transmed pipeline bringing piped North African gas to the European continent for the first time.[8]

While European countries were the first in the world to turn to LNG, their main source of piped gas would come to be the Soviet Union, and later Russia. Their pipeline development was slow, but eventually resulted in four key transportation routes outlined in the previous chapter. In brief, the first was Brotherhood, an export transmission system comprised of multiple pipelines going via Ukraine to Slovakia (then Czechoslovakia) and beyond, developed in different stages from the 1960s to 1980s. The second, built in the 1970s, was the Northern Lights, bringing Russian, rather than Ukrainian, gas via Belarus to the West. The third, the Yamal-Europe pipeline to Germany via Russia, Belarus and Poland, was gradually completed between 1994 and 2006 after the dissolution of the Soviet Union. The fourth and the newest, the Nord Stream gas pipeline, which supplies Germany via the Baltic Sea, launched in 2011 with plans for expansion announced in 2015. The impact of early Soviet pipeline systems was evident by 1980 when Norway, the Soviet Union, and Algeria accounted for 97–98 percent of the gas imports of the European Communities, totaling 52 percent, 40 percent, and 6 percent of the imports, respectively.[9] The dramatic growth of Soviet exports is even better illustrated when compared to the 1970s, when the largest exporter of gas to European countries belonging to the Organisation of Economic Co-Operation and Development (OECD) was the Netherlands (31.4 BCM), compared with the Soviet Union (1 BCM), Algeria (1.6 BCM), and Norway (still 0 BCM at the time, but set to rise to 23.7 BCM in 1980). The picture changed dramatically in 1988 when the total trade of gas to OECD Europe almost quadrupled, to 131.8 BCM, and the largest exporter became the Soviet Union (46.8 BCM). In 1990, Russian gas imports to the European states that would later form the EU-28 accounted for 56 percent of total imports. Such high dependency on a small number of energy suppliers continued until the late 1990s, when apart from imports from Russia, Algeria, Norway, and the Netherlands, the share of other gas suppliers such as United Kingdom or Qatar started to increase. By the 2000s, energy security and diversification became watchwords in the EU. At the same time, although the share of Russian gas in EU imports decreased in comparison to the Cold War era, in absolute numbers EU imports of Russian gas increased also in part due to EU enlargement. Subsequently, Russia's share of EU-28 natural gas imports declined from about 44 percent to just under 30 percent between 2003 (125 BCM) and 2010 (110 BCM).[10] The EU's gas import volumes from Russia peaked before the global financial crisis of 2007–2008, which led to a reduction in the EU's gas demand. For instance, in 2005, the European states that would later form the EU-28 imported 136 BCM from Russia, or around 34 percent of their total gas imports.[11] In 2013, the EU-28 imported nearly 126 BCM of Russian gas, which accounted for 39 percent of EU's gas imports, while Norway trailed with 30 percent and Algeria with nearly 14 percent share of imports.[12]

The Contemporary EU Gas Sector

Since the 2000s, the EU's gas balance reflects a combination of increasing import dependency, declining domestic production, high consumption needs, and the rising popularity of natural gas as a cleaner fossil fuel. The share of natural gas in energy consumption of the EU-15—the fifteen member states that made up the EU before enlargement in 2004—expanded between 1990 and 2000 from 17 percent to 23 percent. The growth was primarily at the expense of petroleum products and solid fuel, largely coal and peat, but was also partially offset by a rapid growth of renewable energy, which went from a nearly 5 percent share in gross inland energy consumption in 1990 to nearly 16 percent in the EU-28 in 2014. Since the 2000s, the relative share of gas in the EU's gross inland consumption mix has remained fairly constant at about 24 percent (even though EU enlargement in 2004 increased gas consumption in absolute numbers), and in 2014 it remained at 21 percent for the EU-28. Besides gas, only petroleum products held a greater share (33 percent) in the EU's gross inland consumption, while other sources of energy such as solid fuels including coal and peat, among others (17 percent), nuclear (14 percent), and renewables, including wood (12 percent), trailed behind.[13]

Domestic Reserves and Production

Compared to the proved gas reserves of the Middle East or Russia, Europe is not well endowed with natural gas. Europe's conventional natural gas is found (in order of largest to smallest producers) in Norway, the Netherlands, the United Kingdom, Romania, Germany, Italy, Denmark, and Poland, the first three member states being the dominant sources of domestically produced gas to the Continent. In 2015, Norway had 2.547 trillion cubic meters (TCM) of proved natural gas reserves, which was considerably more than the entire EU's proved reserves of 1.722 TCM.[14] Still, Norway's proved natural gas reserves comprised just 1 percent of total global reserves, compared to Russia (17.4 percent) and the Middle East, including Iran (42.7 percent).[15] Nearly half of EU reserves are in the Netherlands (864 BCM) and around a quarter are in the United Kingdom (407 BCM); its third largest proved natural gas reserves are in Romania (109 BCM) as of 2015.[16] But those with the greatest gas reserves are not necessarily the greatest producers. For instance, even though Germany is only the seventh in the EU with regard to proved natural gas reserves, it stands among top three producers, along with the Netherlands and the United Kingdom. Other EU gas producers include Italy, Hungary, Poland, Romania, Croatia, and Bulgaria.

Despite some recent discoveries in the Eastern Mediterranean, such as in Cyprus and Israel, natural gas production in the EU has been on the decline.[17] European gas production (if including all twenty-eight future EU member states) peaked in 1996 at 272 BCM, with production in the Netherlands accounting for 35 percent of the peak amount while UK production accounted for 33 percent. Since then, the EU's own gas production halved to around 136 BCM in 2015. Overall, the biggest drop in natural gas production occurred between 2003 and 2013, totaling a nearly 35 percent decrease. This was largely due to depleting reserves, which more than halved from 1990 to 2015.[18] In 2016, the EU's largest gas-producing country, the Netherlands, capped its production at 24 BCM annually, a figure that is unlikely to increase in the future.[19] At the same time, the EU's depleting natural gas reserves have been perceived by Russian Gazprom and other suppliers as a major opportunity. As Gazprom observed in its 2014 annual report: "Given a decline in Europe's domestic gas production due to gradual depletion of major fields and an anticipated gas demand growth in the near future, the market environment can be expected to favor Gazprom's plans of growing gas exports to European countries in the mid-term."[20]

The EU's declining gas production has necessitated an even greater reliance on imports. Yet, since the 2000s, and especially after the Ukraine crisis in 2014, the EU's relationship with its most important gas supplier, Russia, has been increasingly politically charged, reaching levels unseen since the Cold War. In this context, the EU has increasingly looked to LNG as a means to meet its appetite for gas, to diversify its gas imports, and to gain greater flexibility in the global gas markets.

Europe's LNG Infrastructure

Europe is no newcomer to LNG imports. Some states have been importing North African LNG since the 1960s. As of 2016, there were a total of thirty operational LNG import terminals in Europe (including Turkey and Israel) of which twenty-five were large-scale terminals for a total annual capacity of over 220 BCM. EU countries with LNG import capacity included Belgium, Finland, France, Greece, Italy, Lithuania, the Netherlands, Poland, Portugal, Spain, Sweden, and the United Kingdom.[21] The terminals are distributed relatively unevenly across the EU, with the vast majority of them concentrated in southern Europe and Western Europe, followed by northern Europe and Eastern Europe, where they are sparse.

In southern Europe, in 2016 there were seven operating LNG import terminals on Spain's Mediterranean and Atlantic coasts (Gijón, Mugardos, Sagunto, Bilbao, Cartagena, Huelva, and Barcelona), three in Italy (Panigaglia, Porto Levante, and Toscana), one in Greece (Revithoussa), and one in Portugal (Sines).[22] These countries all possess extensive shore lines that can accommodate LNG terminals. Thus, it is hardly surprising that southern Europe is home to the Continent's oldest LNG import terminals, such as the Barcelona LNG terminal, built in 1968, or the Panigaglia LNG terminal, built in 1971.[23] When fully operational, the combined regasification capacity of all of the southern European LNG terminals equals about 97 BCM of gas per year and could satisfy nearly a quarter of the EU's annual natural gas consumption needs.[24] In Spain two small LNG terminals (Arico-Granadilla and Gran Canaria) are under construction while the Revithoussa terminal in Greece is currently adding a storage tank in order to be able to import more gas.[25]

With the launch of a new, medium-sized LNG terminal in Dunkerque in July 2016, France became the country with the highest concentration of LNG infrastructure in Western Europe, with four import terminals (Fos-Tonkin, Montoir-de-Bretagne, Fos Cavaou, and Dunkerque). The United Kingdom, the pioneer of LNG imports, follows France with three terminals (Isle of Grain, Dragon, and South Hook), and is also home to the largest LNG import terminal in Europe, South Hook, which has a maximum regasification capacity of 21 BCM. Belgium is about to finish a significant expansion of its only LNG terminal in Zeebrugge in 2018. Meanwhile, at the end of August 2016, the Netherlands expanded its Gate LNG Terminal in Rotterdam. On the whole, the current maximum combined regasification capacity of Western European LNG import terminals is nearly 116 BCM per year, which could satisfy more than a quarter of EU natural gas demand. Thus, together, southern and Western European LNG import terminals could theoretically meet half of the EU's gas demand.

Northern Europe's LNG import capacity is dominated by small-scale terminals such as those in Finland, Sweden, and Norway, which could not accommodate large tankers and thus could not import LNG from the United States and other big global suppliers. Meanwhile, Poland and Lithuania have recently completed larger-scale terminals that will allow them to take part in the global gas markets. Lithuania has one new, medium-sized LNG import terminal (FSRU Independence).[26] The FRSU Independence is of strategic importance not only for its host country, but also for Latvia and Estonia, because it will enable the three Baltic states to access alternatives to Russian gas for the first time ever.[27] In 2016, the share of LNG in gas supply in Lithuania was expected to reach 60 percent. The region's LNG importing capacity was similarly reinforced in 2015 with the completion of Poland's Świnoujście LNG terminal. What makes northern Europe stand out on the European continent, however, is that it is also home to the only LNG *exporting* terminals in Europe. Norway currently has five such operational liquefaction terminals, although four of them are small scale and are used only to meet local demand. The largest is the Snohvit terminal, which has a 4.2 BCM annual export capacity.[28]

Ambitious LNG Plans

While Europe's LNG import capacity appears relatively modest in comparison to Asia, a number of EU countries have ambitious plans to significantly bolster their LNG infrastructure. Optimism, however, should be tempered because it is uncertain when some of these planned projects will be realized, or if at all. Whether they will be or not depends on market conditions, political will, and commercial priorities, which must balance risk in a changing global LNG market with placating vested interest groups. Moreover, many of these plans have been around for a while and are not perceived as viable by energy industry experts, who point to the current underutilization of existing European LNG terminals.[29] At the same time, after a number of years of decline, European net LNG imports went up by 15.8 percent in 2015, in comparison to 2014, returning to the level they had reached in 2005.[30] This was in part due to Europe's absorption of Asia's unwanted cargoes where demand slowed. Ultimately, the success of these costly long-term import projects and the ability to guide them from start to finish will require sizable investment, bureaucratic and institutional capacity, and stability within each member country.

As of 2016, there are plans to build twenty-one new LNG regasification facilities in the EU, with most of them located in southern European countries, including Italy (Falconara Marittima, Porto Empedocle, Gioia Tauro, Trieste), Greece (Aegean Sea, Alexandroupolis), Croatia (Krk), and Malta (Malta LNG), followed by Western European countries, including the United Kingdom (Port Meridian, Anglesey), France (Fos Faster), Germany (Rostock), and Ireland (Shannon).[31] A significant number of regasification terminals are also planned for Sweden (Göteborg, Gävle) and Finland (Pansio Harbour, Finngulf), followed in the East by Estonia (Muuga, Padalski), Latvia (Riga LNG), and Romania (Constanta).[32] Many of these planned LNG terminals, such as those in Finland, Estonia, Latvia, Romania, and Croatia, would further reduce Russian gas dominance in the region, especially in countries where Gazprom has 100 percent market share.[33]

If these terminals are ever completed, at full capacity they would provide additional regasification capacity of around 100 BCM per year, and together with the existing LNG import capacity, would meet more than half of the EU's gas needs. Total European LNG import capacity (including border non-EU states like Morocco, Ukraine, and Albania) is expected to reach around 280 BCM by 2019.[34] Whether LNG infrastructure will develop based on the optimistic or conservative scenarios will depend on market conditions and geopolitical developments.

The EU-Russia Relationship: The Politics of Dependence

The relationship between the world's largest importer of gas and the world's largest exporter plays a central role in the current geopolitics of gas. Russian gas made up nearly one-third of the EU's gas imports (30 percent) in 2014.[35] During the preceding decades, despite being a huge gas importer, the EU failed to leverage its power of demand, largely because it failed to negotiate with one voice for its gas supplies. As a result, large suppliers like Gazprom were able to extract concessions from various EU markets and leverage their power of supply over the divided individual countries.

Differences among EU members will likely persist in terms of their dependence on Russian gas, as will differences in degrees of energy security and energy vulnerability, complicating efforts to develop a common approach. Such differences became acute after the EU's post–Cold War rounds of enlargement in 1995, 2004, 2007, and 2013, when it took in former border states, former Soviet satellite states, and former Soviet republics into its fold, all of which had much higher rates of dependency on Russia gas given their long-standing energy ties to Soviet-era infrastructure. This created a new dividing line between "old" and "new" EU member states and significantly changed the energy security posture of the EU. At the same time, EU gas dependence on Russia, highlighted by the vulnerability of new EU member states, has gradually strengthened the political will in Brussels and across European capitals to focus on energy security and diversification. It also reiterated the somewhat divergent goals of functioning markets and hopes for negotiations with "one voice."

Several flashpoints in the 2000s contributed to a reevaluation of the EU's perception of Russia as a reliable gas supplier and marked a new stage in EU-Russian gas relations. The first wakeup call came at the end of 2005 and the first days of 2006, when tensions between Russia and Ukraine over gas prices flared. The dispute resulted in Russia decreasing gas supply to Ukraine in early January and thus momentarily halting EU gas deliveries transiting through pipelines on Ukrainian territory.

Although no EU countries were completely cut off from supplies, many that received gas via Ukraine, such as Austria, Croatia (candidate member at the time), Germany, Hungary, Italy, Poland, Romania, and Slovakia, reported a drop of gas volumes.[36] When Kyiv and Moscow finally reached a gas agreement in January 2006, the Austrian minister for the economy and president-in-office of the EU Energy Council, Martin Bartenstein, reflected on the fragility of European gas supplies: "We must discuss what has happened over recent days very seriously, as many European countries suffered a reduction in their gas supply, for the first time in forty years."[37] And although most of the EU gas importers did not feel the consequences of this mainly Russo-Ukrainian dispute, Brussels's perceptions of the reliability of gas deliveries from Russia via Ukraine were irrevocably transformed.

The late 2000s also saw Russia pursue a strategy of eliminating energy transit states and raising energy prices to market levels for the Baltic states, Ukraine, and Belarus, which were gravitating away from Russia. The resulting tensions further reinforced Moscow's reputation for abusing its position as a supplier and using energy as a political tool. For instance, later in 2006, after the Polish oil company PKN Orlen beat out competing Russian investors and acquired the Lithuanian oil refinery Mažeikių Nafta, Moscow permanently shut off crude oil supplies via the Druzhba ("friendship" in Russian) pipeline's extension to Lithuania, citing pipeline malfunctions and other technical reasons. Though Russia had commercial and arguably political motives, the incident was perceived as energy bullying.[38] The following winter, in January 2007, another dispute followed, this time between Gazprom and the Russian state oil transport company, Transneft, on the one hand and Minsk on the other hand.[39] In the midst of the argument, Belarus itself cut oil supplies to the EU via Druzhba for three days, impacting Poland and Germany.[40] Energy tensions among Russia, the new EU member states, and the transit countries of Ukraine and Belarus were becoming increasingly difficult to ignore in Brussels.

January 2009 was marked by the notorious second Russian-Ukrainian gas crisis, which lasted two weeks and impacted the supplies to the Czech Republic, Romania, Austria, Poland, Croatia, and Slovakia.[41] This time the dispute had deadly consequences. Because gas is used for household heating, a cutoff in the midst of winter resulted in at least eleven people freezing to death, including ten in Poland, where temperatures reached –20°C.[42] While the EU had not become involved in the 2006 dispute, in 2009 its tone was much harsher, and energy security rose to the top of the agenda when the Czech Republic took over the presidency of the Council of the EU—an EU institution led by each member state on a rotating basis, with that state being in charge of coordinating EU-wide policy and adopting EU law, as well as serving as the voice of EU member governments. On January 6, in a joint statement, the European Commission (EC) and the Czech presidency described the situation as utterly unacceptable. Calling for "gas supplies to be restored immediately to the EU," they demanded "that the two parties resume negotiations at once with a view to a definitive settlement of their bilateral commercial dispute."[43] Coincidentally, but largely irrespective of the crisis, later that year the EU adopted possibly its most important energy regulation, the Third Energy Package (to be discussed later in this chapter), which had been in the works for years.[44]

Still, energy tensions with Russia persisted, and almost every winter some conflict would arise. In early 2010, another dispute between Russia and Belarus over natural gas prices and transit fees reignited. In June 2010, Russia partially cut gas supplies to Belarus, which resulted in a short-term decrease of 40 percent in gas deliveries to Lithuania and Russia's own territory of Kaliningrad.[45]

With Russia's annexation of Crimea in March 2014 and the subsequent war in eastern Ukraine, gas transit via Ukraine once again became a critical issue and one in which Brussels increasingly took a more active role. After the fall of former Ukrainian president Viktor Yanukovych, Russia dramatically increased the price of gas sold to Ukraine, and then, in June 2014, when Kyiv had been unable to pay debts for several months, it cut off supplies.[46] Russia only resumed gas supplies to Ukraine in December 2014, after an agreement mediated by the EU was made with Ukraine.[47]

The 2014 crisis resulted in the EU states taking on a new role as *suppliers* of gas to Ukraine. During and even after the gas cutoff, Slovakia, Poland, and Hungary supplied gas to Ukraine via eastward reverse flows of the pipelines that feed into the Brotherhood gas pipeline system.[48] Reverse flows via Slovakia were ensured on the reconstructed Voyany-Uzhgorod pipeline, launched by Slovakia in September 2014. In response, Gazprom Export blocked reverse flows at the Velké Kapušany (Slovakia) and Uzhgorod connection points of the Brotherhood system.[49] Moreover, after the visit of Gazprom CEO Alexei Miller to Budapest, Hungary temporarily stopped reverse flows to Ukraine in September 2014.[50] From that time to March 2015 Gazprom Export made further attempts to discourage reverse flows to Ukraine by reducing gas supplies to Poland.[51] Nonetheless, EU efforts to stand by Ukraine marked a new level of engagement and concern, which went beyond securing supply for its member states.

In sum, then, the enlargement of the EU to incorporate energy-vulnerable Central and Eastern European states and the repeated halts in Russian energy exports to European states, particularly in gas supplies via Ukraine, fundamentally contributed to Brussels's increased focus on energy security and countering Russia's gas dominance.

The EU's Counter to Gazprom's Dominance: Regulation, Diversification, and Integration

In 2000, the EC's Green Paper, "Towards a European Strategy for the Security of Energy Supply," described the EU's energy partnership with Russia in words that would be difficult to imagine a decade later, stating that "the continuity of supplies from the former Soviet Union, and then Russia, over the last 25 years is testimony to an exemplary stability. A long term strategy in the framework of a partnership with Russia would be an important step to the benefit of supply security."[52] The gas wars of the 2000s proved this confidence wrong, and in the coming years Brussels launched a new regulatory agenda with a seemingly political undertone to counter Russia's and Gazprom's gas dominance in the EU, even if this aim was not explicitly stated. Starting in the late 2000s, a flurry of such initiatives were adopted, including the Third Energy Package (2009), new internal energy market legislation (2009), Security of Gas Supply Regulation (2010), and the launch of the Gazprom antitrust case (2011). The EU, above all a legal construct and a "regulatory state," relied on regulation to pursue diversification and integration of its gas markets.[53]

Overall, the EU's natural gas market regulations have two main aspects: first, they include directives, decisions, regulations, and infringement procedures that are specifically tailored for the energy sector; second, they invoke general competition law. Additionally, the EU has a broad set of policies and appropriate tools ranging from reducing demand for fossil fuels to diversifying supply sources and routes, breaking up energy monopolies and enforcing transparency of bilateral relations between member states with outside suppliers, coordinating mechanisms in case of short-term supply disruptions, and finally, creating a single energy market. Five of the most important EU regulatory and diversification policies as they pertain to the future developments of European gas markets are discussed below.

20/20/20: Renewables and Efficiency

One of the EU's first objectives, to reduce its reliance on fossil fuels, including natural gas, was related to its focus on renewable energy and efficiency. Although this initiative did not stem primarily from energy security concerns, it did become one of the reasons (in addition to cheap coal prices) for the EU's reduced demand for gas, and it had direct consequences for Gazprom. In 2008, the European Parliament adopted the European Union Climate and Energy Package 2009, also known as "20/20/20." According to the package, by 2020, the EU plans to decrease greenhouse gas emissions by 20 percent compared to 1990 emission levels, to produce 20 percent of energy from renewables, and to increase energy efficiency by 20 percent.[54] Germany emerged as the preeminent executor of this initiative, which is particularly significant because it is also the largest energy consumer within the EU. In June 2011, the German Parliament adopted an ambitious plan, known as the Energiewende ("energy transition" in German), to transform its power sector from nuclear and coal to renewables within the next four decades.[55] Indeed, within only a few years, Germany was able to produce more than 50 percent and sometimes more than 70 percent of electricity needs solely from renewables during some time periods.[56] Nonetheless, Germany relies extensively on coal for electricity when its solar and wind farms do not generate enough power. Other frontrunners in use of renewables include EU member states that traditionally relied on hydropower, such as Sweden and Latvia.

The EU's energy efficiency programs directly translate to less consumption of energy, gas included. From 1995 to 2014, the EU's energy intensity of economy indicator of all fuels, which shows an economy's energy consumption and its overall energy efficiency, improved from 173 kilograms of oil equivalent per €1,000 in 1995 to 122 kilograms in 2014.[57] In addition, in 2014 the EU agreed on new targets for 2030 consisting of a 40 percent cut in greenhouse gas emissions compared to 1990 levels, as well as at least a 27 percent share of renewables in energy consumption and at least a 27 percent energy savings compared with the business-as-usual scenario.[58] EU renewable and efficiency policies have been a significant contributor to reducing its overall demand for natural gas.

The Single EU Energy Market

Possibly the greatest regulatory and policy undertaking of EU institutions to date is the attempt to merge the separate national energy markets into a single EU market. In it, market forces would determine gas demand and supply as driven by transactions between companies in hubs and other trading platforms rather than by the long-term contracts that were increasingly perceived as inflexible at best and politicized at worst. While the EC did not go as far as directly challenging the existing system of long-term contracts, it did start to encourage non-oil-linked pricing and objected to long-term capacity reservations in cross-border pipelines. In some regards, unification of energy markets is a natural extension of the EU's origins as the European Coal and Steel Community, founded in 1952. Although the EU aims to build an internal market first, rather than to diversify from Russia, the successful implementation of this energy market will reduce the scope of the Russian gas giant's influence on the European continent by reducing the dependence and the vulnerability of isolated energy markets, particularly those of Central and Eastern European states.[59] The single energy market plan envisaged by the EU is a step-by-step process made of up three legislative packages—the First (1996–1998), Second (2003), and Third (2009)—to gradually "liberalize" energy markets by allowing a larger proportion of consumers to become eligible to choose among different energy suppliers.

An important part of the single EU energy market project is investment in infrastructure and interconnectors between countries, as well as enabling of bidirectional flow of gas in the pipelines, as was used for the first time in 2014 to supply gas to Ukraine. To encourage investments, the EU assigned funding to infrastructure deemed to be "projects of common interest" (PCI), and by November 2015, 195 such energy projects had been selected and prioritized.[60] PCIs may receive accelerated licensing procedures and access to financial support up to €5.35 billion from the EU's Connecting Europe Facility (CEF) between 2014 and 2020. As energy researcher Tim Boersma argues, the creation of an internal energy market also largely depends on the cooperation of the member states. However, such cooperation remains institutionally challenging, as countries often wait for Brussels's payouts rather than invest in such energy projects without any such financial assistance.[61]

Because most of the major pipelines run from east to west, the EC proposed the construction of north-south interconnections in Central and southeastern Europe to link gas supply sources from the Baltic, Adriatic, and Black Seas to the rest of the EU.[62] These plans award special attention to weakly interconnected or isolated EU regions, so-called "energy islands," addressing the concern that "no EU Member State should remain isolated from the European gas and electricity networks after 2015 or see its energy security jeopardized by lack of the appropriate connections."[63] A special plan for the interconnection of the Baltic states, the Baltic Energy Market Interconnection Plan (BEMIP), coordinates and financially supports such projects, including the expansion and improvement of storage facilities.[64] Such CEF cofinanced projects will enable Lithuanian and Polish natural gas systems to link by 2019, thereby giving Lithuania a third source of gas supply, in addition to its new LNG terminal and piped gas from Russia.[65] Once the Baltic energy markets are integrated, they will have the option to no longer import any gas directly from Russia. EU financing from the temporary program that preceded CEF, the European Energy Programme for Recovery (since 2009), helped to complete thirty-four gas and electricity infrastructure, storage, and reverse flow projects by the end of June 2015. To note just one instance, this funding enabled Hungary to establish gas pipeline links with Romania and thus increased connections between different Central European states of the EU.[66]

The buildup of interconnecting infrastructure will have lasting consequences for gas trade. As of 2017, there is a visible gap in the energy trade levels between the old and new EU states. Even heavily import-dependent Western European states tend to export and trade gas, whereas the majority of Central and Eastern European states, often less connected to EU gas markets, tend to keep a larger share of natural gas for themselves. Such differences partially stem from the lack of interconnecting infrastructure in new EU member states, partially from the lack of a gas trading culture and partially from territorial restrictions in the contracts with Gazprom. Previously, these trade differences also propagated a cycle of stimulating investments in cross-border interconnections between the old EU member states and thus furthering trade, and possibly discouraging them between the "new" EU member states, thereby leaving them path dependent and less secure in terms of gas supply.[67] However, this tendency is changing as Central and Eastern European countries are implementing reverse flows and building LNG import terminals such as the one in Lithuania and Poland, which can service a region rather than a single country. Thus, the infrastructural plans of the single EU energy market will not only improve the security of gas supplies, but also encourage more gas trade among the EU countries.

The Third Energy Package

The EU's most far-reaching and important energy legislative measures, referred to as the Third Energy Package of 2009, is a set of key pan-European directives and regulations to make the electricity and gas sectors more competitive. This package, and particularly a component called the Third Natural Gas Directive, is one of the most important examples of EU regulation to date. It emerged from the EC's Inquiry into the Energy Sector of 2005–2007, as did many other EU initiatives regarding antitrust cases or diversification. While the directive was not initially designed to contain or push back Gazprom's influence, in the end its three main provisions were all applicable to Gazprom's business in the EU. These provisions included (1) rules regarding mandatory access to pipelines by competing energy suppliers, referred to as the "third-party access" (TPA); (2) an "ownership unbundling" requirement; and (3) special certification requirements in relation to "third countries" (non-EU).[68]

The TPA rule, or Article 32, requires that, unless a national regulatory authority with EC's approval grants an exemption, various natural gas suppliers and traders can use pipelines that belong to another gas company under fair price and transparency conditions. This has direct implications for Gazprom, which not only supplies gas to the EU markets but also partially owns and operates those pipelines. For example, the inland extensions of Gazprom's Nord Stream pipeline, such as NEL and OPAL in Germany and Gazelle in the Czech Republic, are subject to this requirement. The German and Czech regulatory authorities gave conditional exemptions to OPAL and Gazelle, accordingly, but still have some spare capacity in these pipelines that must be made accessible for potential competitors of Gazprom.[69] Yet, in a blatant and contrary move, against liberalization, in October 2016 the EC allowed Gazprom even greater access to OPAL at about 90 percent of the 36 BCM capacity, thus seemingly defying the intent of the Third Energy Package.[70] This development shows that the TPA rule will require both further clarification at the EU level of policymaking and unified application within all EU member states.

"Ownership unbundling," or Article 9, if applied in its entirety, is possibly the most important clause of the Third Energy Package because it has already forced Gazprom to break up its EU assets. The provision, targeting both electricity and gas sectors, prohibits a single company from simultaneously controlling *production* or supply of energy resources and owning the *transmission* system or pipelines for those resources.[71] The rationale for "ownership unbundling" stemmed from the EC's 2005–2007 inquiry, which concluded that if the same entity both owns energy networks and trades energy, it tends to favor its own affiliates and forecloses access to the network to potential competitors.[72] The proposed remedy was divestment of assets by companies that did not meet the criteria. Instead of full ownership unbundling, a state could still choose two other possible models of organization of the natural gas sector, such as independent transmission operator or independent system operator with strict rules of transparency and "firewalling" of information exchanges between the transmission and supply parts of the business. The ownership unbundling option, if chosen, had to go into effect no later than March 2013.[73]

Among all EU member states, only Finland, Estonia, and Latvia were exempted from ownership unbundling because they depended on a single external supplier, Gazprom.[74] Lithuania also qualified for an exemption, but instead it was one of the first to opt for ownership unbundling in 2010–2011. At the time, the gas transmission system operator and main gas retailer, Lietuvos Dujos, was co-owned by the Lithuanian state (17.7 percent), Gazprom (37.1 percent), and the German energy company E.ON (38.9 percent). When the latter two shareholders resisted unbundling, the Lithuanian government and the foreign shareholders of Lietuvos Dujos resorted to arbitration.[75] At the time, Russian prime minister Vladimir Putin spoke out against the initiative: "Our companies, together with German partners, legally acquired distribution assets in Lithuania. Now they are being thrown out there with reference to the Third Energy Package. What is this? What is this robbery?"[76] After a tedious legal and public relations battle, Gazprom and E.ON sold their shares in Lietuvos Dujos to the Lithuanian state in 2014.[77] Estonia likewise followed suit with unbundling in 2012–2015, despite its official exemption.[78] Meanwhile, Latvia, being 100 percent dependent on Russian gas, held off from unbundling its transmission operator, Latvijas Gāze (34 percent owned by Gazprom), citing its gas contract through 2016, until the Russian giant proposed in early 2016 to sell its shares in order to comply with rather than fight EU regulation. However, in August 2016, Juris Savickis, the vice-chairman of the Latvijas Gāze supervisory board, said that Gazprom might not sell their shares in Latvijas Gāze and remain as an investor.[79] Finland continues to exercise its right to exception from unbundling, but in late 2015 the state acquired Gazprom's quarter share of the national gas operator, Gasum Oy.

The "third countries" requirement, or Article 11, of the directive has unofficially been called the "Gazprom clause."[80] From the beginning, in view of the possibility that ownership unbundling could lead to acquisition of EU energy grids by third countries, the clause targeted natural gas transmission operators from non-EU countries and caused heated discussions in Brussels in 2007–2008, when the EC initially proposed the draft.[81] The directive gave the EC a right to give an opinion on certifying transmission system owners or operators on the domestic level of separate EU member states for companies from "third countries."[82] After the directive was adopted and transposed to different EU member states, such transmission operators had to comply with the same unbundling requirements as the EU operators.[83]

Of the three Third Directive provisions, unbundling has been the most intrusive to Gazprom, though some have argued it could act in its favor. The breakup of larger companies to create smaller, more vulnerable companies has been perceived as potentially risky because it could enable their acquisition by nontransparent entities or Gazprom-linked interests. And there are, in fact, some examples of unsavory practices in Central and Eastern Europe, where Gazprom, being the sole supplier, sold gas to a country's consumers via several "middlemen" instead of directly through Gazprom Export. For instance, in 2009 the Bulgarian government attempted to exclude "middlemen" gas companies Overgas and WIEE (both 50 percent co-owned by Gazprom Export) from the gas supply chain between Russia and Bulgaria.[84] In Lithuania, as well, a number intermediaries previously sold Gazprom's gas to consumers—the privately owned companies Dujotekana, Vikonda, and Stella Vitae.[85] In Hungary, intermediary Emfesz, which benefited from the partially liberalized markets and by 2009 became the biggest player on the country's unregulated part of the gas market only to go bankrupt one year later, is another example. The company was controlled by Ukrainian oligarch Dmytro Firtash and until the end of 2008 received gas from the Ukrainian intermediary RUE, co-owned by Gazprom.[86] These examples show the potential risks of creating small nontransparent "middlemen" companies via unbundling and thus opening companies to potential acquisition by Gazprom via murky intermediaries.

The Third Energy package requirements have also been applied to potential infrastructure projects, such as Gazprom's South Stream pipeline. South Stream, like its northern counterpart Nord Stream, was intended to bypass Ukraine and bring Russian gas via Bulgaria, Serbia, Hungary, and Austria to Western Europe. The EC demanded that all EU member states holding agreements related to the South Stream project renegotiate the terms of their deals to abide by the Third Natural Gas Directive, specifically in respect to provisions concerning unbundling and third-party access.[87] Non-obedience was threatened with "infringement procedures against the government" even for Serbia, a non-EU country.[88] Although there were many other factors going against South Stream, such as questionable profitability and Gazprom's economic difficulties, as well as speculation that Russia never intended for South Stream to be completed and instead used the plans in an effort to foil the competing EU Nabucco pipeline project, the regulatory difficulties posed by the EC contributed to Gazprom's cancellation of the project in 2014. Nonetheless, despite the success of EU regulation regarding South Stream, the acquiescence vis-à-vis OPAL and the lack of unified stance against Nord Stream II shows that Brussels is trying but not always succeeding in transforming its relationship of gas dependence vis-à-vis Russia and in safeguarding its gas markets.

The EC's Antitrust Case against Gazprom

By passing the Third Energy Package, the EU showed it sought to create competition in its energy markets, but in September 2011 the EC reminded the world that it also intended to enforce competition and would not shy away from antitrust regulation vis-à-vis Gazprom, even in Central and Eastern Europe, where the Russian gas giant long held an upper hand. In other words, the EU wanted Gazprom to conform to its regulation like any other EU company or supplier. In September of 2010, after an earlier complaint by Lithuania to the EC Competition Directorate, the EC unexpectedly launched a "dawn raid" of the premises of Gazprom affiliates in Central and Eastern European states, including among possibly others: Gazprom Germania in Germany, EuroPolGas in Poland, Vemex in the Czech Republic, GWH Gashandel in Austria, Overgas Inc. AD in Bulgaria, Panrusgas in Hungary, Latvijas Gāze in Latvia, Eesti Gaas in Estonia, and Lietuvos Dujos in Lithuania.[89]

Based on the information extracted through those searches, in 2012 the EC opened a formal antitrust procedure to investigate a possible abuse of dominant position in Central and Eastern European gas markets by Gazprom.[90] The resolution of the case was unusually long, complicated by the Russian-Ukrainian conflict that started in 2014 and coincided with the end of the term of the previous EC that same year. However, in mid-2015 the EC, already consisting of a new team of twenty-eight commissionaires, charged Gazprom with pursuing monopolistic practices in Central and Eastern Europe, including setting "unfair prices," seeking to partition European gas markets, and making gas supplies conditional upon specific commitments to gas pipeline infrastructure projects such as South Stream and Yamal-Europe. Bulgaria, Estonia, Latvia, Lithuania, and Poland were deemed to be the impacted countries.

Initially many hoped that the case would be groundbreaking. English law scholar Alan Riley labeled the case "the antitrust clash of the decade" and compared it to another prominent antitrust case lodged against Microsoft.[91] Yet in many regards the antitrust case was simply a continuation of the EC's regulatory efforts vis-à-vis its gas markets and Gazprom since the mid-2000s. EU regulators had already forced Gazprom to remove destination clauses from its contracts with Western European companies ENI (2003), ÖMV, and E.ON Ruhrgas (2005). Along these lines, the European commissioner in charge of competition, Margrethe Vestager, stated, "To me, this case is not political. . . . It is pretty straightforward."[92]

After initial efforts to fight back, by the second half of 2015, Gazprom started showing willingness to settle the antitrust case,[93] and subsequently, after meeting with Vestager in early March 2016, Gazprom deputy chief executive Alexander Medvedev said that both parties were moving toward a "mutually acceptable" solution.[94] In the end of October 2016, Brussels prepared a draft settlement with Gazprom that does not include any financial penalties for the Russian gas company.[95] If such an agreement to settle is indeed reached, it may provoke negative reactions of some member states, such as Lithuania and Poland, two countries that have been fighting for Gazprom to face financial responsibility for its historical discriminatory pricing policy in Central and Eastern Europe. As in the case of OPAL and Nord Stream II, the seemingly relatively painless conclusion of the antitrust case has shown that the EU and the EC have pursued their regulatory stick within limits, still leaving room for Gazprom to maneuver.

Diversification of Sources and Routes: The Southern Gas Corridor

In addition to its numerous regulatory efforts, the EU has also sought diversification away from Russian gas and Russian-controlled gas pipelines with plans to create alternative pipeline routes. For various reasons, diversification has been difficult to achieve especially for the new EU member states, which have relatively small markets with monopolist suppliers intertwined with political and business interests. For instance, in 2014, the import dependency on Russia of Mediterranean, Western European, and northern European states was below 30 percent, while that of Central and Eastern European states was above 50 percent.[96] In some countries the dependency was even higher: until 2015, the Baltic states were 100 percent dependent on Russian gas while Slovakia, Bulgaria, and Hungary were from 70 percent to nearly 90 percent dependent. Europe's strategic alternative pipeline plan rests on the Southern Gas Corridor, which covers a set of infrastructure projects that would bring natural gas from the Caspian region to southeastern Europe, mainly from Azerbaijan's Shah Deniz field but potentially even from Turkmenistan and beyond.

The Southern Gas Corridor, along with its predecessor plans, has had several highs and lows in its history—from the initial frenzy and subsequent hibernation in the early 2000s, to its EU priority project status in 2008, to its current downscaled version in the 2010s. It will consist of three interconnecting pipelines: the operating South Caucasus Pipeline from Azerbaijan to Georgia, the Trans-Anatolian Pipeline (TANAP) through Turkey (under construction, to be completed by 2018), and the Trans-Adriatic Pipeline (TAP) between Greece, Albania, and Italy (also under construction, to be completed in 2020).[97] The Southern Gas Corridor pipeline capacity is planned to reach 24 BCM annually by 2023 and 31 BCM by 2026, though it could be scaled up to 60 BCM, particularly if Turkmenistan and potentially even Iran become suppliers.[98]

The Southern Gas Corridor is in many ways a scaled-down version of the earlier, ill-fated Nabucco pipeline project, named after an equally ill-fated Italian opera, which EU politicians promoted for a decade beginning in 2002.[99] Nabucco was intended to bring natural gas from Azerbaijan and possibly Turkmenistan to Europe via a 4,000-kilometer-long pipeline through Georgia, Turkey, Bulgaria, Romania, Hungary, and onward to Austria and the Czech Republic. Despite the highest EU institutional support, the progress of Nabucco stalled for years.[100] From the beginning the problem was that politics outweighed the commercial rationale or the actual gas supplies available for export. Economically Nabucco would have been viable only with additional suppliers such as Iran, which was always more of a future possibility, than a short-term reality.[101] Geopolitical forces were also not favorable: in 2007 Russia proposed the rival South Stream project, which would have followed essentially the same route as Nabucco but went to Russia's traditional ally of Serbia rather than Romania. Engaging the same EU member states that should have been part of Nabucco such as Bulgaria, Hungary, and Austria, in the South Stream project,[102] Moscow created a "divide and rule" strategy that succeeded in slowing down Nabucco, until it was eventually cancelled. The project's failure was also linked to Azerbaijan's choice between TAP and Nabucco to transport Azeri gas from the Shah Deniz Stage 2 field. In June 2013, Baku opted for TAP.[103] Moreover, EU plans to access Turkmen gas never panned out because Ashgabat instead focused on keeping up with Chinese gas demand.

Overall, Nabucco has served as a lesson that large gas diversification infrastructure projects will be difficult to achieve in the EU without significant political will and commercial motivation, as well as highlighting some of the difficulties that the Southern Gas Corridor may face. This is particularly important because southeastern Europe is becoming "a new key battleground" over gas infrastructure between the EU and Gazprom.[104] Here Moscow's on-again, off-again competing Turk Stream project to bring Russian gas to Turkey and southeastern Europe demonstrates Russia's changing strategic approach, which instead of seeking control of pipelines in the EU, would opt to deliver gas to the EU border (Greece in the case of the Turk Stream) without entering into any intergovernmental agreement.[105]

The EU in the 2020s and Beyond

The EU's gas sector is changing due to a confluence of factors. Internally the EU is pursuing regulatory and diversification efforts that seek to constrain Gazprom's role in Europe and at the same time creating a truly pan-European gas market. Externally, global gas markets are experiencing a transformation with a new abundance of natural gas that can now be shipped from more distant markets in the form of LNG. North American gas production is exploding, greatly exceeding domestic demand and launching US LNG exports to Europe in 2016. Closer to home, Iran could open up its gas resources over the coming decades, while new findings in the Eastern Mediterranean may meet the gas demand of Europe's neighbors, such as Egypt and Israel, if not the EU itself. Europe is ready to take advantage of these external developments via its existing and planned LNG import infrastructure and is close to finally accessing Caspian gas. As some energy experts argue, this turn to the Caspian / Central Asian and Mediterranean countries for potential supplies demonstrates an increasingly politically rather than economically minded EU gas policy.[106] The future holds many import and legislative options for the EU's gas market, all of which could contribute to the eventual decline of Gazprom's dominant position on the Continent. Five developments that have the potential to impact the future of Europe's gas sector—plans for an energy union, shale development, American LNG, Iranian and Eastern Mediterranean gas, and Europe's gas demand—are covered below.

The Next Regulatory Step—The Energy Union

The next step in the process toward a single EU energy market is the strategy for an Energy Union, launched in February 2015 by the EC in parallel with Donald Tusks' assuming of the presidency of the European Council. The idea has been long in the making and has had strong political support. In 2010, the EC picked up and modified the idea of the union first introduced by the former president of the EC, Jacques Delors (in office 1985–1995), which was later publicly reinforced by Tusk in 2014 when he was still the prime minister of Poland.[107] "The EU is creating a banking union, a Europe-wide underpinning for its financial institutions. The bloc's 28 members jointly buy uranium for their nuclear power plants through the EU's atomic energy agency, Euratom. They should take the same approach with Russia's gas. I therefore propose an energy union," Tusk wrote in the public appeal published by the *Financial Times* less than a year before he became president of the European Council.[108] Now, the concept is administered by the EC vice president in charge of the Energy Union, Maroš Šefčovič, who called the union "undoubtedly the most ambitious European project since the formation of the coal and steel community." Its main goals are "continuing diversification of Europe's natural gas market, modernization of the electricity market and the development of energy efficiency through the use of renewables and other alternative forms of energy."[109]

In essence, the Energy Union can be seen as the next logical step of the single market agenda, but it goes even further than that. The Energy Union strategy consolidates five energy issues under one umbrella: supply security, a fully integrated internal energy market, energy efficiency, emissions reduction, and research and innovation. The Energy Union plan, when implemented, could authorize the EC to have oversight and involvement in intergovernmental agreements on energy issues between member states and third countries like Russia that supply external gas.[110] Member states, however, would still decide their energy mixes, which could impede the formation of the union. The Energy Union proposal also stirred discussions that several EU member states could voluntarily consolidate their gas demand into joint purchase agreements especially "during a crisis and where member states are dependent on a single supplier."[111] This idea emerged in the context of motivating possible suppliers from the Caspian Sea and the Middle East, but could be applied to negotiations with Iran or Russia.[112] According to energy expert Nikos Tsafos, the model of consolidating gas demand is not without issue—including unresolved legal questions and the risk that it could inadvertently quell competition in EU markets, thus potentially acting as a force to raise gas prices. Likewise "negotiating as one" will remain problematic for the EU because doing so not only clashes with development of a fully functioning market in which gas deals depend on market players rather than on monopoly purchasing agreements, but also contradicts the EU's historical experience of how important liberal markets and non-incumbent commercial players in Italy, Spain Greece, and the United Kingdom have been in delivering supply diversification.[113]

A year after the Energy Union project was launched, in February 2016, the EC proposed more specific steps for its implementation. European commissioner for climate action and energy Miguel Arias Cañete noted the urgency: "After the gas crises of 2006 and 2009 that left many millions out in the cold, we said: 'Never again.' But the stress tests of 2014 showed we are still far too vulnerable to major disruption of gas supplies. And the political tensions on our borders are a sharp reminder that this problem will not just go away."[114] The energy security package presented consists of four main dimensions: security of gas supply regulation, heating and cooling strategy, a decision on intergovernmental agreements in energy, and LNG and gas storage strategy. Out of these four, the LNG strategy component is of particular importance because, according to the EC, it should complete the internal energy market and end single-source gas dependency for many EU countries.

To what extent the Energy Union will succeed remains uncertain. For one, the EC's wish to expand its oversight of energy deals between member states and third countries such as Russia, the so-called ex-ante compatibility check, may cause objections from some EU members.[115] Also, some analysts argue that the new proposed framework to implement the Energy Union in fact only stipulates activities already present in various existing policies and measures, but does not point to the most pressing issues, nor prioritize them, and does not offer plans for concrete policy implementation.[116] Others argue that if one of the goals of the Energy Union is a liberalized gas market, it may bring a politically unwelcome consequence in that free market rules would dictate that as the most competitive source, Russian gas would contribute to a substantial share of the EU's supply.[117] At the same time, developments with OPAL, Nord Stream II, and the antitrust case against Gazprom show that for Brussels there is still much work to be done in terms of enforcing its existing regulation. Despite these concerns, and even if it takes decades to fully implement, the Energy Union is symbolic of the EU's vision and political will to transform its gas import dependence into the power of gas demand.

Iranian and Eastern Mediterranean Gas

Searching for ways to diversify its sources of gas, Europe may eventually be able to import gas from Iran, which borders Turkey and Azerbaijan, both of which have plans for pipeline infrastructure connecting with Europe. Depending on various estimates, Iran competes with Russia for the first place in the world in terms of the size of its natural gas reserves.[118] Moreover, as energy analyst Brenda Shaffer noted, "Iran is the only country with the volume and location to pose any major threat to Russia's dominance in Europe and some of the markets of the in the former Soviet Union."[119] Yet, because of the lack of pipeline infrastructure and years of Western sanctions that only ended in 2016, Iran has never supplied gas to Europe; instead, it has limited its exports to its immediate neighbors, such as Turkey, Armenia and Azerbaijan.[120] Even before international sanctions were officially lifted, numerous European companies began negotiating billions of dollars' worth of new energy (including gas) deals with Iran.

Going forward, Iran could in theory supply gas to Europe by connecting to the Azeri pipeline system or to the TANAP-TAP network starting in Turkey, but it will take time to complete the gas pipeline infrastructure and thus even longer for Iranian gas exports to make their way to Europe. It seems more likely that Iran would first pipe gas to neighboring countries, including Turkey, Kuwait, Oman, and the United Arab Emirates, which also do not have enough gas to meet domestic demand.

The more likely and preferable option for Iranian gas exports in the near term would be in the form of LNG, a project that was abandoned in 2012 due to international sanctions[121] but that now has a number of possibilities going forward. First is an onshore export terminal with a planned capacity of 10.5 million tons per annum (MTPA)—scheduled to become operational in 2018—that is currently pursued by the National Iranian Gas Export Company and a number of foreign investors.[122] Although the licenses and some of the plant's facilities were already in place, sanctions until 2016 restricted the delivery of liquefaction technology to Iran to complete this project. Other options would be a few small to mid-size LNG terminals. Likewise, Iran's National Iranian Oil Company is also considering construction of a floating LNG facility rather than a land-based plant, which would save both time and money. Alternatively, Iran may opt to build a pipeline under the Persian Gulf to Oman—which could be done within two years—and utilize Oman's already existing Qalhat LNG facility. Yet another option would be the use of the Das Island LNG facilities in Abu Dhabi. Greece was one of the first European countries to express interest in Iranian LNG, which could help it become a regional gas hub and help southern Europe diversify its gas sources. It is estimated that Europe may import between 25 and 35 BCM of gas if Iranian LNG does in fact become reality. Nonetheless, optimism should be tempered because it will take Iran time to rebuild its gas industry, and LNG projects may be difficult to finance given the global gas glut and fallen global oil prices since 2014.[123]

Gas discoveries in the Eastern Mediterranean, specifically off the coast of Israel, Cyprus, and Egypt, have also raised hopes for new sources of supply for Europe, but optimism should be tempered. In the late 2000s, Israel discovered several vast offshore gas fields: the Tamar in 2009 of around 10 trillion cubic feet (TCF), the Leviathan in 2010 of 22 TCF, and, together with Cyprus, Aphrodite (5.7 TCF) in Cypriot territorial waters.[124] These discoveries have transformed Israel's gas security as the previously energy-poor country can now rely on its own domestic gas resources when supply is otherwise unavailable or its relations with regional energy-rich states of the Middle East are tense. However, the potential for Israeli exports to Europe (via Cyprus, Greece, and Turkey) remains largely theoretical. Natural gas exports from Israel are subject to a complex web of regulations that require fields of various sizes to earmark up to 50 percent of their resources for domestic consumption. These constraints are combined with the security risks of laying pipelines or building LNG export terminals in a region fraught with mostly unfriendly states and territories facing societal unrest, war, and terrorism, as in the case of Egypt, Syria, Lebanon, and the Gaza Strip. For instance, the American energy company Noble's offshore gas drilling platforms and onshore facilities are within firing range from Gaza Strip. Meanwhile, the Leviathan field is in close proximity to Lebanon, where the US-designated terrorist group Hezbollah claims without substantiation that the field extends into Lebanese waters. Although the security considerations do not significantly increase the costs of Israeli gas exploration, the market conditions of low gas prices and the gas glut of mid-2010s do not encourage sizable export infrastructure projects. Meanwhile, Egypt's gas discoveries in the Mediterranean, which date back to 2003, have enabled the country to emerge as an LNG exporter to regional markets, but rising domestic demand forced it turn to imports from Israel in 2015. The 2015 discovery of the "supergiant" 30 TCF field of Zohr, as well others such as the Nooros in 2016, will help meet Egypt's needs going forward, but this domestic demand makes it unlikely that sizable volumes will be available for exports to Europe in the near term. Instead, the most realistic if still theoretical prospect of Eastern Mediterranean gas going to Europe would be an Israeli (and potentially Cypriot) project to send its gas to the Egyptian LNG facilities via subsea pipeline, from where it could be exported to European markets.[125]

Domestic Gas and Shale Production

The future of Europe's domestic gas production is uncertain. Conventional gas production is expected to fall as gas reserves in the North Sea dwindle beyond 2020 and as production from the Netherland's Groningen field is expected to continue declining after a series of earthquakes heightened public safety concerns over the environmental consequences of gas extraction. Norwegian gas is not expected to compensate for domestic decline in production.[126] Europe's shale gas reserves could be the long-term answer to boost domestic production, though near-term prospects remain dim. Western and Eastern Europe (including Ukraine and Russia's Kaliningrad territory) have an estimated 598 TCF of technically recoverable shale gas reserves, the largest volumes of which are estimated to be in Poland (148 TCF), France (137 TCF), Romania (51 TCF), Denmark (32 TCF), and the United Kingdom and the Netherlands (both at 26 TCF).[127] However, up to the present time there has been no shale gas production in Europe. Moreover, in my view, exploitation of shale gas in the EU will be slow in the near to medium term.

In Europe, greater population density and greater concern for the environment than in the United States make it unlikely that at least Western European countries would embark on domestic shale production anytime soon. Therefore, the support for exploration of shale has been scarce in many countries where governments have rejected hydraulic-fracturing or "fracking" (as in Bulgaria, France, some regions of Spain, and Germany), while public opposition to fracking has been strong (as in France, Bulgaria, and the United Kingdom). Furthermore, the necessary legislative framework for fracking is lacking in some countries (Lithuania), and environmental NGOs across the Continent actively oppose it.[128] While energy-vulnerable states such as Poland and Lithuania were keen to exploit shale gas opportunities, especially in 2012–2013, when prices were high,[129] in the lower gas price environment and the global gas glut, which have been around since 2014, even such countries have become less eager for domestic shale production and have chosen other means, such as LNG, to ensure their energy security.[130] Nonetheless, in 2015, the United Kingdom adopted some simplifications to its robust shale gas legislation, and in late 2016, the government overturned a previous "fracking" ban in the north of the country, providing a green light for onshore drilling in the near future.[131]

The political environment is more complicated in Germany, where fracking has effectively been banned, especially in areas rich in groundwater. However, the new law, out in June 2016, leaves room for governments on the state level to decide on individual cases and to allow fracking for some noncommercial projects, such as scientific test drilling.[132] Spain (in some regions), Denmark, the United Kingdom, Hungary, and Romania granted research licenses or even authorized shale gas extraction by mid-2014.[133] France and Bulgaria banned fracking in mid-2011 and in January 2012.[134] Moreover, in 2016, France discussed a potential ban on the import of fracked gas from the United States.[135]

Even among those EU member states that are in favor of shale gas, none has achieved commercial levels of shale production. As of 2016, Poland was the only country to have begun exploration, having drilled over sixty exploratory wells. After the initial hype of shale gas by some European and multinational energy companies, the mood turned negative in 2015, with most of the multinational energy companies withdrawing from the EU. For example, the US energy company Chevron withdrew from Lithuania and Bulgaria in mid-2014; then Chevron and fellow American company ConocoPhillips left their shale activities in Poland and Romania in 2015.[136] Romania was the last active shale gas market in the EU for Chevron while ConocoPhillips was the last "global oil firm" to quit Polish shale gas.[137] Analysts have estimated that to reach shale gas production levels of 28 BCM per year in Europe by 2030 (which would still be less than 8 percent of current total gas consumption) 800 to 1,000 wells would have to be drilled annually for ten years, an extremely unlikely task to accomplish.[138]

Going forward, I believe that Europe's shale production prospects are unlikely to be as bright as those of the United States for several reasons. First, with few exceptions such as Latvia, practically no EU country awards landowners mineral rights—which has been a motivating factor in small-scale exploration in the United States. Second, the shale extraction technologies developed to suit North American geology may prove to be less commercially appealing in different European geological conditions. Third, Russian gas interests pursue an information and lobby campaign against shale exploration in Europe.[139] Finally, low global oil and gas prices and the abundance of gas on the global markets have reduced interest in shale gas potential, even in the countries where energy security has been paramount. Nonetheless, shale gas reserves remain an option for some EU member states in the future under more favorable market conditions.

American LNG

LNG imports may prove an easier way for Europe to achieve diversification from Russian gas than building alternative pipeline infrastructure or pursuing shale gas development on the European continent. European LNG imports, some of which will be sourced from the United States, are expected to roughly double between 2014 and 2020.[140] Overall, while American LNG may decrease Europe's dependence on Russian gas, it will not entirely replace it, partly because construction of additional LNG import and export terminals will take many years and partly because the volumes of American exports will fluctuate based on the market conditions. Still, there is evidence of optimism, commercial interest, and political will on both sides of the Atlantic. US ambassador to the EU, Anthony Luzzatto Gardner, has said that

> the United States is deeply committed to working together with the EC, the European Parliament and the private sector on both sides of the Atlantic to address energy security concerns in Europe in a way that keeps the U.S. and EU economies growing, increases investment and business opportunities on both sides of the Atlantic, lowers our collective carbon footprint, and strengths energy security.[141]

In early 2016, the first American company to launch LNG exports from United States to Europe was Houston-based Cheniere Energy, with its first cargo shipped to Portugal and its second to Spain.[142] In mid-2015, Cheniere Energy had signed agreements with the United Kingdom's Centrica; Italy's Enel Group; Spanish energy companies Iberdrola, Gas Natural Fenosa, and Endesa; and French EDF, as well as Total and Engie (formerly GDF Suez); it is also in discussions with Lithuanian importer LITGAS.[143] Due to existing long-term Russian and other gas contracts to Europe, especially to "over-contracted" northwest Europe, it is uncertain how much room there is for additional gas imports and thus for American LNG. American LNG is more likely to make its way to the United Kingdom and the Iberian peninsula, where there are less proximate Russian pipelines. However, if gas demand in the EU does increase, the picture could change. In any case, additional supplies would likely be resold and thus stimulate gas trading and emergence of hubs and exchanges in Europe and beyond.[144] Most importantly, LNG imports from North Africa, Qatar, Norway, the United States, and eventually, possibly even Iran will emerge as new options for European countries that seek to diversify from both Russian piped gas and long-term gas contracts.

EU Gas Demand

While Europe remains the world's largest consumer of gas, demand fell and plateaued in the 2010s. Indeed, demand has fallen across all three major sectors: power, industry, and residential. In 2014, gas use in the EU fell to levels last seen twenty years ago. In comparison to 2014, 2015 marked a 4 percent growth in EU natural gas usage; it was also the first annual increase in four years and was due largely to economic recovery in some EU states like France, the Czech Republic, and Slovakia, as well as to weather patterns. The rise in 2015 was mirrored by a rise in LNG imports as well.[145] Still, in 2015, EU gas demand was 23 percent below its peaks of 2005 and 2010, reflecting the pre– and post– economic recession levels.[146]

A number of economic and political considerations supporting expectations of the EU's falling gas demand trend are expected to continue. First, the economy on the European continent has slowed down, especially since the global economic downturn of 2008. The EU-28 average real GDP growth rates accelerated to 3.1 percent in pre-crisis 2007 but experienced a 4.4 percent decline in 2009. In 2014, after a slow recovery, the average growth rate of the EU-28 GDP reached 1.3 percent compared to 2013.[147] Meanwhile, EU gas usage has also dropped—by around 19 percent, to 422 BCM in 2014 compared to 523 BCM in pre-crisis 2007.[148] In 2013, EU demand fell to the lowest level since 2000.[149]

In addition to economic sluggishness, the EU's long-term focus on renewable and efficiency policies is reaping results and contributing to declining gas demand. In 2015, 80 percent of EU gas demand came from seven western EU states that have strong renewable and energy efficiency policies in place.[150] In addition and in contrast to environmental objectives, since the US shale boom, cheap American coal exports to Europe have also contributed to a decline in gas usage as coal eroded gas's share of the power market.

The future of EU gas demand remains somewhat uncertain. The International Energy Agency projects that in 2016–2021 it will increase only very modestly and then stabilize in the future.[151] It is unlikely that EU states will experience a boom in gas demand like that of Asia's rapidly developing nations, nor is a dramatic drop in EU's gas consumption likely in the near term. European and specifically Dutch gas production is set to decrease, so demand for imports would likely rise. To what extent renewables can replace current gas usage is questionable for the near term. Short- and medium-term gas demand will also depend on the choice between coal and gas. The EU plans to reform the Emission Trading System (ETS)—its "carbon market" wherein companies buy pollution permits in government auctions or trade permits with companies that have a surplus—for the 2021–2030 period by lowering the emission allowances. This could create greater disincentives for coal consumption, and thus, gas demand could rise in response. However, critics maintain that in Europe, domestic policy loopholes exacerbate the consumption of cheap US coal imports. Environmentalists argue that European greenhouse gas reduction targets are not ambitious enough, while EU-wide emission reduction agreements do not necessarily become policy on the member-state level. Moreover, the ETS does not place a high enough price on carbon to cap emissions, given allowances of a mere five euros per ton of carbon burned.[152] Although gas demand may remain stable, imports in the near term may rise as unwanted LNG cargoes in the global gas trade find their way to Europe, which has the capacity to absorb cargoes that cannot find a home in any other market.

Conclusion

The European gas sector of 2020 and beyond will be strikingly different in many ways compared to the latter half of the twentieth century. The EU has already demonstrated that it will be increasingly proactive in its efforts to loosen Gazprom's noose on Europe. Certainly, the EU countries are still far from unified on this issue, and Gazprom will fight hard to hold on to this market by trying to build additional pipelines such as the Nord Stream II and the Turk Stream pipelines, gaining access to European pipelines such as OPAL, or possibly by cutting gas prices. However, the EU's trajectory has been set: regulation, integration, diversification, and a focus on renewables and efficiency. Building on the regulatory successes of the 2010s and learning from failed attempts to build alternative gas pipelines, the EU will continue to set the terms of behavior for external gas suppliers to its market. The EU's most important achievement that would truly allow it to practice the politics of demand rather than the politics of dependence would be to finally speak with one voice and build the much-anticipated energy union. However, this will probably be most difficult to implement.

In the long term, due to internal decisions and the changes in global gas and energy markets, gas will lose some of its strategic importance for Europe. A turn to renewables and efficiency has already slowed EU gas demand, while at the same time gas has become much more abundant in the international markets. In the coming years the effects of the US shale boom, the prominence of LNG, new global gas sources, and possibly even shale production in some European states will start to be felt even more. If the EU stands firm and enforces its regulation, Russian gas will lose its monopolist position in many EU markets and it will also lose the political influence that goes with it. Gas has the potential to become just another commodity on the European continent, and with it the EU's relationship with Gazprom and Russian gas will become more commercially based, less politically tinged, and less geopolitically charged.

5 The Politics of Transit: Ukraine and Belarus

Located at a critical juncture between the world's largest natural gas exporter and the world's largest natural gas importer—Russia and the European Union—Ukraine and Belarus have been the linchpins of the half-century-long gas relationship between Gazprom and Europe, their lands bearing vast networks of pipelines to transport Russian gas and oil products to Europe. As such, the two countries exemplify the politics of gas transit. These two countries have been, and remain, strategically vital for both Russia and the European Union (EU); more than half of Russian gas exports to Europe during the peak winters of the 2010s were piped through Ukraine, and about a quarter through Belarus.[1] This position has awarded Ukraine and Belarus some benefits: negotiating leverage vis-à-vis their supplier, revenues from transit fees, and attention from European states due to Kyiv's and Minsk's capacity to destabilize the transit of supplies through their territory. Overall, though, the costs have probably outweighed the benefits. Being "rentier states," they have been subject to the resulting destabilizing domestic weaknesses. At the same time, Ukraine and Belarus have pursued distinct and complex relations with Moscow, which have been further complicated by relations between the West and Russia. Belarus's and especially Ukraine's strategic roles in gas transit, in addition to their geographic position and history, have placed them squarely in the center of the fluctuations of the balance of power in their regions, as well as in the crossfire of the East-West geopolitical struggle and the EU-Russia gas relationship.

The transit relationship with Russia has been complex for Ukraine and Belarus ever since they gained independence from the Soviet Union. The lack of an overarching regulatory body or any legal oversight of transit between these states has made navigating transit intrinsically complicated and fraught with crises.[2] A further complication is that both countries have been highly dependent on Russia for energy, especially gas. Their gas dependence has been underscored by their overall economic vulnerability, especially in relation to Russia. Economic crises—often magnified by huge, mismanaged energy debts to Russia—have swept dramatically through both countries, resulting in devaluations and inflation. They have been accompanied by political crisis in Kyiv and to some degree also in Minsk. Energy dependence has bred corruption and established a "cycle of rents" in the economic and political life of both countries.[3] It has led to a cycle of loans, debts, and gas price and transit fee wars with Gazprom and political and commercial concessions to Moscow.

This chapter will examine the foundations of the Russo-Ukrainian and Russo-Belarusian gas relationships. These cases will shed light on the role of gas transit states and how this role has been changing since the early 2010s with the changing global landscape of gas markets and deteriorating relations between Russia and the West.

Ukraine

Europe's largest country by landmass, Ukraine serves as the cornerstone of the EU-Russia gas import and export relationship. Ukraine's energy sector, historically marked by a heavy and largely inefficient reliance on Russian gas, has long exemplified the political and economic impact of being an import-dependent gas transit state. As such, its geopolitical shift westward has been fraught with tensions.

The state-owned energy company, NAK Naftogaz Ukrainy (Naftogaz), has dominated Ukraine's gas industry, producing and transporting oil and gas domestically and transiting Russian gas. From the start, the company was linked to the highest echelons of power. President Leonid Kuchma (1994–2005) created Naftogaz in 1998 to counterbalance oligarchs in the oil and gas business and allegedly to raise money for himself and his allies in upcoming elections.[4] By 2014, Naftogaz produced 90 percent of Ukraine's oil and gas and accounted for 8 percent of Ukraine's gross domestic product (GDP).[5] At the same time, Naftogaz received an estimated $2 billion in transit revenues from Gazprom, accounting for another 6 percent of Ukraine's budget revenue.[6] The company wholly owns multiple subsidiaries, such as Ukrtransgaz, its oil and gas transport arm. In 2013, Ukrtransgaz moved 132 billion cubic meters (BCM) of mostly Russian gas through Ukraine's territory.[7] Naftogaz also owns more than 50 percent of Ukrnafta, which controls 11 percent of Ukraine's domestic production of natural gas and 68 percent of its oil production.[8]

Ukraine's underground gas storage facilities, run by Ukrtransgaz, form a robust system, the second-largest in Europe with their 31 BCM capacity, and they offer flexibility in gas storage and consumption between seasons. Nonetheless, due to the conflict with Russia since 2014 and risks of gas cuts, storage is becoming a vital concern. In 2014, Ukraine became the first non-EU state to join the Aggregate Gas Storage Inventory transparency platform of Gas Infrastructure Europe (GIE), in an effort to bring Ukraine's gas storage system up to EU standards.[9]

Import Sources

In 2013, natural gas met 34 percent of Ukraine's energy consumption needs, second only to coal at nearly 36 percent, and followed by nuclear power at 18 percent. That mix remained largely unchanged between 2000 and 2013.[10] Overall, the volume of Ukraine's gas consumption has steadily fallen from 124 BCM in 1990, just before the Soviet Union's collapse, to 50.4 BCM in 2013, to just 33.8 BCM in 2015.[11] The conflict in eastern Ukraine, as well as Russia's annexation of Crimea, have in no small part contributed to the fall in gas consumption since 2014; the loss of Crimea alone may have cut up to 15 percent of Ukraine's annual gas consumption.[12] Ukraine relies heavily on imported Russian gas, but Ukraine-Russia tensions, especially since 2014, as well as Russia's repeated cutoffs of gas to Ukraine in the 2000s, have led Ukraine to reduce its gas consumption and direct imports from Russia. Ukraine cut its imports from Russia from over 28 BCM (over 90 percent of its imports) in 2013 to 6.1 BCM (approximately 37 percent of imports) in 2015.[13] Simultaneously, Ukraine increased its gas imports from neighboring European states—from just over 2 BCM in 2013 to 10 BCM in 2015—through reverse flow of mostly Russian gas via existing pipelines. As a result, by 2015 Gazprom was no longer the dominant supplier of gas to Ukraine, though the country still consumed the same Russian gas transported via other routes.[14]

Domestic Gas Reserves and Production Capabilities

Ukraine's sizable gas import dependence has been largely artificial—a product of both inefficiency and underdeveloped and underfunded domestic gas production. Ukraine has sizeable proved natural gas reserves, up to 1.2 trillion cubic meters (TCM). The US Energy Information Administration (EIA) estimates Ukraine's recoverable shale gas resources at 5.2 TCM, the third highest in Europe behind Poland and France.[15] Meanwhile, Ukraine's scientific research institute, Naukanaftogas, has estimated those resources at a whopping 22 TCM.[16] Prime Minister Arseniy Yatsenyuk (2014–2016) projected in 2015 that Ukraine could shift to using only domestically produced gas by 2025 and reiterated in 2016 that the country will be energy independent in ten years. However, without heavy investments, refurbishments, and improvements of Ukraine's domestic pipeline system and without the participation of foreign companies to develop Ukraine's domestic resources, success is unlikely.[17]

Ukraine has three primary geological formations that are home to various gas fields under development. The Carpathian Foldbelt, a geological formation in western Ukraine, stretches into Poland, Romania, and Moldova. It has 877 BCM of recoverable resources and contains the Olesska shale deposit field, which Chevron was intending to develop. In 2014 Ukraine raised taxes on drilling due its dire financial situation, and

Chevron terminated its product sharing agreement (PSA) without an official explanation.[18] Italy's Eni and Austria's Rohöl-Aufsuchungs Aktiengesellschaft (RAG) also secured a contract in western Ukraine, though drilling plans for 2016 have not progressed.[19] The second geological formation, the Dniepr-Donets Basin in eastern Ukraine, has the highest potential for gas production, estimated at 4 TCM (of which 2.15 TCM is shale gas reserves).[20] In 2013, Royal Dutch Shell won a contract to explore and develop unconventional shale deposits in the Yuzivska field of the basin, and while it started to drill two exploration wells in 2012, in June 2015 it pulled its $10 billion project due the conflict in eastern Ukraine. The war raged just 100 kilometers from Bilijajivska 400, the southern test borehole in Yuzivska.[21]

With Russia's annexation of the Crimean peninsula in March 2014, Ukraine lost the Black Sea and Sea of Azov shelf and its gas fields, where reserves are estimated from 4 to 13 TCM of natural gas resources (45–75 TCM when gas hydrate reserves are included).[22] This spelled the loss of most of Ukraine's offshore reserves. In the annexation, Naftogaz lost one of its subsidiaries, the Crimea-based oil and gas company Chernomornaftogaz, which also operated the local Hlibovske gas storage facility. Naftogaz threatened to sue Russia over $1 billion in "stolen natural gas," but no immediate action was taken.[23]

Prior to Crimea's annexation, Ukraine had been in long and unfruitful negotiations with a consortium of companies, including American ExxonMobil and Shell, and Austrian ÖMV's subsidiary Petrom, to explore and develop natural gas resources off the coast of Crimea. The loss of the Black Sea and Azov Sea shelf could significantly disrupt Ukraine's quest for gas independence because an estimated 30 percent of Ukraine's potential gas resources are located there, where 1.65 BCM of gas was already produced in 2013. Russia began drilling off the shore of Crimea in 2016; Kyiv decried the action as illegal.[24]

Gas Transportation System and Transit Capacity

Ukraine's greatest role in the geopolitics of natural gas has been as a transit country of Russian gas to European markets, and its gas infrastructure reflects this. Ukraine has a vast gas transportation system of some 40,000 kilometers of pipelines running through its territory. The eight different pipelines connecting Russia and Europe through Ukraine can carry a combined flow of 142 BCM of gas and together reach much of Europe: Austria, Bosnia, Bulgaria, the Czech Republic, France, Germany, Greece, Hungary, Italy, Macedonia, Poland, Romania, Serbia, Slovakia, Slovenia, Switzerland, and Turkey. These pipelines have served as Europe's gas lifeline. In 2014, up to 51 percent of Europe's gas imports from Gazprom flowed through Ukraine in the winter season, down from 90 percent in the winter of 2009. Europe received almost 85 BCM in 2013 through Ukrainian pipelines and of the total amount of gas consumed by Europe, regardless of origin, 13 percent flowed through Ukraine in 2014.[25]

Following Kyiv-Moscow tensions over the conflict in eastern Ukraine, in 2015 Russian officials expressed a desire to stop transporting gas through Ukraine by 2019 after the contract between Gazprom and Naftogaz expires that year.[26] Moscow is exploring and developing alternative supply routes, such as additional Nord Stream system lines and other pipelines to access southeastern Europe, though a complete cessation of gas transport through Ukraine is unlikely in the near future. If Russia were to achieve its goal of bypassing Ukraine, the Ukrainian economy would lose gas transit revenues. More importantly, the EU would lose the ability to import gas via Ukraine's transportation system, which, despite political tensions, has certain infrastructural benefits, including a high level of interconnectivity among the main pipelines, which ensures security of supply in face of accidents or other emergencies on individual pipelines.[27]

Ukraine-Russia Gas Relations: Gas Prices, Transit Fees, and Debt

The explosive Ukrainian-Russian gas relationship has been carefully watched in Europe since Ukraine's Orange Revolution and the first gas dispute of 2005–2006, and particularly since 2014, when Ukraine's Maidan revolution and efforts to forge closer ties with the EU and the United States were followed by Russia's annexation of Crimea and the ensuing conflict in eastern Ukraine. Several issues have complicated what could have been a plain commercial relationship. For one, Ukraine is both a significant customer and a transit provider for Gazprom. Thus, gas negotiations between the two parties have included both the price Ukraine pays for gas imports and the transit fees it receives. Only in 2009 were the supply and transit relations contractually separated, though in practice gas deals for supply and transit were still negotiated in tandem. Second, as elsewhere in the post-Soviet space, the countries' gas relationship has been inextricably linked to their political and economic relationship. When Kyiv-Moscow relations were cooperative, Gazprom offered lower prices and at times allowed a run-up of gas debts. Kyiv's attempts to turn toward the West were met with price hikes. At the same time, as a monopolist supplier, Gazprom always held the upper hand, while Ukraine did not have or seek access to alternative gas imports.

Following the fall of the Soviet Union, Ukraine, like all former Soviet republics, had to renegotiate its gas supplies with Gazprom. Moreover, Commonwealth of Independent States (CIS) countries had to renegotiate a new "transit dimension" of gas, a phrase coined by energy researcher Katja Yafimava; the once single, consolidated market had become subject to the regulations, laws, and politics of multiple independent countries.[28] In contrast to oil, which was negotiated on market terms and prices, the prices of Russian gas exports to the CIS remained far below the market level and were determined through bilateral politicized negotiations. For instance, in 1993 Gazprom set the price for gas to Ukraine at $80 per thousand cubic meters (tcm). While that was higher than Gazprom's price of $50 tcm for Russia's closest ally, Belarus, it was still considerably lower than the price Gazprom charged EU states, which averaged about $100 per tcm in the 1990s.[29] At the same time, the transit fee for Russian gas piped through Ukraine to Europe was $1.73 per tcm for each 100 kilometers of transit in 1993. The nature of the Ukrainian-Russian gas relationship continued to be barter-based as in the Soviet era in that Russia paid for transit with gas. In 1998–1999, during the dip in global oil prices and Russia's financial crisis, the transit tariff was marked down to $1.09 for 100 kilometers' transit of a tcm, and the barter price for gas fell to $50 per tcm, which was higher than Belarus paid but considerably lower than what other European former Soviet republics like Moldova or Lithuania paid. Ukraine continued to pay $50 per tcm for gas until 2006.[30]

In 2005, Russian-Ukrainian negotiations over gas prices intensified. Early that year, Ukraine's Orange Revolution ushered in the pro-European government of President Viktor Yushchenko. At the same time, global oil and gas prices were skyrocketing. So Gazprom proposed doubling its sales price to Ukraine to $160 per tcm. The two sides argued for months over a widening web of issues—the gas price, transit fees, allegations that Ukraine was stealing gas, and Russian pressure on Kyiv to let Gazprom buy into Ukraine's pipeline network. The dispute even generated calls within Ukraine to renegotiate the contract that enabled Russia to base its Black Sea Fleet at Sevastopol on Ukraine's Crimean Peninsula. Later that year, Moscow again hiked its proposed price to Ukraine—this time from $160 to $230 per tcm. Ukraine had still not agreed to the original offer of $160, and Yushchenko responded with sarcasm: "Let's start from $500—or $700. This is not a basis for political dialogue. This is an irresponsible approach."[31]

On January 1, 2006, after nearly a year of negotiations, Gazprom declared a cutoff of gas to Ukraine on orders of Russian President Vladimir Putin, who has been intimately involved in the operations of Gazprom.[32] The cutoff reduced Gazprom's supply into Ukraine to the volumes intended for transit to its European customers. At noon that day, the temperature in Kyiv was −4°C, with snow in the forecast. Soon, Naftogaz was diverting some of the Russian gas intended for European markets to provide heat for Ukrainians. European customers bristled; under their contracts, Gazprom had guaranteed delivery of gas through Ukraine to Europe.[33]

By January 4, Ukraine agreed to pay an effective price of $95 tcm for gas from Russia in a complex deal. Officially, the supply was to be a mix of Russian gas at $230 tcm from Gazprom and much cheaper Central Asian and Russian gas supplied via a newly formed company,

RosUkrEnergo (RUE). But the price in the deal was fixed for only six months, and by 2007, it climbed upward to $130, approaching European prices of $160–$230 per tcm.[34]

By the end of 2008, tensions had reemerged, but now they centered on Ukraine's gas debt. Gazprom again sought to raise prices and demanded that Naftogaz pay its debt or face a cutoff. Ukraine had been accumulating a debt from Russian gas imports since 1992. Its weak, corruption-ridden, and energy-inefficient economy, with its low export revenues, could not generate the hard currency needed to meet gas payments, and its political system lacked the will for reform. The domestic gas middlemen traders of the 1990s, such as the United Energy Systems of Ukraine (UESU), were connected to government; they could buy gas from state-owned importers such as Ukrgazprom at low prices and resell it to the industry at high prices. This cannibalized the revenues of state-owned companies, which subsequently could not make payments to Russia. Already by 1993, the unpaid gas import bill was 138 billion rubles (about $110 million at the time).[35]

These gas debts also provoked Gazprom's threats to cut supplies to Ukraine in 1993 and 1995, but the countries came to informal agreements to continue the gas supply after relatively small payments by Ukraine. In exchange, Ukraine's leadership allowed various Ukraine-registered middlemen companies to import gas directly from Russia and Turkmenistan. The beneficiaries of such companies were often allegedly linked to the leadership of the Russian and Ukrainian governments and to persons connected with the managers of Gazprom and Naftogaz.[36] On another concession, President Kuchma and Russian president Boris Yeltsin agreed in 1997 to cancel an accumulated Ukrainian gas debt of $800 million in exchange for Russia's lease of the naval base in Sevastopol and other facilities in Crimea until 2017. Before the 1997 agreement, the Black Sea Fleet was formally under the command of the CIS, though in reality it was directed from Moscow. The deal would have catastrophic consequences for Ukraine in 2014, when the presence of the Russian forces there greatly facilitated Moscow's takeover of Crimea.

Worse, Ukraine quickly returned to accumulating gas debt to Russia. By 2008, in part due to rapidly rising prices, it owed Russia $2.4 billion for gas—and a new bilateral crisis erupted. When Ukraine paid $1 billion of its debt in December 2008, Gazprom demanded the remainder. The crisis escalated on January 1, 2009, when Gazprom reduced exports by 110 million cubic meters (MCM) per day to Ukraine. Hungary, Romania, and Poland soon felt gas shortages, while Ukraine took technical gas from its pipelines, prompting Gazprom to call the diversion of supply illegal. Tensions rose, and on January 7, Russia halted all gas transit through Ukraine to Europe. Ukraine continued to supply gas to the EU from its own storage facilities.[37] While EU officials tried to mediate, only negotiations between the Russian and Ukrainian prime ministers, Vladimir Putin and Yulia Tymoshenko, yielded results. On January 19, the two agreed to a ten-year deal at a basic price of $450 per tcm, with a proviso that prices could change with fluctuations in the price of oil and in transit fees. The resulting average price was reported to be $230–$250 per tcm, though the contract's pricing formula was a commercial secret. While Ukraine's price for gas was still lower than that paid by the EU, it continued to rise rapidly each year.[38]

The 2009 Russo-Ukrainian gas dispute left a lasting imprint on the geopolitics of gas in Europe. For one, it seriously damaged the reputations of both Russia as supplier and Ukraine as a transit country. It also deepened Ukrainian-Russian tensions, which would reach their peak in 2014. More broadly, energy scholar Simon Pirani has argued that the gas price hikes and the disputes between Russia and the CIS countries have contributed to toward an "economization" of Moscow's foreign policy, which has favored greater revenues and accepted the risk of disputes and worsening relations with the CIS countries. Pirani and Yafimava have both argued that the 2009 crisis cemented Russia's decision to build pipelines circumventing Ukraine to increase Russian profits in the gas market.[39]

The gas dispute also jolted Ukraine's politics, leading to the downfall of Prime Minister Tymoshenko—one of the country's most important and visible figures in politics and the gas business. Tymoshenko rose to prominence in Ukraine and earned her nickname "gas princess" as the head of a natural gas trading company and the country's largest gas importer, UESU, which she cofounded in 1995. During her tenure as vice prime minister in 2000–2001 she dedicated herself to reforming the gas sector. She later became a figurehead of the Orange Revolution as Viktor Yuschenko's ally and served as prime minister in 2005.[40] In her second term as prime minister, Tymoshenko seized the initiative in the 2009 dispute with Gazprom by bargaining on gas prices with Putin directly, even though previously only presidents had handled Ukraine's gas negotiations. Tymoshenko eliminated the role of controversial intermediary RUE in gas trade with Russia in the 2009 gas deal, thereby restoring direct negotiations between Gazprom and Naftogaz, which proved to be a positive development for Ukraine in the long term.[41] That step upset Ukraine's oligarchs and ruling class, many of whom were Tymoshenko's political rivals.

When her longtime nemesis and political competitor Viktor Yanukovych (prime minister 2006–2007, president 2010–2014) returned to the presidency in 2010, the 2009 gas deal was used to launch criminal cases against Tymoshenko, including a charge related to her having ordered Naftogaz to sign the 2009 gas contract with Gazprom. Moscow-friendly Yanukovych and Tymoshenko were pitted against each other, given their differing views on how to negotiate gas prices with Russia. Yanukovych disagreed with Tymoshenko's negotiated deal, instead favoring lower prices from Gazprom and fewer European proposed reforms. These tradeoffs remained salient, and Ukrainian leaders have been consistently trying to renegotiate the 2009 contract ever since.[42] In the end, the Yanukovych government charged Tymoshenko with abuse of power in the 2009 gas deal and sentenced her to seven years in prison, where she was incarcerated until Ukraine's Maidan revolution of 2014. Meanwhile, presidential influence over gas contract negotiations with Gazprom never reached previous levels. Yanukovych handed responsibility over gas to the Cabinet of Ministers, leaving only strategic issues demanding his attention. This Ukrainian presidential-ministerial arrangement was further solidified when President Petro Poroshenko's political block selected the minister of energy in 2014, therefore controlling the process of gas contract negotiations between Russia and Ukraine.

The Domestic Implications of Ukraine's Gas Dependence

Ukraine's imports, production, consumption, and transport of gas have been plagued by corruption and political manipulation since the country's independence in 1991. Energy scholar Margarita Balmaceda has demonstrated how Ukraine's highly bureaucratized energy sector was susceptible to manipulation by political elites and oligarchs, who achieved a "de-facto privatization" of the country's energy infrastructure and "capture of state institutions."[43] Indeed, the financial stakes in Ukraine for oligarchs and political elites have always been high, given the country's heavy dependence on gas and its critical role in Russian gas transit guarantee huge sums of money. These factors, plus the general lack of public accountability, have made gas a highly lucrative industry of choice for political elites who use it to maintain economic and political power in the country. According to Balmaceda, the government has even contributed to the "cycle of rent-seeking" where financial rents from the energy sector are used as resources for political campaigns to influence the outcomes of elections, thus further consolidating the power and influence of such interest groups for future rent-seeking.[44] The sheer size of the rents received by oligarchs and politicians contributes to the inability to reform the system.[45] Anders Aslund, a scholar of Ukraine, estimates that between 1996 and 1999, a constantly changing group of Ukrainian oligarchs "netted at least $1 billion a year" through lucrative gas deals.[46] Similarly, economist Chi-Kong Chyong and Russia expert Celeste Wallander have argued that oligarchs in sectors that rely on gas imports (mainly metal and chemical industries) have lobbied the government to keep Russian gas prices to Ukraine low, at the cost of maintaining low transit fees on Russian gas headed to Europe.[47]

A look at Naftogaz and other intermediaries demonstrates the complex and troubling web between gas companies, oligarchs, and political interests. Until 2009, Naftogaz was primarily responsible for domestic pipelines and imports, but imports were dependent on private intermediary companies that bought gas from Gazprom or its subsidiaries in Turkmenistan and resold it to Naftogaz and Ukrainian industrial companies. In addition to Tymoshenko's UESU, a key example of these intermediaries was RUE, which was established in July 2004 in an agreement between Putin and Kuchma.[48] From its inception, RUE served as the primary middleman company in the Russian-Ukrainian gas trade until the crisis of 2009. Gazprom owned half of RUE. Dmitry Firtash—a prominent Ukrainian oligarch who came to symbolize the opacity and wealth of the Ukrainian gas industry—owned 45 percent via a holding company. RUE's operations were clouded in political secrecy, though it is suspected to have made payouts to members of Yanukovych's pro-Russia Party of Regions to secure loyalties and political favors. RUE's operations depended on working the system to purchase gas from Russia and Central Asia and reselling it to Naftogaz, as well as on taking advantage of low transit costs through Ukraine and profiting from re-exporting gas to Europe and CIS countries.[49] Between 2004 and 2007, RUE sold a total of $21 billion worth of gas and made a profit of $2.3 billion, while Naftogaz accrued over $500 million in debts to RUE.[50] Most importantly, with this arrangement, Naftogaz lost its monopoly over the reexportation of gas to Europe and CIS countries.

The election of Yanukovych to the presidency in 2010 gave a boost to Firtash and a new "family" of oligarchs, as Firtash's allies and associates were appointed to government positions overseeing the gas industry. The intricate web of gas traders, local businessmen, politicians, and Gazprom interest groups created a particular political environment in Ukraine, where involvement in the gas sector and political favors were connected, and corruption rampant.[51] Not only was the country highly dependent on Russia for natural gas, but its elites were also dependent on Russian gas contracts for their wealth and power. The corruption, lack of transparency, and influence of gas interests in the domestic and foreign policy of the country became endemic to Ukraine regardless of who was in power in Kyiv. Its domestic, political, and economic systems became subject to destabilizing forces as a result of the vast flows of resources through its territories and the broader geopolitical tensions. While Belarus, the Baltic states, and other EU and CIS countries exhibited similar conditions to various degrees, in Ukraine these conditions were entrenched and their scale largely unprecedented.

The Foreign Policy Implications of Ukraine-Russia Gas Relations

Russia's dependence on Ukrainian territory for the transit of its gas exports has been one of Moscow's main motives for keeping Ukraine in its sphere of influence. In many regards this is not surprising, because a country so dependent on revenues from oil and gas exports bears even greater risk by being dependent on a foreign territory where there is no overarching body to resolve regulatory, legal, and contractual risks.[52] Russia's dependence on Ukraine for transit and Ukraine's gas dependency on Russia have had seismic consequences for the two countries and for Europe. The ramifications eventually challenged Europe's post–Cold War order in the Russo-Ukrainian conflict of 2014.

The conflict originated in Ukraine's gas debt of the 1990s. Its earliest turning point was Ukraine's "debt-for-fleet" lease agreement of 1997, when it let Russia station its Black Sea Fleet, together with ground forces and aircraft, in the Crimean city of Sevastopol until 2017.[53] In exchange, Russia agreed to write off a Ukrainian debt of $726 million, owed mostly to Gazprom. Kyiv probably understood the security risks of accepting the Russian fleet, but at the time, Moscow was perceived as more of an ally than a threat, and Ukraine's government chose to compromise with Moscow over the fleet to avoid further tensions. The continued presence of the Russian navy on the peninsula influenced the political situation in Sevastopol and Crimea. The Black Sea Fleet directly owned a popular TV and radio channel, a newspaper, and a printing establishment, all of which contributed to a propaganda campaign promoting separatism among the Crimean population.[54]

Starting in 2010, the Ukrainian government under the leadership of Yanukovych prioritized Russian gas price discounts rather than security considerations. The Kharkiv Agreements of 2010 extended the 1997 agreement, giving Ukraine a ten-year discount of $100 per tcm for Russian gas in exchange for a twenty-five-year extension of the base from 2017 to 2042. This agreement was already highly criticized by the opposition and society at large and was one of the lesser-known reasons for the Maidan revolution.[55] After Russia annexed Crimea, it reneged on the agreement, scrapping the payments to Ukraine. Although Russia then offered to apply the gas discount through export duty relief, Ukraine, not recognizing Crimea's annexation, has insisted on upholding the agreement and asking Russia to implement the lease payment and the gas discount.[56] Overall, the looming gas debt and desire to gain discounts on Russian gas have proved disastrous to the Ukrainian state and its sovereignty. Not only has the country suffered from entrenched gas interest groups and accompanying corruption and political weakness, but it has also lost territory to Russia and since 2014 is facing a military conflict in eastern Ukraine.

Ukrainian policy vis-à-vis the EU has also been deeply affected by its gas dependence on Russia. Most significantly, this dependence and consideration for gas prices contributed to President Yanukovych's refusal to sign the EU-Ukraine Association Agreement, which ignited the Maidan revolution. During the lead-up in December 2013, Russia had reportedly offered to buy roughly $15 billion of Ukraine's government bond debt and slash gas prices from more than $400 to $268 tcm, seemingly in return for Ukraine's choosing closer cooperation with Russia rather than the EU.[57] At the same time, the EU's interest in the future and stability of Ukraine and potentially even the offer of the Association Agreement was related to the country's role as a transit territory for Russian gas to European markets. Since the crisis in Ukrainian-Russian relations of 2014, all gas negotiations between Ukraine and Russia have included the EU as a mediator.[58] Ukraine has also appealed to the European Energy Charter Treaty, which provides norms for cross-border cooperation in the energy sector, to persuade EU countries to reexport Russian gas to Ukraine, which they have been doing since 2014. Overall the EU has largely supported Ukraine's role as its main gas transit country even after the Moscow-Kyiv crisis in relations.

The Transformation of Ukraine's Gas Sector

The ongoing transformation in Ukraine since the 2014 Maidan revolution, change in government, and conflict with Russia has resulted in important though slow shifts in Ukraine's gas sector. These have also demonstrated the extent to which conditions systemic to gas transit states and import-dependent states can be changed via reform, and to what extent they are sticky. The country's dependence on and links to Russia are gradually weakening, but a notable transformation will be determined by the success of four actions: (1) tackling corruption, (2) reforming the gas sector and complying with EU regulations, (3) diversifying gas imports with reverse flows and LNG, and, possibly most important, (4) ramping up domestic gas production. In turn, all of these actions depend on the stability of the government and the sustainability of Ukraine's broader domestic energy and political reforms. A true transformation will also depend on external actors and changes in the global gas markets, as well as the possibility of a resolution to, or de-escalation of, the conflict in eastern Ukraine. Uncertainty remains high as to whether these conditions can be met.

Tackling Corruption

Seeking to tackle corruption, President Petro Poroshenko's government has pursued a gradual reduction of oligarchic power over Ukraine's gas and oil business since 2014. One such target was the political "family" of oligarchs associated with Yanukovych's government and with the oligarch Firtash. Firtash himself was detained in Austria on an FBI warrant in March 2014 and accused of bribery, but was released after an Austrian court refused to extradite him to the United States.[59] He reportedly held onto large stakes in the Ukrainian gas market, although his companies have not imported gas from Russia and Gazprom since Yanukovych left power.

Poroshenko has also sought changes in the Ukrainian gas sector's key companies, such as Naftogaz subsidiary Ukrnafta, which sometimes meant moving against former allies. For example, in the spring of 2015, Poroshenko dismissed Ihor Kolomoisky, a Ukrainian billionaire whom Poroshenko had appointed governor of Dnipropetrovsk and who owns 42 percent of Ukrnafta. The government continued to put pressure on Kolomoisky's energy assets and influence in 2015, especially over Ukrnafta, by increasing the state's shares of that company through Naftogaz.[60]

Beyond tackling oligarchs, the new government has also tried to reform and reduce corruption in private regional gas supply companies—"oblgaz companies"—which rent gas supply infrastructure from state and local governments and have been accused of supplying low quality gas.[61] In addition, the government lowered the unmeasured levels of gas supply for consumers and made local gas supply companies install gas meters in order to prevent fraudulent sales by local gas companies to the industry.[62]

Despite the apparent efforts by Poroshenko's government to reform the energy sector and address corruption, real change will be slow. The majority of Ukraine's elites have no incentive to reduce gas dependence on Russia or improve the transparency and functionality of Ukraine's domestic gas sector, because they profit from the system.[63] Thus it is likely that vested interest groups will remain lobbyists for Russian gas and hinder real change in the gas sector. The catch-22 is that Ukraine's reforms and diversification from Russian gas would reduce corruption and the power of vested interest groups in the gas sector, but while these conditions exist, they make it difficult to break the cycle of rents or pursue long-term investment and strategies toward gas import diversification and domestic production.

Reforms and Complying with EU Energy Regulations

Among the most important changes to Ukraine's gas sector will be the extent to which Kyiv chooses to adopt EU regulations to help transform its energy sector for the long term. Since the 2000s, reforms of the gas sector have been introduced in the Verkhovna Rada, but little progress has been made due to a lack of political will.[64] Since 2013, the International Monetary Fund (IMF) and the World Bank have stipulated gas sector reforms as preconditions for Ukraine to receive further economic assistance.[65] In 2015, Ukraine received $17.5 billion from the IMF out of a potential $40 billion total loan package, but EU and IMF leadership stipulated that the next portion of assistance will not be disbursed without continued reforms, especially in the energy sector. The EU, the IMF, and World Bank have repeatedly urged Ukraine to adopt laws reforming its gas sector in line with the European market standards of the Third Energy Package to ensure Europe and Ukraine's energy security. The main tasks at hand are unbundling the assets of monopoly energy companies and introducing an independent regulatory authority.[66] In 2016, the World Bank agreed to provide Naftogaz with a $500 million loan to purchase gas and get Ukraine through the winter without turning to Gazprom.[67]

Ukraine's sweeping 2015 Gas Market Law seeks to align the country with the Third Energy Package and has "de-oligarchised and demonopolised" Ukraine's gas sector, according to then–Prime Minister Arseniy Yatsenyuk.[68] The law raised gas prices for the general population, ending government subsidization of gas. It levied a fee on regional gas distribution companies (often owned by oligarchs); they must now pay for access to gas transport networks, whereas previously access was free. Additionally, the Gas Market Law sought to create an impartial and apolitical regulatory body from the existing National Energy and Utilities Regulatory Commission of Ukraine (NEURC) to oversee the domestic gas industry. Previously, the president could appoint the head of NEURC while the State Committee for Regulatory Policy could overturn its decisions and findings in provisions that invited political influence if not outright corruption. The new law replaced presidential appointment with more transparent procedures to determine appropriate candidates and removed NEURC from the State Committee's review powers.[69] However, NEURC is still subordinate to the president and is accountable to the parliament, limiting its real independence.

The law also included an addendum promising further parliamentary debate on additional laws, which may include further cuts of gas subsidies. In July 2016, Ukraine approved a resolution that will "unbundle" or split Naftogaz into different operating companies (in charge of transport, production, storage, and supply) to reduce Naftogaz's financial troubles and break its monopoly of the market. Unbundling would make Ukrtransnafta, the transit subsidiary of Naftogaz, an independent company.[70] The aim of creating separate depoliticized companies for different operations is to reduce bureaucracy and facilitate a move toward market-driven rather than government-subsidized pricing of gas.

Eliminating gas subsidies will be a high cost to bear for the industry, but even more for the Ukrainian population, whose annual GDP per capita in 2014 was just over $3,000 and yet who must rely on gas for heating in the winter. Over six million Ukrainians receive energy subsidies, making eliminating them a difficult change politically.[71] On the other hand, if Ukraine creates a more liberal gas market aligned with EU policies, it will boost its potential to complete gas deals with European companies and secure foreign investment to bolster its domestic industry.[72] Reformed gas pricing and a liberalized gas sector will mean the reform of a highly inefficient energy sector and underperforming economy. While the crisis in eastern Ukraine since 2014 has been used as an excuse for the country's slow progress, reforms in the energy sector had failed for years prior.[73] Ukraine's conflict with Russia, the near-collapsing economy, and pressure from international organizations in fact probably provide the greatest incentive and opportunity Kyiv will ever have to push through change.

Reverse-Flow Gas and LNG Imports

After Gazprom cut off supplies to Ukraine in July 2015, Kyiv's need for diverse gas sources led to a critical search for alternative routes for gas imports. The easiest solution has been "reverse-flow" gas—Russian gas pumped from west to east, from Ukraine's European neighbors into Ukraine.[74] Moreover, in 2015, Gazprom officials expressed their goal to bypass Ukraine completely for gas flows by 2020, though this is dependent on the construction of the Nord Stream II pipeline.[75] Ukrainian officials also expressed a desire—verbally, at least—to be completely free of Russian gas imports, and many Ukrainian political and energy analysts such as Mikhailo Gonchar say that Ukraine can subsist on reverse-flow imports alone, without Russia's direct supplies.[76]

Ukraine has the pipeline capacity to import up to 15 BCM per year in reverse flow.[77] Predating the 2014 crisis, reverse flow of imports started from Poland in November 2012, followed by Slovakia and Hungary in December 2013.[78] However, the amounts were still small in 2014, totaling only about 5.1 BCM. Statoil, EON, GDF Suez, and TrailStone provided the bulk of this supply.[79] In 2015, Ukraine imported 10.3 BCM through reverse flow. The same year, Slovak operator Eustream increased reverse capacities to Ukraine by more than 60 percent in comparison to 2014. Bulgaria's Bulgartransgaz and Romania's Transgaz signed a deal with Ukrtransgaz to test reverse flow routes by the end of 2016. France's Engie provided 3.5 BCM of gas, and in November 2016, signed an agreement with Kyiv to increase reverse flow and begin storing its natural gas in Ukraine's gas storage system, a first for European gas companies.[80] But in order to secure significant reverse gas flows from EU countries, the cross-border pipeline network with the EU has to be modernized. With improved utilization of Ukrainian and European pipeline capacity and continued commitment to increasing Ukraine's non-Russian gas routes, Ukraine could theoretically increase its reverse flow capacity to 24 BCM of gas annually, well beyond its 2015 gas import total of 16.4 BCM.[81]

The primary constraint on Ukraine's reverse-flow imports is financial—the need to reduce gas consumption when its economy is facing a severe downturn. In October 2015, the European Bank for Reconstruction and Development loaned Ukraine $300 million over three years to purchase 1.1 BCM of gas from Europe, but the loan is contingent on the ongoing restructuring of Ukraine's gas sector.[82] Yet it is important to realize that reverse flow of gas imports is only a partial solution: it does not improve Russo-Ukrainian gas relations, it accepts that Ukraine will no longer be a gas transit state, and most important, it does little to reform the country's inefficient gas utilization and economy. At most, reverse flows reduce the influence of interest groups that would benefit from direct Russian gas imports.[83]

As part of its efforts to buy gas outside of Russia, Kyiv has also discussed the prospects of building a LNG import terminal, despite its significant price tag. In mid-2014, Ukraine announced plans for an LNG terminal on the Black Sea coast that could import up to 5 BCM, though the plan has seen no progress since then.[84] In 2015, a plan was floated for the US-based company Frontera to help build a terminal in Ukraine to receive LNG from the Republic of Georgia's gas fields.[85] Such exports are unlikely in the near term, because Georgia's gas reserves are still speculative, and it is uncertain whether or when Georgia will gear up its gas development and build an LNG liquefaction and export terminal. In the long term, if Ukraine fails to develop its own gas resources, it could reasonably tap into the growing LNG market, though the prohibition of LNG shipments through the Turkish Straits will limit Ukraine's LNG trade partners, as it will for other Black Sea littoral states.

Increasing Domestic Production

Ukraine's domestic gas production has underperformed and slowed in recent decades, and turning this around will require significant investment and reforms. Historically, Ukraine's gas production peaked in the mid-1970s, reaching 68.5 BCM in 1975; it then dropped to 41 BCM a decade later and fell further to about 20 BCM in the 2000s. By 2013, more than three-quarters of Ukraine's explored gas reserves were depleted, and in 2015 domestic production totaled 19.9 BCM.[86] Since Ukraine's independence, gas exploration and production have also been discouraged by Naftogaz's monopoly over the domestic market. Private investors have found it unprofitable to drill for gas because they have to sell it all to Naftogaz at a low, fixed price linked to the rate that households pay for gas.[87] State enterprises have been unable to afford the greater expense of exploring for unconventional or offshore gas reserves.

Ukraine also did not find reliable local or foreign investors for offshore exploration outside of a single attempt to explore the Prykerchenska area in the Black Sea, south of Crimea.[88] Since Crimea's occupation, most of Ukraine's offshore gas reserves have fallen under Moscow's control, while foreign investors such as Chevron and Shell have pulled out of their projects for onshore explorations because of the increased risks raised by the conflict in eastern Ukraine.

Ukraine could counter these challenges with reforms to boost its domestic production in the long term. First, analysts have suggested that Ukraine reduce its taxes and change its tax laws on new gas drilling projects to encourage investors. Eliminating gas price subsidies for households and letting households purchase gas directly from both private and state companies are other alternatives.[89] Despite the conflict in eastern Ukraine, Kyiv could try to attract private gas drilling companies in the western Olesska shale deposit field, where it is secure and government and regional authorities have already granted permission to drill to companies like Chevron. However, achieving these reforms will be problematic in the near term. Ukraine's financial troubles and overstretched state budget will make reducing taxes difficult, and earnings are still too low for all households to pay market prices for gas. The drop in global oil and gas prices since 2014 likewise decreases incentives for new exploration projects. Finally, Ukraine's reputation for corruption and inefficiency discourages investors from operating in the country. Such factors constrain Ukraine's domestic production and hinder Kyiv's future ability to reduce its dependence on Russia for transit revenues and gas imports.

Belarus

Neighboring Belarus is another key transit state for Russian gas to Europe, one that is almost completely dependent on Russian gas and oil. This stems from its lack of indigenous resources, historical legacies, and Soviet-era infrastructure, and it is the cause of Minsk's frequent troubles with Moscow over energy supply, pricing, and transit fees. Like

Ukraine, Belarus is a highly inefficient gas consumer—it has been described as a "gas addict"—and it, too, has experienced a cycle of gas debts as well as Moscow's price hike threats and offers of loans and assistance.[90] In contrast to Kyiv, however, Minsk has progressively sold off its energy infrastructure to Russian energy companies, primarily Gazprom.

Natural gas accounts for an astounding 60 percent of Belarus's total energy consumption as of 2015.[91] Oil provides about 33 percent, with coal and hydroelectric power contributing less than 1 percent each.[92] From 1995 to 2015, Belarus consistently imported at least 86 percent of its gas needs, and from these imports, Russia provided 95–98 percent, or 19 BCM in 2015.[93] In 2015, Belarus met only 1 percent of its gas needs domestically through its small and underperforming gas production industry and tiny natural gas reserves, estimated at less than 3 BCM.[94] The Pripyat Basin, which stretches through Ukraine to the southeast corner of Belarus, is estimated to have significant unproven gas reserves, although it has yet to be explored beyond initial ventures by Belorusneft, and the bulk of the gas resources are estimated to be in Ukrainian, rather than Belarusian, territory.[95] Thus, for the foreseeable future Belarus will remain a gas importer.

Belarus's vast network of gas pipelines, at 7,870 kilometers, is second only to that of Ukraine and is a second crucial route for Russian gas to European consumers. This gas transit depends on two pipeline systems—Northern Lights and Yamal-Europe—both of which are owned and controlled by Gazprom. The Northern Lights pipeline is one of Russia's largest gas export systems to Europe, carrying up to 51 BCM of gas per year to Belarus and then onward to Poland, Lithuania, and Ukraine via five smaller branches.[96] Gazprom acquired its Belarusian section in 2011, when it gained full ownership of the Belarus state pipeline operator, Beltransgaz, which was renamed Gazprom Transgaz Belarus. The Yamal-Europe pipeline was built by Gazprom, which also wholly owns its Belarusian section via Gazprom Transgaz Belarus. With an annual capacity of 33 BCM, Yamal-Europe passes through Russia, Belarus, Poland, and Germany, and accounts for a fifth of Russian gas exports to Europe and 63 percent of all Russian gas transported through Belarus.[97]

Via Gazprom Transgaz Belarus, Gazprom also owns Belarus's small underground gas storage infrastructure of combined 1 BCM capacity.[98] Amid Ukraine-Russia tensions and Kyiv's efforts to turn westward, Moscow has favored gas transit via Belarus through the Gazprom-owned pipeline infrastructure rather than via Ukraine. Transit volumes of Russian gas via Belarus increased from nearly 25 BCM in 2001 to 45 BCM in 2014.[99] Gazprom would like transit via Belarus to increase further, though plans to build a second branch, Yamal-Europe II, have seemingly been postponed indefinitely.

Gas Prices, Debts, and Tensions

Gazprom's acquisition of Belarus's gas infrastructure has largely been a story of frequent tensions between Minsk and Moscow over gas prices, transit fees, and debts. It illuminates Gazprom's overall investment strategy in gas-vulnerable states, where allowing an accumulation of gas debts and threatening price hikes facilitates its eventual acquisition of assets. Moreover, it reflects Gazprom's broader concerns about the security of its exports and its efforts to gain control over transit infrastructure for its gas and oil.

Gazprom and Minsk have been in a nonstop cycle of gas price and debt disagreements since 2004. In a first significant attempt to ratchet up pressure on Minsk, Gazprom suspended gas shipments to Belarus in 2004, accusing it of siphoning gas from pipelines and breaking agreements on the privatization of Beltransgaz.[100] Earlier, Belarus had agreed to sell 50 percent of Beltransgaz to Gazprom in exchange for a five-year plan of gas sales to Belarus at Russian domestic gas prices. But the two parties did not agree on a price for Beltransgaz in 2004, when Minsk asked for $5–$6 billion while Russia offered $500–$600 million.[101] The crisis was resolved after an agreement increasing Belarus's price for gas from $35.60 per tcm to the still very low $46.68 per tcm.

Just two years later, in late 2006, Gazprom threatened Beltransgaz with a more than four-fold increase in price, to $200 per tcm. This was in no small part due to rising global energy prices, but tensions escalated.

The confrontation was followed by a new wave of Russian import restrictions on Belarusian goods, including sugar, confectionery and candy products, and alcohol.[102] In January 2007, Russia and Belarus clashed over oil transit and oil export taxes, with Russia accusing Belarus of siphoning oil from the Druzhba pipeline to resell for a profit. Oil transit and sales were briefly halted between the two countries.[103]

The oil crisis of 2007 spilled over into ongoing tensions over gas prices and helped to finally secure Gazprom's first round of acquisition of Beltransgaz. That purchase became part of a larger Minsk-Moscow deal to avoid a longer shutoff of oil transit through Belarusian pipelines. On January 1, 2007, Belarus and Russia reached an agreement stipulating a gas price of $100 per tcm for Belarus and the sale of 50 percent of Beltransgaz for $2.5 billion. In addition, Belarus agreed that prices for natural gas would be slowly increased to reach European levels, and in return, Minsk could charge Russia a higher gas transit fee.[104] However, by August 1, 2007, Gazprom was threatening to cut off gas supplies by 45 percent after Belarus failed to pay off its $456 million debt for gas, among other charges. Belarus scrambled to pay $190 million of its debt by August 3, thereby avoiding an immediate gas shutoff.[105] Through its management (and manipulation) of the issue, Belarus found ways to delay and soften the effects of its gas price increase throughout 2007. Belarus won an agreement in which, for the first half of the year, it would pay only 55 percent of the new price of $100, with the rest put on credit to be paid by the end of the period. In December 2007, Russia extended a $1.5 billion stabilization loan on favorable conditions to help Belarus pay for the increased price of gas.[106]

However, by June 2010, the dispute flared once again. Gazprom cut Belarus's gas supplies by 15 percent and demanded payment of a $192 million debt. President Alexander Lukashenko claimed that Gazprom owed Belarus $260 million in unpaid transit fees and called Russia's treatment of its neighbor a "humiliation" of the Belarusian people.[107] At that, Gazprom cut its supply by an additional 15 percent. By the end of the month, Belarusian officials stated that the country had paid its debts to Gazprom in the amount of $187 million, and gas flows were normalized.[108] In 2011, Gazprom bought the rest of the Beltransgaz, including the Belarusian section of the Yamal-Europe pipeline, as part of a loan and discounted gas package worth $14 billion. In that deal, Belarus, desperate for economic assistance amid a financial crisis, received a $10 billion twenty-five-year loan toward its planned nuclear power plant and $2.5 billion cash for half of Beltransgaz.[109]

The early 2010s were relatively calm, and Russia charged "comfortable" rather than market gas prices to Belarus.[110] However, another dispute erupted in July 2016, when Belarus asked for a gas price cut from $132 tcm to $73 tcm to reflect the drop in global energy prices. Meanwhile, Moscow pointed to Belarus's unpaid gas debt of some $200 million for the first half of 2016, and in the absence of remaining gas infrastructure to use as a bargaining chip, Belarus may have to bargain with military alliances or its other businesses to lower its debt.[111] Indeed, tensions over gas supplies, debts, prices, and transit fees can erupt any day between Minsk and Moscow.

Overall, the lower prices that Belarus traditionally enjoyed from Gazprom offered some short-term benefits, yet left it both vulnerable in its dependence on Russian gas and able to avoid reforms, diversification, and energy-efficiency programs. As Belarusian energy expert Tatiana Manenok explained, "Belarus does not need to look for Russian energy alternatives because it already receives the cheapest oil and gas in the post-Soviet world."[112] These Russian energy subsidies have been substantial, totaling about $14–$15 billion between 2004 and 2008.[113] According to IMF estimates, in 2004 specifically, the subsidies amounted to 10 percent of Belarus's GDP, with 6–7 percent being the result of subsidized gas prices and 3 percent being the result of lowered oil prices in comparison to those paid by European customers.[114] Even while Russia sought to increase gas export prices in the 2010s, Belarus consistently paid the lowest prices. For instance, in 2014, Gazprom delivered gas to Belarus at an average price of $164 per tcm, while Germany paid $323, and Poland $379.[115]

At the same time, its dependence on cheap Russian gas left Minsk with little choice but to acquiesce to the conditions proposed by Moscow for the various gas deals, making Belarusian domestic and foreign policy further dependent on Russia. Balmaceda noted that Lukashenko managed to score short-term gains vis-à-vis Russia more successfully than states like Ukraine, but has not been able to reverse Belarus's role as a "client" country to Russia or secure other long-term energy supply alternatives.[116] Others have pointed out that unlike Ukraine, which views energy dependence on Russia as a threat to its security, Belarus has approached dependence on Russian energy as a "fact of post-Soviet life" and thus will remain a "captive market . . . to a significant degree" for the foreseeable future.[117] By acquiring the majority of the Belarusian pipeline system, Moscow has reduced some of the uncertainties related to transporting gas through Belarusian territory, and Minsk's largely cooperative stance toward Moscow makes it likely that Russia will increasingly favor Belarus rather than Ukraine as its key transit state. However, the fact that Belarus has already sold off most of its infrastructure to Gazprom raises the question what future concessions Minsk will have to make to Moscow to settle its ever-mounting gas debts.

New Opportunities Not Taken

The geopolitical developments in Europe and the changes in the gas markets in the 2010s have offered Belarus some new opportunities to diversify its gas imports and evolve from its role as a transit territory for Russia. In 2015, Kyiv asked Belarus to transit 1.5 BCM of gas from Lithuania's new LNG terminal to Ukraine in reverse flow. However, Gazprom Transgaz Belarus refused to serve as a transit operator for this contract, a decision it declined to explain publicly.[118] Belarus reportedly has shown some interest in importing gas from the Lithuanian LNG terminal for domestic consumption, but as of 2014, former Lithuanian energy minister Arvydas Sekmokas said only "very insignificant volumes" would be exportable unless the existing infrastructure is improved.[119] Overall, as a landlocked country, Belarus is unable to directly import LNG, and if in the future it seeks to diversify its gas sources, it will have to rely on reverse flows of piped gas from neighboring countries such as Lithuania or Ukraine.

Belarus instead has pursued deeper energy relations with Russia, offering in August 2015 to increase its transit of Russian gas to Europe.[120] The offer coincides with Gazprom's construction of its new headquarters in Belarus, a project worth approximately $250 million and slated to be finished by 2018.[121] Overall, the Lukashenko regime prefers to maintain the dividends of cheap oil and gas that come from being a loyal ally to Moscow, rather than pursue gas diversification policies or new opportunities.

Domestic Political and Foreign Policy Implications

Belarus's choices in the gas sector have had serious implications for its domestic politics; the country's dependence on Russian gas and Lukashenko's reputation as the Kremlin's ally are intrinsically linked. Often called "the last dictator of Europe," Lukashenko has been the primary decision-maker for Belarus during his more than two decades in power, interacting directly with the Kremlin and Gazprom leadership to ensure gas and oil deals in exchange for political support. He has said that Russia and Belarus are "equally dependent on each other."[122] Indeed, Minsk is an important customer of Gazprom; Belarusian consumption of natural gas constituted 9 percent of Russia's total gas exports in 2014.[123]

On the other hand, Belarus's gas deals with Moscow are often coupled with economic aid packages that impact the domestic politics of Belarus and tie Minsk's foreign relations to Russia. For instance, Belarus's sale of Beltransgaz resulted in Russian approval of a $3 billion loan from the Stabilization Fund of the Eurasian Economic Community (EurAsEc) and a reduction in the 2012 price of gas to $165 per tcm, a yearly saving of $3 billion over Belarus's 2011 gas bill.[124] Taken together, the injection of $2.5 billion for the purchase of Beltransgaz shares and $1.24 billion for the first two installments of the EurAsEC loan contributed to "a short-term economic respite" for Belarus during one of the most challenging economic downturns the country had experienced in the last decade.[125] Likewise, it eased pressure on Lukashenko's regime. Belarus's economy has not been able to function without foreign economic aid, and Russian economic assistance in the form of gas subsidies and linked economic aid reached approximately $28 billion between 2012 and 2015, in comparison to $5 billion from European and Western financial institutions. This difference alone reinforces Belarus's ties to Russia.[126]

Gas prices have also frequently been a source of contention between the relatively poor Belarusian population, which relies on gas for heating in the winter months, and the government. Thus, Lukashenko has used cheap gas deals to stifle domestic discontent. At the same time, the lower gas prices have meant a further delay in the reform of Belarus's economy away from reliance on large, energy-inefficient industries.[127] Lukashenko is supported domestically by a network of powerful and wealthy oligarchs, many of whom benefit from lower gas prices, as in Ukraine; however, the oligarchs have not been as influential in the national gas sector because Gazprom itself owns the gas transit system, unlike in Ukraine.[128] Thus, according to Balmaceda, Lukashenko has established a vertical line of control over foreign and energy policies toward Russia, taking advantage of a lack of transparency in the domestic gas sector to use bilateral energy deals for personal political gain.[129] His 2015 "reelection" for another five-year term makes significant changes unlikely in the near to medium term. Only a dramatic turn of events, such as a popular uprising, Moscow's territorial incursions vis-à-vis Belarus, or turbulence in the Kremlin itself, might precipitate some changes.

At the same time, the 2016 gas price dispute demonstrated that Minsk will point to changing global gas market conditions to try to negotiate a gas price discount, even if it is not necessarily interested in making long-term investments to diversify its gas sources. Minsk's efforts to free-ride on lower gas prices might be perceived as a clever strategy. However, these temporary gains should be considered a Pyrrhic victory in the long term. Minsk has become fully dependent, and has very few options to escape the straightjacket of Russian gas imports and transit and Gazprom's infrastructure.

Conclusion

Despite being important to both Europe and Russia for their gas flows, Ukraine and Belarus hold questionable leverage in their energy relationships with Moscow. Certainly, they have benefited for years from transit revenues and lower gas prices, but these benefits have accrued to each country's elites and vested gas interests rather than to the economy or the population at large. Depending on such rents and subsidies has hindered any energy diversification, efficiency, or domestic production efforts, entrenching the status quo of transit and import dependence and gas debts. Indeed, Gazprom's flexibility on gas prices has been limited. Since the 2000s, Gazprom sought to raise the prices of its gas exports to market levels even for countries of the "near abroad," making Minsk's or Kyiv's long-term hopes for lower prices in exchange for concessions unrealistic without the diversification that allows imports from elsewhere and improves their bargaining positions.

In the past, Gazprom also issued threats to increase gas prices if Ukraine or Belarus did not agree to sell off their energy assets, repay their debts, or follow Moscow's political lead. The massive run-up of debts by both countries in the end served Russia's interests; these were used to facilitate not only Gazprom's acquisition of local gas infrastructure but also, in Ukraine's case, Moscow's lease of the bases in and around Crimea. In the end, Ukraine's gas debts indirectly cost the country its territorial integrity, with Russia's annexation of the peninsula. The Russo-Ukrainian conflict, spilling over into a separatist war in the Donbas region, has spurred some changes in the country since 2014. Among Ukraine's most significant achievements as a result of this crisis is increased reverse flow of gas from European states and the start of energy-sector reforms. However, it is uncertain whether Ukraine will be able to stay the course. Its ability to complete a critical mass of reforms to secure its gas sector through diversification, domestic production, and efficiency programs remains at the whim of a number of domestic, regional, and international political and economic factors.

This established status quo of gas transit and import dependence and the accompanying corruption and entrenched interest groups will constrict the abilities of both Ukraine and Belarus to seize opportunities that arise from changes in global gas markets, such as the shale boom or greater interconnection of gas markets in Europe. Yet Kyiv has at least made an effort to alter its gas predicament, being forced to do so in part by the crisis in relations with Russia and threats to its state since 2014. Although Ukraine currently seems to be in a much more difficult situation in facing conflict and the Kremlin's aim to reduce if not eliminate transit via the country, it still has more options than Belarus in the long term.

Looking forward and considering the transformation of the global sector, we can expect that the role of transit states will diminish. As European countries seek access to LNG and greater pipeline interconnection with their neighbors, enabling different supply routes, the importance of gas transit via traditional routes such as Ukraine and Belarus will decrease. As a result, in the eyes of the EU, the strategic importance of Ukraine and the Moscow-Kyiv energy relationship may decrease. If Russia's strategy to turn to Asian gas markets is eventually successful, the effect for the EU will be the same. As a result, while Belarus may be satisfied serving as the fallback route for Russian gas exports and Moscow's backyard, Ukraine will have to carve a new identity, not only in the context of the gas markets but also in the politics of Eurasia.

6 The Politics of Isolated Suppliers: The Caucasus and Central Asia

The regions of the South Caucasus and Central Asia lie at the crossroads between Europe, Russia, and Asia and are continuously shaped by the political, economic, and energy developments of their powerful neighbors. This is a region of contrasts, of both energy-rich and energy-poor countries. The energy poorest countries of these regions—Armenia, Georgia, Tajikistan, and Kyrgyzstan—have been exposed to even more vulnerability and pressure than Ukraine and Belarus in their energy relationship with Russia. Meanwhile, the isolated gas producers of Central Asia and the South Caucasus of Azerbaijan, Turkmenistan, Uzbekistan, and Kazakhstan had difficulty competing with Russia's gas riches in the post–Cold War era and were dependent on Russian pipeline infrastructure for exports in the 1990s. As former Soviet republics and subsequently largely cooperative neighbors, these countries have long been considered to be in Moscow's "near abroad" backyard, but since the 2000s, due to geopolitical shifts, changes in global gas markets, access to Western technologies, and evolving gas prices, they have had some success in stepping out of Moscow's shadow and reaching new gas customers such as China and European states.

The study of the gas producers of the South Caucasus and Central Asia will serve to exemplify the politics of isolated suppliers. Their case demonstrates how even gas-producing states can be disempowered in the geopolitics of gas when they do not have access to diverse markets or control of their own routes and infrastructure for exports. Besides earlier political and infrastructural constrains, the export opportunities of these producers have been greatly constrained by their geography. All five Central Asian countries of Kazakhstan, Kyrgyzstan, Tajikistan, Turkmenistan, and Uzbekistan, and two of the South Caucasus states, Azerbaijan and Armenia, are landlocked. The region's false "sea"—the gas-rich Caspian Sea—bordered by Azerbaijan, Kazakhstan, Turkmenistan, and Iran—connects the three distinct regions of the Caucasus, Central Asia, and the Middle East. The Caspian Sea is the largest enclosed inland body of water with no outflows and it offers no routes for exports by sea beyond the Caspian-bordering states.

The landlocked nature of these gas producers and their region limits exports to land-based or trans-Caspian pipelines and excludes the possibility of entering the growing global liquefied natural gas (LNG) trade. Indeed their landlocked position limits the extent to which they can become global rather than regional gas suppliers.

At the same time, despite their inherent geographic limitations, the production and exports of countries bordering the Caspian Sea have been influenced by broader global political and economic developments, especially evolving technology and prices in the gas markets. During the Soviet era, Moscow controlled the Caspian Sea energy reserves of the Soviet republics, though officially the sea resources belonged to all the bordering countries equally. Nonetheless, this made little difference, because the Soviet Union did not have the technology to exploit offshore oil and gas. The Soviet Union's demise opened the region to external actors and allowed foreign companies to develop energy reserves and construct alternative pipeline routes. New Western technology accelerated the exploitation of offshore oil and gas fields in the Caspian Sea. The rise of oil and in turn gas prices in the 2000s also created greater incentives for large scale investments in gas pipeline infrastructure to enable exports to other markets. Thus by 2009, the Central Asian gas producers broke out of isolation, ended their dependence on Russia, and launched gas exports to China. Meanwhile, Azerbaijan launched exports to Georgia and Turkey in the mid- to late 2000s and has plans to reach European markets by 2019.

However, since the 2010s, changes in the global gas markets, such as competition from LNG and the shale gas boom, have also reduced the significance of Caspian or Central Asian gas for Russia, Europe, and China. Both Europe and China now have a number of alternatives in their gas diversification and supply strategy, especially in the form of LNG. Unlike goliaths such as Russia or the United States, which can engage in the politics of gas supply, these isolated producers demonstrate the more limited role that smaller, newcomer suppliers could play in gas trade and the geopolitics of gas in their region and beyond.

The South Caucasus

Nestled between the Black Sea and the Caspian Sea, the South Caucasus is where Europe meets Asia. It includes the former Soviet republics of Armenia, Azerbaijan, and Georgia, as well as part of Turkey and Iran. Because it lies between the European Union (EU), Russia, Central Asia, and the Middle East, for decades the South Caucasus has been considered as a potential transit corridor of energy resources between the landlocked Caspian basin and European and Turkish consumer markets. However, each of the three countries of the South Caucasus is distinct, and their complex relations with each other and with other states of the region, particularly Russia, complicates many matters, including their energy policies.

Azerbaijan stands out among the three states of the South Caucasus due to its notable oil and gas reserves and access to the Caspian Sea. Azerbaijan and Armenia do not have diplomatic relations because of an ongoing bloody conflict in and around Azerbaijan's separatist region of Nagorno-Karabakh. Meanwhile, Baku maintains economic ties with Russia while simultaneously doing its best to assert its independence. For the energy poor and regionally vulnerable Armenia, Russia has become an unlikely friend, security guarantor, and seeming protector of Armenia in its regional tensions with Azerbaijan and Turkey. With Black Sea access, Georgia is also the only non-landlocked country in the region and is emerging as an important player in gas diplomacy, particularly as a transit state for Caspian gas.

Azerbaijan

Azerbaijan is the most important gas player in the South Caucasus and a historic energy powerhouse. However, to date the present-day capital of Baku has had limited ability to parlay its vast gas resources into significant exports or influence. Azerbaijan has only recently emerged as a gas producer, though historically it was one of the world's oldest oil producers.[1] In 1847, during Russian tsarist rule, the first oil well was drilled on the Absheron peninsula, where Baku is located; the well preceded by eleven years the first American oil well in Pennsylvania. The second half of the nineteenth century saw an explosion of oil production, so that by 1901 Baku, or the "Black Gold Capital," had more than 3,000 oil wells and contributed to half of the world's output.[2] During the Soviet era, Azerbaijan's oil production was decimated first after the Bolsheviks occupied the country in 1920 and nationalized its oil assets, and later in the 1980s because of discovery and development of other oil fields in the Soviet Union.[3] Following its independence, in 1994, Azerbaijan made a comeback as an important regional energy player when it signed the "Contract of the Century" for oil development with more than a dozen international companies from the United States, Great Britain, Russia, Turkey, Norway, Japan, and Saudi Arabia.[4]

Azerbaijan's gas industry was a late bloomer in comparison to the oil industry. Its eventual success owed much to access to Western technology in the 1990s, which allowed for accelerated exploitation of offshore oil and gas fields.[5] Although tensions between Azerbaijan and Turkmenistan over demarcations of Caspian reserves initially posed difficulties, in 1996 an international consortium of companies, including lead operator British Petroleum (BP, 25.5 percent) and Statoil (25.5 percent), State Oil Company of the Azerbaijan Republic (SOCAR),

Lukoil, Naftiran Intertrade Company (NICO), Total (10 percent each), and Turkish Petroleum Corporation (TPAO, 9 percent), signed a production sharing agreement to prospect Azerbaijan's section of the Caspian Sea.[6] The discovery of the offshore Shah Deniz gas field in the Caspian Sea just southeast of Baku in 1999 seemed to promise another heyday for the country, considering that its overall reserves are estimated at over 1 trillion cubic meters (TCM).[7] The start of exploration of Shah Deniz also coincided with the rise of global energy prices from its lows of the 1990s to the highs of 2000s, which further incentivized investment and exploration. In 2007, BP announced it had discovered a new, high-pressure reservoir beneath the northern flank of the Shah Deniz field. This additional development—the Shah Deniz Stage II—is a giant project that will add a further 16 billion cubic meters (BCM) of gas production per year to the nearly 10 BCM annual production of Shah Deniz I.[8] Gas from this second stage of development is expected to supply Turkey in late 2018 and subsequently Europe.[9]

Shah Deniz may play a significant role in the EU's gas efforts to diversify away from Russian gas. The Southern Gas Corridor pipeline project, with a planned annual capacity of 31 BCM by 2026, will deliver gas from Shah Deniz to Italy via Azerbaijan, Georgia, Turkey, Greece, Albania, and Italy.[10] Moreover, the first of the three pipelines that will make up the Southern Gas Corridor, the South Caucasus Pipeline (SCP), is already in operation and has already served to improve the energy security of Georgia. Also known as the Baku-Tbilisi-Erzurum (BTE) pipeline, it runs parallel to the Baku-Tbilisi-Ceyhan (BTC) oil pipeline for 429 miles through Azerbaijan and Georgia before landing in Erzurum, Turkey. In late 2006, the SCP started exporting gas to Georgia and in 2007 to Turkey from Shah Deniz. In 2008, the pipeline exported some 8.4 BCM annually but there are plans to eventually expand its capacity to about 25 BCM by the end of Shah Deniz Stage II development in 2017. What is most unique in the post-Soviet space of gas infrastructure is that Russian companies do not dominate the SCP or the other Southern Gas Corridor pipeline ownership. While Russian Lukoil holds a 10 percent share, the main shareholders of the SCP are BP (28.8 percent),

Turkish TPAO (19 percent), and Malaysia's Petronas (15.5 percent), as well as the National Iranian Oil Company's subsidiary Naftiran Intertrade Company (NICO) and Azeri companies AzSCP and SGC Midstream.[11]

The SCP pipeline will link to two other key pipelines of the Southern Gas Corridor. At the Georgia-Turkey border it will connect to the Trans Anatolian Natural Gas Pipeline (TANAP), which will continue to the Greek border.[12] TANAP's construction started in March 2015 in Turkey and is scheduled for completion by the end of 2019.[13] TANAP's shareholders include Azeri national energy company SOCAR (58 percent), Turkey's BOTAŞ (30 percent), and BP (12 percent).[14] At the Greek border TANAP will link to the planned Trans Adriatic Pipeline (TAP) to transport Azerbaijan's natural gas exports through Greece and Albania to Italy. TAP's shareholders include BP (20 percent), SOCAR (20 percent), Snam (20 percent), Fluxys (19 percent), Enagás (16 percent), and Axpo (5 percent).[15] In July 2015, the first construction work began on Albanian roads and bridges to access future pipeline construction sites,[16] and TAP's construction inauguration ceremony was held in Thessaloniki, Greece on May, 17, 2016. Deliveries to the EU via TAP are planned for early 2020, and before the pipeline's completion were already contracted at about 10 BCM of gas annually.[17] Future plans also include a potential fourth leg to the project—a Trans-Caspian pipeline to bring additional gas from Turkmenistan to Azerbaijan and fill the SCP, TANAP, and TAP with additional resources.[18]

When completed, the Southern Gas Corridor has the potential to diversify gas imports of southeast Europe, primarily in Greece and Italy, where in 2014 national gas consumption was 63 percent and 37.5 percent dependent on Russian imports, respectively.[19]

However, by the time Azerbaijan launches significant exports, the strategic importance of natural gas and Azeri gas may have diminished for Europe. Before the shale boom, the growth of LNG trade, and US LNG exports, Europe's options to diversify its gas imports away from Russia were much more limited. Changes in the global gas markets since the 2010s have created an abundance of gas and low gas prices, while at the same time the EU has also become increasingly interested in efficiency and renewables rather than just diversification. The Southern Gas Corridor will thus be just one of the many new tools and sources that the EU will have at its disposal to reduce the influence of Gazprom and Russian gas on the European continent. What most sets the Southern Gas Corridor apart is the fact that it will bring non-Russian gas by pipelines not controlled by Russia to Europe. It is also significant that the capacity of the Southern Gas Corridor pipelines can be expanded in the future to meet Europe's demand. Meanwhile, Moscow has worked hard to make the Southern Gas Corridor obsolete by forging its own plans for the Turk Stream pipeline to send Russian gas to southeastern Europe via Turkey. Russia will continue to woo Turkey and Azerbaijan—both key states in Europe's plans to access Caspian gas. Azerbaijan already started exporting small amounts of gas (under 1 BCM annually) from Shah Deniz to Russia in 2010.[20]

Due to its status as a landlocked gas producer, Azerbaijan has limited options for becoming a player in the global gas markets. Plans have been floated for an Azerbaijan-Georgia-Romania Interconnector (AGRI) that would include building an LNG terminal on neighboring Georgia's Black Sea coast to export Azerbaijani natural gas onward to Romania and farther to Central Europe. To date however, the project has not moved past the feasibility study stage. Among similar distant opportunities, Azerbaijan also has several shale gas fields in the territory of Gobustan, Shemakha, and other regions, though when production will start, considering the low gas price environment since 2014, remains highly uncertain.[21] In sum, Azerbaijan's greatest role in the gas markets still rests on maximizing pipeline exports via the Southern Gas Corridor to help European states and its more immediate neighbors like Georgia and Turkey diversify away from Russian gas.

Georgia

The Southern Gas Corridor project and Azeri gas have made a great impact on neighboring Georgia. As a historically energy-poor country that had fought a war with Russia in 2008, Georgia long sought to mitigate its former 100 percent dependence on Russian gas. Since 1988, Georgia and Armenia have received Russian gas via the North Caucasus–Trans-Caucasus pipeline. However, by the 2000s, the pipeline was a frequent target of sabotage by separatist movements in the northern Caucasus.[22] The SCP pipeline from Azerbaijan had been under construction since 2004 when, amid political tensions between Tbilisi and Moscow, in the winter of 2006, the coldest in some twenty years, unknown saboteurs bombed the North Caucasus–Trans-Caucasus pipeline, interrupting Georgia's supply.[23] At the time, president of Georgia Mikheil Saakashvili contended that this represented Moscow's political and economic pressure on Georgia, which had been pursuing a strongly pro-Western course since the country's 2003 Rose Revolution. He also argued that the bombing was part of Moscow's effort to pressure Georgia into selling its gas pipeline network to Russian companies.[24] After the attack, Georgia could still import Russian gas through a pipeline in Azerbaijan, but a few days later Gazprom announced that this pipeline was also damaged and could no longer supply Georgia.[25] Tbilisi resorted to importing Iranian gas via Azerbaijan for ten to fifteen days before the pipelines returned to operation.[26] When the SCP was completed at the end of 2006, Tbilisi turned to Azeri gas.[27] By 2015, Georgia depended nearly 90 percent on Azerbaijan for its gas imports and only about 3 percent on Russia, with an additional option to seek Iranian supplies in case of emergency.[28] However, in 2015 and 2016 the Georgian government started controversial discussions of diversification in the other direction—that is, increasing once again its gas imports from Russia.[29] This was met with a public protest movement urging the state to say "No to Gazprom," and eventually the government decided to increase their gas imports from Azerbaijan.[30]

Georgia also holds promise as a gas producer and potentially even as a net exporter. In 2015, an independent American company, Frontera Resources, announced that it discovered massive gas resources in Georgia's region of Kakheti. Estimated at 3.8 TCM, these could be comparable to the Azerbaijan's Shah Deniz 2 field, and would theoretically enable Georgia to become a gas exporter.[31] However, these findings have not been broadly confirmed, and, on the contrary, the Georgian government has maintained that Frontera's estimates are vastly optimistic and that all of the gas and oil produced by the company come from existing Georgian fields.[32] In the near term, Georgia's most important role in the gas markets will be that of a transit country for Caspian gas to European markets via the Southern Gas Corridor project. As the only country out of the South Caucasus and Central Asian states to have access to the Black Sea, Georgia may also eventually build an LNG export facility serving landlocked Azerbaijan and Turkmenistan, and potentially even its own gas production if estimates of its reserves are confirmed and production is successful.

Central Asia

In comparison to the South Caucasus, Central Asia's natural gas reserves are much greater—estimated from around 14 to 20 TCM—and have already made a significant impact on the region's development and foreign policy trajectory. Of the five Central Asian states, Turkmenistan is by far the richest in gas resources, but Uzbekistan and Kazakhstan are also resource-endowed. Since the 1990s, Western technologies and foreign investment have accelerated the region's energy exploitation. The rising appetite for gas in China and rising gas prices in the 2000s gave momentum to gas exports. This newly harnessed energy wealth was used to try to unshackle the countries from Russia and to pursue more independent foreign political and economic policies, including forging new alliances and trade routes to the West and to Asia.

Nevertheless, geopolitical realities have forced the Central Asian states to take into account interests of great powers involved in the region—first and foremost, Russia, followed by nearby China, then the United States and to some extent the EU. The energy-rich and authoritarian states of Kazakhstan, Uzbekistan, and Turkmenistan have been able to pursue more balanced and independent foreign policies, while the less-endowed and more economically vulnerable states of Tajikistan and Kyrgyzstan have grown to be much more dependent on Moscow's financial aid, as well as remissions from their labor migrants in Russia.[33] Finally, despite sharing a common region and many common interests, the level of integration and cooperation among the five regimes has remained extremely low; some scholars see this as a reflection of the dominance of national elite interests across the region, which benefit from the preservation of border restrictions and national regulations.[34] Indeed, Central Asian states are more likely to cooperate with outside players than among themselves, as is evident in their gas and overall energy sectors.

Turkmenistan

The story of Central Asian gas must start with Turkmenistan, which has by far the most resources in the region. However, to understand Turkmenistan and its role as a gas developer and exporter it is crucial to understand the country's unique regime and recent history. After it gained its independence twenty-five years ago, following the dissolution of the Soviet Union, Turkmenistan emerged as the most autocratic, closed, and self-isolated nation of Central Asia. Even among autocratic Central Asian regimes, Turkmenistan has been infamous for its leadership's cult of personality pursued by both presidents Saparmurat Niyazov (1990–2006) and Gurbanguly Berdymukhamedov (2006–present). This gas-rich country of just over 5 million people has failed to become what could be the "Kuwait of Central Asia" because its resources have benefited mostly Turkmen leaders and elites known for their affinity for luxuries and symbols of absolute power, such as the ostentatious Turkmenbashi Palace (Oguzkhan Presidential Palace) built in 1997 in the capital of Ashgabat during Niyazov's rule.[35] In foreign policy, Turkmenistan has officially declared neutrality. Described as "domestic-oriented," this foreign policy has included periods of engagement and of disengagement with Russia, both of which have contributed to the survival of regime.[36] The launch of the Kazakhstan–Turkmenistan–Iran railroad in December 2014, which is expected to boost trade between Central Asian producers and Iran, was one of Ashgabat's most important cooperative ventures of the early 2010s.[37] There have also been minor instances of cooperation with other powers, as in Turkmenistan's agreement to act as a transportation hub for US and NATO (North Atlantic Treaty Organization) forces during their operations in Afghanistan and Iraq.[38] Meanwhile Beijing has emerged as an important trade partner, largely due to Turkmen gas exports to China.

The extent of Turkmenistan's gas riches is still contested. Global energy companies Eni and BP estimate the total natural gas reserves of the country at 10 TCM or 17.5 TCM, respectively.[39] The Turkmen government is more optimistic, suggesting reserves of 24 TCM.[40] Even with the lower estimates, Turkmenistan places about fourth globally after Iran (34 TCM), Russia (32.6 TCM), and Qatar (24.5 TCM); it is followed by the United States (9.8 TCM) and Saudi Arabia (8.2 TCM).[41] Geologists have estimated that 99.5 percent of Turkmenistan's territory of mainly vast deserts is conducive to oil and gas prospecting.[42] To date, Turkmenistan has not considered shale gas exploration and has instead focused on increasing its conventional production.

Turkmenistan launched gas production in the 1970s, when it was still the Turkmen Soviet Socialist Republic.[43] By the 1980s, the country's total production level reached around 80 BCM.[44] While the fall of the Soviet Union resulted in a drop in gas export volumes to the former Soviet republics due to delays in payment in the 1990s, gas exports resumed by the end of that decade.[45] As of the mid-2010s, over 30 gas fields and over 1,000 gas wells are being developed in the country, including Galkynysh—the world's second largest gas field after South Pars in Iran. Galkynysh was discovered in 2006, with reserves estimated from 13.1 TCM to as high as 26.2 TCM by the government, and has been in production since 2013.[46]

Modern Turkmenistan's gas production efforts have been largely underwhelming. In 2014, the country's production still lagged behind Soviet-era levels, reaching between 69.3 and 76 BCM, according to BP and state numbers, respectively.[47] Despite having the fourth largest gas reserves in the world, the volume of Turkmen gas production is almost nine to ten times lower than that of the United States (728.3 BCM) and considerably lower than that of Russia (578.7 BCM).[48]

According to the "Oil and Gas Industry Development Programme of Turkmenistan for the Period till 2030," adopted in 2006 under Berdymukhamedov, Turkmenistan is targeting to reach the production level of 230 BCM annually by 2030 and is counting on the giant Galkynysh field to achieve this.[49] However, optimism should be tempered, and not only because of Turkmenistan's regime peculiarities. Gas available for future production in Turkmenistan is high in hydrogen sulfide and carbon dioxide and has greater pressure and temperature, which increase capital costs for exploration and development of gas fields. Technical challenges, investment constraints, and geographical obstacles may all affect gas production.[50]

The country is also a highly inefficient consumer of its own gas: the tiny domestic market of just 5 million people consumed the vast sum of around 30 BCM of gas in 2014, or nearly half of total production, which is ten times more than Finland, a county with a similar population, consumed in 2014.[51] This is largely because the energy sector of Turkmenistan, including electric power and heating, is completely dependent on natural gas.[52] Due to subsidized lower prices and abundance of supplies, both the population and the local industry use gas highly inefficiently. As a result, despite its gas riches, Turkmenistan's gas exports of around 40 BCM are still relatively small in relation to its production and certainly in relation to its reserves.

Kazakhstan

The largest and wealthiest country in the region, Kazakhstan is often considered to be the regional leader, although neighboring Uzbekistan with its vibrant historical legacy might have some objections to this claim. Despite its huge territory—it is the second largest former Soviet republic after Russia—Kazakhstan is sparsely populated, with the majority of its approximately 17 million inhabitants living in the country's few urban areas. Kazakhstan's economic indicators have been impressive—its gross domestic product (GDP) per capita of $13,610 in 2013 was almost ten times higher than that of Uzbekistan, Tajikistan, and Kyrgyzstan and nearly twice that of Turkmenistan.[53] Even though the economy of Kazakhstan was hit by falling oil prices and devaluation of the local currency *(tenge)* in 2015, its GDP per capita indicators are still relatively high (around $11,000) especially for the region.[54] Its political model is described as a "benevolent dictatorship" or authoritarian government that increasingly limits civil freedoms but is taken to have the consent of the majority of population.[55] Kazakhstan is firmly associated with its longtime political leader, Nursultan Nazarbayev (1991–present), who enjoys the image of a wise and experienced politician whose main motto is "economy first," implying political developments are secondary to economic growth.[56] Despite Kazakhstan's reputation as being one of Russia's closest allies and having the largest ethnic Russian minority in the region (accounting for 21 percent of the Kazakhstan's population), Nazarbayev's position toward Russia is more cautious, especially after the annexation of Crimea in 2014.[57] Meanwhile, he has also consistently pursued relationships with the United States and China.

Kazakhstan, while rich in oil reserves, falls far behind Turkmenistan in its gas endowment. Gas reserves of Kazakhstan are under 2 TCM, not including an additional 779 BCM of wet shale gas. Some forty gas fields have been discovered there, though their distribution throughout the country is uneven.[58] Most of the reserves are located in central Kazakhstan, including 27 percent in the Atyrau region, and in western Kazakhstan (50 percent); both are sparsely inhabited and far from industrial centers.[59] More than 70 percent of all natural gas reserves in Kazakhstan are located in the Karachaganak gas field in the western region near the town of Aksai.[60] Unlike the rest of the region, Aksai has a large Russian minority (around 40–50 percent of the population), which raises security concerns, given the town's and the gas field's proximity to the Russian border and Moscow's previous territorial expansion in Ukraine and Georgia on the pretext of serving the interests of Russian minorities.[61]

Despite the fact that Kazakh gas reserves are about a tenth of Turkmenistan's, its production is catching up, having increased by nearly 40 percent from 2005 to 2014. In fact, Kazakhstan's gas production grew only in the 2000s, in contrast to its position in the Soviet era, when it was only a transit country for Turkmen and Uzbek gas.[62] In 2014, it produced over 45 BCM of gas, out of which 21.7 BCM was market gas in the gasified form and 23.8 BCM was associated gas, which is pumped back into oil wells to enhance oil production.[63] Due to lack of countrywide gas transportation infrastructure, the government was not able to deliver Kazakh gas to all of its population. Thus, Kazakhstan consumed about half of its produced market gas and exported the other half to Russia.

It imported roughly the same amount from neighboring Uzbekistan and Russia to be used domestically. With the extraction level of 8.5–9 BCM per year, the Karachaganak deposit has been the largest contributor to the gas production in Kazakhstan. Ninety percent of gas produced in Karachaganak, however, is delivered to the Orenburg gas processing plant some 130 kilometers away in Russia, and then more than half of that is returned to Kazakhstani customers, with the rest remaining in Russia.[64] While there have been efforts by the Kazakh authorities to develop their own gas processing capacity, not much progress has been achieved so far.[65]

Likewise, Kazakh authorities have made several statements regarding their interest in developing shale gas in the country for both domestic and export purposes. In 2014, the prime minister, Karim Massimov, said that a special program will be developed for the extraction of shale gas in the country.[66] Even though no comprehensive studies have been conducted on the shale gas potential of Kazakhstan, according to some very rough estimates Kumkol and Amangeldy fields in the southern part of the country may possess some shale gas reserves.[67] Government officials claim that some shale gas exploration work has already started and that if production turns out to be expensive, gas will most likely be directed to external markets.[68] However, these claims have been unsubstantiated, and no details of the study have been revealed. Besides, the Kazakh authorities are known for promoting certain projects to improve country's international image. Shale gas exploration may turn into one of those projects.

Uzbekistan

Uzbekistan, the most populous country of the region with 30 million people, is also resource endowed. Its long-serving president, Islam Karimov (1990–2016), was a typical example of the region's Soviet-bred politician who managed to preserve political power after the collapse of the USSR.[69] Following Karimov's death in September 2016, Prime Minister Shovkat Mirziyaev was appointed as interim head of state until he was elected president in December 2016.[70] Mirziyaev promised that Uzbekistan would stay devoted to Karimov's legacy, and there is little reason to expect any radical changes in the country's closed political system in the post-Karimov period.

Uzbekistan's natural gas reserves, estimated to be above 1 TCM, are lower than its neighbors.[71] Nonetheless, while lacking the vast gas riches of Turkmenistan and the economic development of Kazakhstan, Uzbekistan has been a major producer of gas in the region since well before the disintegration of the Soviet Union. According to Uzbekneftgas, a 100 percent state-owned oil and gas company, 60 percent of the country's territory has a potential for gas extraction. There are 108 gas fields discovered in the country thus far, out of which around 50 percent are currently being exploited.[72] More than two-thirds of the total annual gas production comes from the Bukhara-Khiva region on the border with Turkmenistan.[73] This includes Gazly, the largest gas field in Uzbekistan, with proved gas reserves of 500 BCM.[74] Uzbekistan can also refine most of its gas domestically, unlike other Central Asian states, particularly Kazakhstan, Kyrgyzstan, and Tajikistan, which rely heavily on the imported gas products and the refining plants of Russia and China.[75]

In fact, Uzbekistan produces almost as much gas as the vastly more endowed Turkmenistan, but as the largest consumer of gas in Central Asia, the country is not a large exporter. In 2014, the Uzbek domestic market consumed up to 85 percent of all produced gas—or nearly 49 out of 57 BCM produced—while the rest was exported to Russia, China, Kazakhstan, and Kyrgyzstan.[76] As in Turkmenistan, natural gas is the main source of energy in the domestic energy mix in Uzbekistan,[77] and while its high rate of consumption is explained in part by its large population, it is also due to highly inefficient energy processing and delivery facilities, which account for 60 percent of primary energy loss. The World Bank estimated that between 2013 and 2022, the country could increase energy efficiency and reduce the consumption of industrial electricity (mostly produced in gas-fired thermal power plants) by 15 percent with an investment of $170 million in energy efficiency programs. The World Bank even provided $100 million of the financing.[78] If Uzbekistan along with Turkmenistan and Kazakhstan could succeed in increasing their energy efficiency, they would be able to increase their gas exports further even without additional investment going into boosting production.

Uzbekistan has also shown some interest in shale exploration, but to date the focus has been on shale oil rather than shale gas. The estimated shale oil reserves of Uzbekistan are around 340 billion barrels, and Uzbekneftegaz started the development of the Sangruntau shale oil field in 2013.[79] However, the future of Uzbekistan's gas sector will also depend on the power struggles that could ensue in the post-Karimov period, both domestically and among outside powers such as Russia, China, and the United States.

Regional Gas Infrastructure

The landlocked positions of the Central Asian countries make pipeline infrastructure essentially the sole means of fossil fuel transportation. In Central Asia, unlike in many other parts of the world, this is the case for both oil and natural gas transport, both of which are similarly constrained by geography and vulnerable to the vagaries of regional political relationships. The inability of Central Asian producers to directly reach global energy markets prevents these countries from enjoying full control over exports of their resources.[80] The region's transnational gas pipeline networks, established in the Soviet era, served two purposes: exporting gas to Russia via the Central Asia Center and Bukhara-Ural gas pipelines, and supplying gas through the Bukhara-Tashkent-Bishkek-

Almaty pipelines to southern Kazakhstan and northern Kyrgyzstan, and through smaller pipelines connecting Uzbekistan with southern Kyrgyzstan and Tajikistan.[81]

In an attempt to establish integrated national gas pipeline systems and to increase gas export capacity to external markets, Turkmenistan and Kazakhstan have in recent years started to prioritize both export and local gas pipeline infrastructure. The East-West gas pipeline is designed to integrate all major gas fields in Turkmenistan into a single system capable of increasing export capacity in all directions, including to the Caspian Sea coast, with the future potential of further exports via the Caspian Sea to Turkish and European markets.[82] The pipeline, with an annual capacity of 30 BCM, was financed by Turkmengaz and completed in December 2015.[83] Another new Beineu-Bozoi-Shymkent pipeline is designed to supply southern regions of Kazakhstan with their own gas. The first section of the pipeline, Bozoi-Shymkent, was launched in September 2013, and the second section was completed in late 2015. However, the Beineu-Bozoi-Shymkent is also expected to feed the region's major exporting Central Asia–China gas pipeline.[84] In the end, some fear that Chinese gas demand may overshadow Kazakhstan's desire to supply a sufficient amount of gas to its own southern regions because of Astana's focus on export revenues.

Relations with Russia: Exports and Imports

Gas relations between Russia and the Central Asian states have never operated in a vacuum of commerce or supply-and-demand forces. Just as politics have shaped EU-Russian and Ukrainian-Russian gas relations, so has this been the case in Central Asia. During the dissolution of the Soviet Union, Central Asian states were on the sidelines of decision-making. Subsequently, their independence was a result not of nationalist liberating movements but of decisions taken in Moscow and elsewhere.[85] But whereas Moscow was indifferent to Central Asia and saw subsidizing the region as an unnecessary cost under the Soviet regime, in the mid-1990s Russian foreign policy began recognizing its strategic importance and focusing on reestablishing Russian interests there.[86] This trend continued under Vladimir Putin from 2000 onward, when Russia reinvigorated its efforts to maintain Central Asia firmly in its sphere of interest, especially in view of the emergence of new global players in the region, such as the United States and China. For Russia, the strategic importance of Central Asia is mainly that it serves as a buffer against the more volatile states of Afghanistan and Pakistan. Moreover, Kazakhstan is crucial for the security of Russia's southern regions as the two countries share a border of more than 4,000 miles. It is within these regional and international political developments, as well as changing conditions in the gas markets, that the Central Asian–Russian gas relationship evolved.

Decades of Dependence: The 1990s and 2000s

In the 1990s and 2000s, Central Asian gas exports to Russia largely depended on factors beyond the producing countries' control. First, Russia was never a true importer of Central Asian gas but rather was a reexporter of these resources to other markets. The Russian economy was never dependent on Central Asian gas, and it needed this gas only to help fulfill its commitments to European customers. Second, until the late 2000s, Central Asian exporters were almost completely dependent on Russian pipeline infrastructure to move their resources to any external markets and were thus extremely vulnerable to Russia's whims and shifts in its political and economic policies. Third, demand for Central Asian gas was largely tied to broader economic and political developments in the post-Soviet region. The breakdown of the Soviet Union and the decline of the economies and industries of its member states decreased the demand for gas in the former Soviet countries and resulted in the first gas production decline in Central Asia, mainly in Turkmenistan, which lasted until 1998.[87] Whereas in 1990 Turkmenistan produced 86 BCM, out of which it consumed only 8 BCM and exported the rest to Russia and other former Soviet countries, including Ukraine and Georgia, by 1997 production fell to 17 BCM.[88] Similarly, while Uzbekistan's gas exports to Russia were 11 BCM in 1990, they fell to less than 1 BCM in the end of 1990s before rebounding to more than 15 BCM in 2009.[89]

By the beginning of the 2000s, increasing demand for natural gas in Europe and the post-Soviet states, coupled with rising energy prices, enabled Russia to use its gas infrastructure to profit from reexporting Central Asian gas to other markets. At this time, gas supply relations with Russia were slowly changing in favor of Central Asian exporters as Moscow slowly introduced payment in hard currency for Central Asian gas. However, by using its transit infrastructure leverage, Russia did not allow Central Asian producers of gas to sign direct contracts with European customers and continued to keep the Central Asian gas markets isolated. Not surprisingly, Central Asian leaders were unhappy about the situation, but did not have the wherewithal to change it.

The unsatisfactory predicament in which the Central Asian states found themselves in the 2000s, despite the rising demand for their gas, is best exemplified by Turkmenistan, the region's largest producer. Beginning in 2000, when growing external demand boosted its gas production, President Niyazov insisted that Russia pay $42–$45 per thousand cubic meters (tcm) of gas.[90] Russia, in turn, proposed the price of $36 per tcm, to which Niyazov replied, "Let's calculate: You sell gas to Europe for US$85 per tcm and you want to buy from us for US$36."[91] Moreover, to pay for the gas, Russia used a barter system in which Russia delivered 50–60 percent of gas payments in goods.[92] Nonetheless, the arrangement persisted, and export volumes to Russia continued to grow.

In 2003, Russia signed an agreement quadrupling its planned gas imports from Turkmenistan from 20 BCM in 2000 up to 70–80 BCM annually.[93] Experts labeled it "a deal of the century."[94] At that time, the largest existing gas pipeline network in the region, Central Asia Center (CAC), with a capacity of up to 50 BCM, was supposed to carry this gas to Russia.[95] To increase the capacity of gas imports, Russia was planning to build the Pre-Caspian pipeline to run parallel to the CAC along the Caspian Sea coastline from Turkmenistan through Kazakhstan to Russia but this plan was later cancelled, the pipeline was never built, and the deal would eventually fall through.[96] As Russia was making promises, Turkmenistan's leadership was getting restless. In 2006, when Russia was purchasing Turkmen gas for $60 per tcm, and then for $100 per tcm during the last quarter of that year,[97] Gazprom was reselling that gas via RosUkrEnergo to Ukraine and Europe for at least $250 per tcm.[98] At that point, as will be discussed later in the chapter, with global energy prices soaring, Niyazov had had enough and traveled to Beijing to strike a better deal.

In many ways Moscow's effort to secure Central Asian and especially Turkmen gas in the 2000s reflected a strategic effort to gain a first-mover advantage over the resources in a market of rising demand and rising gas prices. Turkmenistan's commitment to fulfill its contractual obligations to Russia would secure Turkmen dependence on Russian pipelines for its gas sales, remove the EU's planned Trans-Caspian pipeline to transport Turkmen gas to Europe from the negotiating table, and constrain Turkmenistan's exports to China. Without Central Asian gas Russia would have more trouble fulfilling its commitments to keep increasing gas supply to Europe, through which Moscow could exert influence over the European continent. However despite Gazprom's healthy profits from reexports of Central Asian gas, the "deal of the century" with Turkmenistan never delivered on its potential as the market and political winds became increasingly volatile in the late 2000s.

Russo-Ukrainian Tensions and the 2008 Financial Crisis

The first Russian-Ukrainian gas dispute of 2006 and the 2008–2009 Russia-Ukraine gas crisis, which occurred simultaneously with the 2008 global financial crisis, reduced European demand for Russian gas exports and thus Gazprom's demand for Central Asian gas. Moreover, ever since 2006, unstable gas transit via Ukraine revived European leaders' interest in directly accessing Turkmen and Azeri gas without depending on Russian infrastructure. But had the EU waited too long? The year 2006 was also when Turkmenistan started talks to export gas to China. Would it have enough gas to supply both China and the EU?

In this context, Gazprom sought to secure for itself the volumes of Turkmen gas production and thus ensure Europe's dependence on Russian-sourced gas by offering Turkmen gas producers the long desired "European price" for gas, even at the expense of larger profits. Thus, in 2008 Gazprom agreed again to purchase 70–80 BCM Turkmen gas each year after 2009 for the "European price" of $350 per tcm and to cover transit fees on its own.[99]

In the mid- to late 2000s, most expected an unending rise of oil prices driving up gas prices, but then came the financial crisis of 2008, which negatively impacted the demand for oil and gas and resulted in a five-fold drop in prices (from $147 per barrel in July 2008 to as little as $32 in early 2009).[100] It was evident that the Russian-Turkmen deal of 2008 would no longer reap huge profits for Russia from Turkmen gas reexports to Europe. Instead of the agreed-upon 70 BCM of gas, in 2009 Russia started purchasing only around 10 BCM per year.[101]

Then events took a more sinister turn. The mysterious explosion on one of the lines of the CAC gas pipeline on April 9, 2009, was perceived by many in the region to be Russia's doing because it provided Gazprom an excuse to reduce the amount of gas imported from Turkmenistan.[102] According to the Turkmen Foreign Ministry, "This accident was caused by a gross unilateral violation by Gazprom Export of the norms and rules of the natural gas sales agreement."[103] Thereafter, Russian–Central Asian gas relations became more frayed. Russia also abandoned the Pre-Caspian pipeline project.

Moscow's annexation of Crimea in 2014 and the Western sanctions against the Russian economy that followed further reinforced the trend of declining Russian imports of Central Asian gas. For instance, whereas in 2007 Gazprom imported 42.6 BCM of Turkmen gas, the amount fell to 11 BCM in 2014 and then down to only 4.5 BCM in 2015.[104] Likewise, imports of Uzbek gas dropped to just 1 BCM in 2015. Following Russian-Turkmen disagreements over gas prices and Gazprom's nonpayment for Turkmen gas imports during 2015, Ashgabat ended all gas exports to Russia.[105] By 2016, the Russian–Central Asian gas relationship was largely history.

Exports to China

Well before the explosion on the CAC pipeline and collapsing energy prices made Gazprom renege on its contracts with Turkmenistan, Niyazov had already made a decision to seek new markets in China. Traveling to Beijing in 2006 to launch negotiations, he signed an agreement with the Chinese president, Hu Jintao, according to which Turkmenistan would supply 30 BCM of gas via a new gas pipeline to China.[106] By the end of 2009, the first line of the Central Asia–China gas pipeline came into operation. While Turkmenistan became the most important gas supplier to China, Uzbekistan and Kazakhstan also signed on.

The rapid construction of the first line (A-Line) of the pipeline, which was built in less than a year and a half, demonstrated China's firm commitment to obtaining Central Asian gas. This was no easy feat: the A-Line extends 1,830 kilometers, starting from the city of Gedaim on the border of Turkmenistan and Uzbekistan, then crossing central Uzbekistan and continuing through southern Kazakhstan until it reaches Horgos, a city in China's Xinjiang Uygur Autonomous Region. The Chinese national energy company, China National Petroleum Corporation, financed the pipeline.[107] The second B-Line became operational in 2010. Both lines reached a delivery capacity of 30 BCM per year by 2011. The same year China also loaned $4.1 billion to Ashgabat in exchange for future gas supplies.[108]

The C-Line, completed in late 2015, increased the delivery capacity of all three lines to 55 BCM annually. The construction of the 1,000-kilometer D-Line, intended to go through Uzbekistan, Tajikistan, and Kyrgyzstan and to add 30 BCM of annual gas delivery capacity, started in September 2013, but its completion has been delayed to 2020 or even later.[109] Nonetheless, China plans to achieve a total annual delivery capacity of 85 BCM (65 BCM from Turkmenistan, 10 BCM from Uzbekistan, and around 10 BCM from Kazakhstan) by 2025, making the Central Asia–China pipeline network the largest not only in Central Asia, but in the world.

The growing gas trade between Central Asia and China has translated into closer economic and even political ties between the two parties. Beijing's Silk Road Economic Belt Initiative, announced in 2013, is a keystone project intended to create an economic and transport corridor that would increase regional connectivity in Eurasia.[110] This initiative would build on a series of bilateral agreements between China and the Central Asian countries, as well as on the work of the Shanghai Cooperation Organization, established in 2001 to provide mutual Eurasian political, economic, and military aid by the leaders of China, Kazakhstan, Kyrgyzstan, Russia, Tajikistan, and Uzbekistan, and which in 2017 is expected to include India and Pakistan.[111] These developments will only continue to strengthen China's influence in Central Asia at the expense of Russia.

In just a decade after Niyazov's first visit to Beijing, Central Asia's complete dependence on Russian pipelines and gas markets turned into a heavy dependence on those of China. However, according to energy scholar and Central Asia expert Farkhod Aminjonov, the nature of these dependences differs. The first was dictated by infrastructural limitations—lack of alternative pipelines to transport Central Asia gas to external markets—while the second is a product of temporary obligations according to gas sales agreements. Now having at least two export routes, the Central Asian exporters could, in theory, use the Russian route to improve their bargaining power vis-à-vis China and vice versa.

In addition, as Aminjonov argues, the Central Asia–China gas relationship also differs from that of Central Asia and Russia both economically and politically. For one thing, the Central Asian–Russian relationship was legacy based. Russia did not invest in pipeline networks, which were inherited from the Soviet Union. China, on the other hand, has almost entirely financed the new pipeline infrastructure and will need a long-term return on its investment.[112] Moreover, Russia can meet its own gas demand and needs Central Asian gas only to help fulfill its export commitments to Europe. Meanwhile, Central Asian gas has become important to meeting China's domestic energy needs.

While China's demand for gas is forecasted to increase and its concern over pollution to intensify, the changed market conditions of gas abundance and lower prices may somewhat dampen Beijing's determination to invest in additional expensive gas import infrastructure. Moreover, China has also turned to Russia in hopes of meeting its gas needs, thereby creating a unique triangle of competition in the region. On the one hand, Russia and China both compete for influence and access to gas reserves in Central Asia, while on the other hand Central Asia and Russia are potential competitors to supply China with gas.

Instead, Central Asia, and particularly Turkmenistan, looks to Iran, South Asia, and Europe as additional export markets, especially after the postponement of Line D to China. In 2014, the Turkmen deputy minister of oil and gas industry and mineral resources, Kurganguly Yaziv, ranked the country's priority annual gas export destinations as China (65 BCM), South Asia (33 BCM), Europe (30 BCM), Iran (20 BCM), and Russia (10 BCM).[113] Export plans to China, Russia, and Iran are more realistic, while ambitions for South Asia and Europe are still little more than ambitions not backed up by a detailed plan on how to increase gas production capacity or build gas pipelines in those directions.

Iran: Old Partner with New Possibilities

Outside of Russia, Iran was the first country to receive Turkmen gas exports starting in 1997; by 2014 it had become Ashgabat's second largest customer after China.[114] While Iran itself enjoys an abundance of gas reserves and production, due to its lack of domestic gas transport infrastructure it has been cheaper to import gas from Turkmenistan to supply some of its regions. Turkmenistan supplies natural gas to northern Iran via the Korpezhe–Kurt Kai pipeline, built in 1997, with a capacity of 8 BCM, as well as through the Dovletabad-Serahs-Hangeran pipeline, completed in 2010, with a capacity of 12 BCM.[115] While Turkmenistan could thus transport up to 20 BCM of gas to Iran, the actual volume reached only 9 BCM in 2014, largely because Iran has been undertaking a major gas pipeline expansion program to make it possible to transport its own gas to northern Iran.[116] Meanwhile, whereas the Iranian gas market was important for Turkmenistan in the late 1990s and early 2000s, when it had limited access to others, with the opening of the Chinese market, Iran has lost some of its strategic importance for Ashgabat.

Going forward, it is likely that gas exports from Turkmenistan to Iran will dwindle as Iran gears up its own production and builds up its infrastructure. There is also a possibility that the two countries may eventually end up competing for the European gas market, and the Iranians further believe that the existing infrastructure could be used to strengthen gas swap relations within the Turkmenistan–Iran–Europe triangle. There is no specifically designed solid pipeline network connecting Turkmenistan to Europe via Iran. There are, however, two separate pipelines connecting Turkmenistan and Iran and one pipeline to transport gas from Iran to Turkey and on to Europe. In mid-2016, the Tabriz-Ankara line, which has a total capacity of 14 BCM, transported about 10 BCM from northwest Iran to Turkey.[117] The extra capacity can be used by Iran to swap Turkmen gas to Turkey and potentially onward to European consumers.[118]

Exports to European Markets

Before Turkmenistan turned to Chinese markets, Europe looked to Central Asia gas to diversify its gas imports away from Russian sources. Although the long-planned Trans-Caspian gas pipeline has failed to materialize to date, it still remains an option and one that would be in the interest of both Ashgabat and Brussels to restart. This 30 BCM capacity, 300-kilometer pipeline would transport Turkmen gas beneath the Caspian Sea, connect to the Southern Gas Corridor's TANAP pipeline to supply gas through Azerbaijan, Georgia, and Turkey to Greece, and eventually link with the TAP to reach mainland Europe.[119] If and when the pipeline is built, the planned 30 BCM exports of Turkmen gas pale in comparison to Russia's 150 BCM exports to Europe; thus, Turkmen gas would serve only a diversification purpose for southeastern Europe.[120] Nonetheless, Russia would prefer that Turkmen gas never gets to Europe, and in any case, these plans have not moved much beyond political proclamations and memorandums since the 1990s.

At that time, discussions to export Turkmen gas to Turkey and Europe began among European and Turkish leaders and then-President Niyazov, but due to disagreements over the terms of gas supply and investment obstacles, no detailed plan was worked out. Interest revived under the new Turkmen president, Berdymukhamedov, who signed a number of memorandums and agreements with European, Turkish, and Azeri leaders, including the 2008 Memorandum of Understanding and Cooperation in the field of energy between Turkmenistan and the EU and the 2013 Framework Agreement on the cooperation on transportation of natural gas from Turkmenistan to Turkey.[121] Following Crimea's annexation and increasingly deteriorating European-Russian relations, the EU's energy commissioner, Maros Sefcovic, visited Turkmenistan in May 2015 to give new impetus to the negotiations and accelerate the construction of the Trans-Caspian gas pipeline. He was reported as saying, "Europe expects supplies of Turkmen gas to begin by 2019."[122] Germany has also expressed continuous political support for the Trans-Caspian project.[123] Presidents of Turkey, Turkmenistan, and Azerbaijan plan to hold a formal summit in the end of 2016 devoted to the construction of the Trans-Caspian pipeline.[124] Although Iran and Russia would almost certainly object to this project, Turkmenistan believes that consent from Turkey and Azerbaijan should be sufficient to start construction.[125]

The main hindrance to the project remains the undetermined legal status of the Caspian Sea.[126] Resources of the Caspian Sea can be divided and controlled differently by the littoral states depending on whether the body of water has the status of a lake or a sea, which is defined individually by each of littoral states. If the Caspian Sea is recognized as a lake, its resources are equally divided, and in that case littoral states enjoy exclusive rights to develop national sectors of 16 kilometers from the shoreline. Interpreting the Caspian Sea to be a lake, Turkmenistan and Azerbaijan claim that the pipeline will pass through only their territorial sector and can thus be built without the consent of other states. While Iran also insists on the Caspian Sea's having the legal status of a lake, it has suggested that a land route for the pipeline from Turkmenistan to Europe via Iran and then Turkey would be more cost-effective. Because Western sanctions against Iran were lifted in 2016, this could emerge as an alternative, but overall the EU supports the Trans-Caspian pipeline and its connection to the planned Southern Gas Corridor.[127] Unlike other Caspian littoral states, Russia is interested in keeping the issue of undetermined Caspian Sea status open because that way construction of the Trans-Caspian pipeline would require consent from all littoral states, including Russia—a major adversary to the project.

Beyond these demarcation issues, building the Trans-Caspian pipeline and connecting it to Europe's Southern Gas Corridor project is technologically and financially challenging. Since 2014, the abundance of gas in the markets and low energy prices have also contributed to slowing down the project. Going forward, Russian meddling can be expected as Moscow already emphasized environmental concerns; generally, though, it does not want to see Europe less dependent on Russian gas. Moscow's strengthening of its Caspian fleet, which was used in the operation in Syria in 2015 and 2016, could imply that Russia could use more than political pressure to derail the project. Likewise Beijing may use its political and economic influence to postpone the construction of this pipeline, because Turkmen exports of gas to Europe without fulfilling its commitments to China directly threaten energy interests of the latter. It is also still unclear where the financing will come from and whether the EU will contribute. Although Turkmenistan is ready to cover the cost of onshore section of the pipeline, it may not be able to build the offshore section due to lack of funds and technical expertise.

Untapped South Asian Markets

Outside of Europe, another means for Turkmenistan to diversify its exports is to access the Indian market by building the Turkmenistan-Afghanistan-Pakistan-India (TAPI) gas pipeline, with a capacity of 33 BCM.[128] While no significant progress had been made since the mid-1990s, when the idea was first introduced and supported by the United States, in early 2016 it gained further momentum as Afghanistan, India, Turkmenistan, and Pakistan have signed an initial investment agreement of $200 million and agreed on the overall price of the project at $10 billion.[129] The ownership rights of the project will be concentrated with Turkmenistan holding an 85 percent share and Afghanistan, Pakistan, and India each holding 5 percent shares.[130] In October 2016, Turkmenistan signed an agreement with the Islamic Development Bank to finance the construction of the Turkmen section of the TAPI gas pipeline for a total amount of $700 million.[131] While the TAPI gas pipeline is planned for completion by 2021, instability in Afghanistan remains a significant challenge.[132] Though Afghanistan has said it will raise a 7,000-member force to guard the pipeline within its territory, these security realities will likely increase the costs and the timetable of the project's implementation.[133]

Conclusion

The long-term growth of global gas demand will ensure continued interest in Caspian gas in China, Europe, India, and beyond. The extent of such interest, however, may increasingly be challenged by the emergence of American LNG on the global market as well as the longer-term potential of unconventional gas production in other countries. Not only do Europe and China possess considerable estimated amounts of shale gas—some 16 TCM and 36 TCM, respectively—but China possesses the largest shale reserves in the world.[134] When and if China's production reaches significant levels, this could free up Central Asian gas for European markets, where development of shale reserves will be slower. At the same time, piped Caspian and Central Asian gas from Azerbaijan, Turkmenistan, and beyond may be commercially more attractive than American LNG for both Europe and China, depending on fluctuations in the global prices.

The global gas markets have shifted tremendously since the EU first considered Caspian gas imports and China signed up for Turkmen gas in the mid-2000s. In the early to mid-2010s, the abundance of gas on the markets, the growth in global LNG trade, and low gas prices have all been pointing to a buyer's rather than a seller's market and thus further disadvantaging landlocked and isolated gas-producing states. Overall, Caspian and Central Asian gas is not a game changer for any of its current and potential customers like China, Russia, or even Europe. Because Central Asian gas can only partially meet major consumers' energy needs, it serves its customers mainly as a means to diversify away from excessive gas dependence on Russian or other sources, or on LNG imports. European gas demand is forecasted to rise to 618 BCM per year by 2030, out of which Central Asia's share, if any, would not likely exceed 14 BCM.[135] The planned 10 BCM of Azeri gas exports to Europe by 2019 is equally underwhelming. Meanwhile, China is forecasted to consume around 420 BCM of gas annually by 2020, but even if Central Asian gas exports to China reach 80 BCM annually, they will meet less than 20 percent of China's total needs.[136]

The export capacity of Caspian and Central Asian gas will remain relatively constrained largely due to limited investment in production and export capabilities in face of low global gas prices since 2014. This makes it unlikely that the Central Asian countries will soon significantly increase the level of their conventional natural gas production or launch shale gas exploration. In sum, while Caspian and Central Asian gas will face more competition on the global gas markets, these countries will remain regional suppliers, and their role in the global gas markets will be more limited than for larger gas producers with access to both sea and land routes for exports.

7 The Politics of Demand: China and Beyond

No continent forces us to reassess the future of geopolitics, energy, and gas markets more than Asia. The energy demand of the developing world is outpacing that of the developed world, and much of that demand is in Asia. Gas markets in developing Asia are expected to contribute nearly half of the increase in global gas demand by 2040.[1] Already the world's top liquefied natural gas (LNG) buyers are all in Asia—Japan, South Korea, China, India, and Taiwan; together they accounted for almost two-thirds of global LNG demand in 2015. Moreover, Japan, South Korea, India, Taiwan, and Singapore are expected to import more than half of US liquefaction capacity between 2016 and 2019.[2] In this new global gas reality, Asia—the key global center of gas demand—is an essential player in determining the market's development.

There, while Japan and South Korea represent traditional import markets, China and India are the new demand centers. As rising economic and political powers and the most populous nations in the world, China and India have a vast appetite for energy, which will potentially have the greatest impact on the status quo of global gas markets and their new geopolitics. In comparison to China, South Korea, or Japan, India is much less integrated in the international gas markets. In contrast, China's ambitions, successes, and increasing interconnectedness with the global gas markets make it unique. China's gas demand has already resulted in Central Asia's shift of its export routes away from

Russia to supply China in the 2000s, and it has led to plans for a Sino-Russian gas relationship, which has been gaining momentum in the 2010s. American LNG exporters launched their first deliveries to China and India in 2016, while historic suppliers such as Australia, Qatar, and others want to hold on to these important markets.

This focus on China in this chapter is for two reasons. First, China's story of the "power of demand" reflects the changing dynamics between gas-importing and gas-exporting states in the new geopolitics of gas. Second, China's diversification strategy stands as a stark contrast to historic European trends. Unlike many European states, China has not been weakened by its gas demand and reliance on imports. On the contrary, China has leveraged its huge demand in negotiations with suppliers and boosted its influence vis-à-vis supplier states without falling into the trap of undiversified import dependency. China's strategy rests on diversification of its gas import sources in parallel to pursuit of energy independence. The country's own reserves will further strengthen its position—it is already only one of four countries in the world (and the only one in Asia) with an active (though still underperforming) shale gas development program. In addition to considering China, this chapter will also look at how India, Japan, and South Korea impact Asia's gas demand and energy mix, shifting trade flows, and regional competition for influence and resources.

Meeting Chinese Demand: The Power of a Diversified Energy Mix

Decades of population and energy-intensive economic growth resulted in China replacing the United States as the world's top energy consumer in 2011. Historically, coal has been the leading resource in China's energy mix and supplied nearly 66 percent of China's primary energy needs as late as 2014. China is the world's largest producer and consumer of coal, accounting for nearly half of the world's consumption, and it is facing increasing criticism at home and abroad for that reason. For instance, despite government censorship, a 2015 Chinese documentary entitled *Under the Dome,* which vividly depicted coal as the perpetrator of the country's notorious smog, became a viral sensation, thereby demonstrating the importance of the issue of pollution to the public. So have the so-called airpocalypse days, or pollution emergencies, in which students were ordered to stay indoors and which became routine in the 2010s.

After coal, oil came second in China's energy mix, accounting for 20 percent in 2012. At that time natural gas contributed only 5 percent to China's energy mix and hydroelectric power 8 percent, while nuclear and renewable sources contributed less than 1 percent each.[3]

As the world's largest carbon dioxide emitter facing dual international and domestic pressure, China is increasingly looking to natural gas and renewables to meet domestic energy demand. In 2014, China struck a landmark agreement with the United States whereby the United States would reduce emissions and China would stop emission growth by 2030.[4] That same year, China's National Development and Reform Commission (NDRC) released the government's official plan for addressing climate change between 2015 and 2020,[5] affirming again that it would promote cleaner fossil fuels—namely, natural gas—and limit nationwide coal consumption. According to this plan, natural gas will account for 10 percent of the country's primary energy consumption (or about 360 billion cubic meters [BCM]) by 2020. China's special envoy on climate change, Xie Zhenhua, further reaffirmed this shift toward low-carbon gas and renewable energies at the 2015 Paris Climate Conference, although he fell short of making further commitments that would compromise China's economic growth and development.[6] To meet its energy appetite and sustainable development goals, China has been looking to the gas markets for imports and placing special emphasis on diversification of gas sources and routes.

China in the Gas Markets

China's natural gas consumption has exploded in the last fifteen years, rising from 25 BCM in 2000 to over 197 BCM in 2015.[7] Before 2007,

Chinese consumption of natural gas was met by its own domestic production, but subsequently the country became a net natural gas importer for the first time. Gas imports have continued to increase dramatically in tandem with rapidly developing gas pipeline projects and LNG terminals. In 2015, about a third of China's gas consumption was met with imports.[8] Between 2005 and 2015, domestic natural gas production also skyrocketed from 51 BCM to 138 BCM, while in contrast coal production only rose by 47 percent, from 1,242 to 1,827 million tons of oil equivalent (MTOE), during the same time period.[9] These were the early signals of China's slow shift away from coal to natural gas, and they fueled high expectations about the country's gas demand. Thus far, though, the outcome is not quite what was expected.

In 2015, China made headlines for its slow economic growth and for its subsequent lower-than-expected gas demand growth. Experts expressed doubt about the rise of a "global age of gas" because Chinese gas consumption grew only 4 percent, its slowest pace in seventeen years, as compared to the average growth rate of 15 percent between 2009 and 2014. As Fatih Birol, executive director of the International Energy Agency (IEA), observed in 2016, "This is one of the factors that stalled the golden age of gas."[10] Unlike any other country, China, with its huge demand, seems to be able to sway the rate of development of a global gas market and the growth of global LNG trade on its own. Although low gas prices should have spurred demand in China, gas demand growth in 2015 was slower than expected, raising concerns that gas may not be able to compete against alternative fuels like coal and oil after all.

Simultaneously, China's domestic natural gas price reform, first discussed in the early 2000s and then launched in 2006 and 2010, has been an important ongoing factor directly impacting the country's gas demand and supply.[11] This reform, which aimed to gradually lower the domestic gas price to reflect that of international gas, officially ended the two-tier pricing system as of 2015. As a result, wholesale gas prices in the country have gone down and have thus also arguably made LNG imports even less profitable.[12] Even before the price reform was finalized, state-owned companies such as Sinopec had already planned to resell some LNG volumes as China's economic growth began to slow down.[13] While the reforms can hurt the dominant national oil companies and LNG importers in the near term, they will eventually introduce more market competition and provide smaller companies with more market share.

Chinese gas adoption is also marked by significant uncertainties. The use of gas in its energy mix is still quite low, and a slowing economy may limit investments in the exploration, production, delivery infrastructure, and promotion of consumption of natural gas to make it the main fuel of choice for the country. It is also unclear to what extent and how quickly the government will put in place a policy to prioritize natural gas as a way to solve the country's air pollution problem, especially considering that the influence of the coal industry interests in Chinese energy policy can be seen at all levels of government.[14]

China remains "the main wildcard" in global gas markets, and when considering its efforts to promote natural gas as a key resource, it is important to remember that the key word is "potential."[15] Nonetheless, I argue that this potential is huge and that natural gas will play an increasingly significant role in China's energy mix in the long term. In 2015, China remained the world's largest energy growth market for a fifteenth consecutive year, and its gas demand growth was still higher than the global average of 1.7 percent in 2015 and the ten-year average of 2.3 percent. The IEA predicts that China will emerge as a key gas buyer with gas demand expected to rebound between 2016 and 2021. Indeed, China's latest Five-Year Plan (2016–2020) calls for the replacement of coal with natural gas or electricity in the non-power sectors as a way to help solve its pollution crisis.[16] To meet this gas demand, China can turn to a variety of options, the first of which is LNG.

Already by 2012, as the LNG markets boomed, China became the world's third-highest LNG importer in the world, behind Japan and South Korea, and relied largely on Qatar, Australia, Malaysia, and Indonesia for imports.[17] In 2014, China imported 19.83 metric tonnes per annum of LNG and consumed about 8 percent of the global LNG trade, and it is set to continue raising LNG imports over the coming years. As of the start of 2016, China had twelve major LNG import terminals with another ten terminals under construction, all concentrated on the seacoast in the provinces of Guangdong, Zhejiang, and Jiangsu.[18] In August 2016, China received its first cargo of American LNG from Cheniere's Sabine Pass, and more imports are on the horizon.

As part of its strategy to secure diverse sources and routes of gas supply, China maintains relations with its traditional LNG suppliers of Qatar and Australia. Qatar, which is the world's largest LNG exporter, supplied about 25 percent of China's LNG imports in 2015. The China-Qatar gas relationship dates back to 2008, when a twenty-five-year LNG Sales and Purchase Agreement was signed by the China National Petroleum Corporation (CNPC), the Chinese state-owned oil and gas company and the country's largest integrated energy company; Qatar's LNG company, Qatargas; and the Shell Group. In 2009, Qatar sent its first cargo of LNG to China,[19] and in 2014 China imported 9.2 BCM of Qatari gas, close to the 10 BCM target.[20]

Australia is China's oldest LNG supply partner. Gas trade began in 2006 and entered a new era with increasing investment in the LNG sector. China's three biggest national oil and gas companies—China National Offshore Oil Corporation (CNOOC), the state-owned Sinopec, and PetroChina (the publicly listed oil and gas arm of CNPC)—have all invested in Australia's LNG sector, acquiring stakes in upstream energy assets in return for financial investment in producing these natural gas and coalbed methane deposits—something that is more difficult to negotiate in Qatar. CNOOC was the first to form a joint share company in 2004 in the North West Shelf project, Australia's first and biggest offshore conventional gas project; in 2009, it also joined in the Queensland Curtis LNG (QCLNG) project, an unconventional gas field. PetroChina invested in three Australian LNG projects—Arrow Energy, Browse LNG, and Fisherman's Landing—though the first two were canceled in 2015 and 2016, respectively, after the plunge in oil and gas prices. Meanwhile, Sinopec is involved in the Australia Pacific LNG (APLNG) project, which is supplied by coalbed methane.[21] China's LNG imports from Australia are likely to further increase and drop in price when the

Australia-China free trade agreement is finalized and 93 percent of China's total imports from Australia will benefit from zero tariffs starting in 2019.[22]

Piped Central Asian gas has emerged as an equally (if not more) important source of supply as LNG imports. As discussed in Chapter 6, in 2009 the Central Asian Gas Pipeline (CAGP) from Turkmenistan was completed, allowing China to receive pipeline gas imports for the first time ever. When the planned fourth branch of the pipeline is eventually completed, the CAGP capacity will reach 85 BCM of gas per year.[23] Since 2010, CAGP and other pipeline buildup have changed China's gas import mix, tilting it more toward piped gas rather than LNG imports. Thus, whereas in 2010 China received under 13 BCM in the form of LNG and over 3 BCM of natural gas through pipeline, with piped gas making up 22 percent of China's total gas imports, in 2015 China imported 59.8 BCM (26.2 BCM of LNG and 33.6 BCM of pipeline gas), making it the world's fourth largest natural gas importer, with piped gas making up just over 56 percent of total gas imports.[24]

China has also turned to neighboring Myanmar to meet its energy demand and diversification aims. Pipeline plans were first put forward in 2004, and after years of negotiation over both oil and gas pipelines, in 2013, the Sino-Myanmar Gas pipeline, which runs across Burma from the Bay of Bengal to Southwest China, became operational, with an annual capacity of 12 BCM.[25] However, China received only about 4 BCM of gas from this pipeline between February 2015 and October 2015 due to poor local distribution in China and a slowdown in investment from the CNPC, which is preoccupied with a domestic anticorruption movement. Iran is also eyeing the Chinese gas market and wants China to invest in its gas pipeline and LNG projects in exchange for energy cooperation.[26] These projects carry risks, though, including a lack of access to the technology needed for higher oil and gas recovery rates, potential conflict with Saudi Arabia or Israel, a lack of clarity on investment returns and terms of operation, and even the return of sanctions.[27] Other Chinese energy projects may involve similar risks as China continues to seek out diverse regional sources of gas—such as from Russia, as discussed below—to maintain its strategy of diversification in face of long-term rise in gas demand. Beyond imports, China will also turn to domestic conventional and unconventional gas production to meet its demand going forward.

Domestic Conventional Gas Resources and Coalbed Methane

Fortuitously, in the face of its high gas demand and ambitions of gas independence, China boasts significant conventional natural gas reserves, ranging between 3.8 and 11.4 trillion cubic meters (TCM).[28] Even the low end of the range constitutes about 2 percent of total world gas reserves. According to China's Ministry of Land and Resources, there are 251 proved natural gas fields in the country, with 80 percent of the gas reserves located in nine major basins. Six of these basins are located onshore: Ordos, Tarim, and Qaidam in the northwest part of the country; Sichuan in the southwest; Songliao in northwest; and Bohai Bay in the north, near Beijing. Three of the basins—Pearl River Mouth, Qiongdongnan, and Yinggehai—are located offshore in China's territory of the South China Sea close to Hong Kong.[29] The other promising offshore fields are located mostly in disputed territories in the South China and East China Seas.

In 2014, the CNPC announced the discovery of the Anyue gas field, a major field in Sichuan Basin with an estimated 0.3 TCM of technically recoverable reserves, equivalent to the total gas reserves of Myanmar.[30] The CNOOC also made two significant gas discoveries (Lingshui 17-2 and Lingshui 25-1) in Qiongdongnan Basin.[31] China's conventional domestic gas production has doubled over the last ten years, yet it has not been able to keep up with the country's growing demand, nor has it been able it achieve a gas production boom like that of the United States. In 2015, China's total natural gas production, which was mostly from conventional gas, totaled 138 BCM, or just over one-sixth of US production in the same year.[32]

While China is blessed with sizable conventional gas resources, its unconventional resources are even larger. Shale gas and coalbed methane

(CBM) are the two most promising sources, with China's total reserves of CBM standing at around 40 TCM, out of which approximately 10 TCM are recoverable.[33] Already back in 1995, China started commercial production and utilization of CBM, and in the 2010s CBM development centered in Qinshui Basin, Shanxi Province, where the world's largest CBM power plant and a large-scale liquefaction plant were built.[34] Twenty years after the start of exploration, however, CBM has proven extremely difficult to produce, and China has thus come to focus more on shale gas development. The success of the shale gas industry and the difficulty China faces in meeting energy demand may affect the extent to which the government prioritizes CBM production.[35] What is clear is that the exploration of conventional and unconventional gas resources domestically is of utmost importance for China, or rather its NDRC.

Shale Gas Development in China

In terms of unconventional gas, the US Energy Information Administration (EIA) estimates China's total shale gas reserves as the largest in the world, at 31 TCM.[36] China has already launched its efforts to be a player rather than a bystander in the global shale gas revolution. Since Beijing's primary goal is not simply to manage import dependence but to progress toward greater energy independence, shale is expected to serve these ends. In 2009, PetroChina drilled China's first vertical exploration well in Weiyuan shale gas field in Sichuan Basin.[37] In August 2011, the first fractured horizontal well—Wei201-H1—was put into production in Sichuan Basin, making China the third country after the United States and Canada to commercially recover shale gas. PetroChina had many subsequent breakthroughs in the Weiyuan, Changning, and Zhaodong gas fields of the Sichuan Basin, where 125 wells were drilled by 2014 and 74 were brought into production, though actual output was modest.[38] In 2012, Sinopec fractured its first well, Jiaoye1-HF, in the Fulin gas field, which is likewise located in the Sichuan Basin. By the beginning of 2015, Sinopec had drilled 200 wells but production volumes were likewise underwhelming.[39] In 2016, the China Geological Survey announced the discovery of a huge 13.54 TCM shale field in Guizhou Province, of which 1.95 TCM are believed to be recoverable.[40]

China still has a long march ahead to keep up with the US shale boom. Compared to China's 0.2 BCM shale gas output in 2013, the United States produced an estimated 336 BCM of shale gas in the same year.[41] By the end of 2014, China had invested over $3 billion into shale gas, drilling 400 wells and producing a meager 1.25 BCM from fracking and 3.69 BCM from coalbed methane extraction. In 2016, China's Ministry of Land and Resources announced that shale output in 2015 reached 4.47 BCM, which, although it missed the government target of 6.5 BCM, was a 259 percent increase in production from 2014.[42] Indeed, China is not discouraged. As Zou Caineng, vice president of the Research Institute of Petroleum Exploration and Development, stated in 2015, "Compared with decades of exploitation history in North America, China has just started shale gas work since around 2005, and had substantial progress."[43] Overall, shale production in China is more of a political mission than a commercial project, and state-owned companies will aim to meet their production targets despite costs or setbacks.

At the same time, China has had to revise some of its early targets regarding shale production. For instance, the National Shale Gas Development Plan of 2011–2015 had ambitious long-term aims to boost shale gas output from near zero to 40–60 BCM by 2020 (or about one-fifth of the country's total annual gas production).[44] By 2014, however, the director of China's National Energy Administration, Wu Xinxiong, along with the Ministry of Land and Resources, both admitted that this figure would realistically reach only around 15 BCM in 2017 and 30 BCM in 2020 due to the many challenges in the country's shale gas development.[45]

Many analysts have questioned whether China can achieve a boom in shale production in the short to medium term.[46] They cite tough geological conditions that make extraction much more difficult than in the United States; lack of access to sufficient quantities of water, especially in Western China; high project costs that will make some shale plays less commercially attractive; and insufficient technology transfer for shale gas success. They argue that many of the conditions that made US shale production successful, such as attractive tax laws, a vibrant services industry, deregulated prices, and a large number of small- to medium-size shale gas exploration and production companies, are still lacking in China. Thus some believe that China will first focus on exploring its conventional gas sources—which are plentiful and cheaper to produce than shale—before it takes on the task of producing unconventional sources of energy.[47]

Other industry experts, however, debate the extent of China's disadvantages in shale development. Some point out that China identifies and targets "sweet spots" for shale development and that Sichuan, a name that in Mandarin means "the four rivers," has satisfactory water volumes for development. But then others are less optimistic, saying that water scarcity in underdeveloped regions is a serious obstacle to shale exploration. Project finances, to some, are not an issue, either, with production costs actually falling (though they are still considerably higher than in the United States[48]) and the Chinese government offering subsidies for production up until 2020. On the other hand, lack of interest from private investors, even as recently as 2015 and 2016, may disprove such theory. For its part, China has more than $15 billion worth of acquisitions in Canadian companies and shale plays, a $4–$5 billion investment in the United States, and a joint venture with two US drilling technology companies that will focus on Chinese shale development for the next fifteen years. The Chinese also invested $3 billion for shale stakes in US companies Devon Energy and Chesapeake Energy for the purpose of gaining know-how that can be later applied at home, particularly to lower development costs.[49] Thus, technology and know-how are being transferred to the Chinese shale industry. Furthermore, despite some setbacks such as ConocoPhillips's termination of its contract with Sinopec in 2015 and Shell's revaluation of its cooperation with PetroChina, international companies are not shying away from taking part in China's shale development.[50]

History has shown that while international companies may back out of energy projects in China for financial reasons, all of the country's energy plans were or are in the process of being realized.[51] But the cost effectiveness of production of wells as far as five kilometers deep in remote areas of the country is debatable. Domestic production of conventional and eventually unconventional gas can help Beijing meet its long-term strategy of energy security, as imports alone cannot. In China's shale program, as in the country's gas demand in general, "potential" is the key word.

South China Sea Tensions

China's gas ambitions have not been free of controversy, as efforts at domestic production and diversification of imports have also led to tensions over resources within Asia itself. The future of the geopolitics of gas in Asia, and particularly Chinese energy relations with other regional powers, will be greatly impacted by relations in the South China Sea. China, Taiwan, Malaysia, the Philippines, Vietnam, and Brunei have disputed over the territory and sovereignty of the South China Sea, surrounding waters, and two groups of mostly uninhabitable islands, reefs, and banks including the Paracels (Xisha) and the Spratlys (Nansha). Several major shipping routes and fishing areas and the discovery of oil and gas reserves of the islands have added another dimension to the dispute. These tensions could unsteady the broader Asian-Pacific region, which accounted for over 20 percent of global natural gas consumption and nearly 16 percent of global production in 2015.[52]

Estimates of the volume of the oil and gas reserves of the South China Sea differ significantly due to insufficient exploration. China's National Energy Administration claims that petroleum resources under China's jurisdiction in the South China Sea equal roughly 168 to 220 billion barrels and natural gas resources total 16 TCM, accounting for one-third of China's total oil and gas reserves.[53] In contrast, in 2013, the EIA estimated that there were approximately 11 billion barrels of oil reserves and approximately 5.32 TCM of natural gas reserves in the South China Sea.[54] Similarly in 2012, the US Geological Survey (USGS) estimated that about 12 billion barrels of oil and 4.53 TCM of natural gas might exist in the South China Sea region, excluding the Gulf of Thailand and other adjacent areas. About one-fifth of these resources are in contested areas, particularly in the Reed Bank at the northeast end of the Spratly Islands.[55] Besides proved and probable reserves, the South China Sea may have additional hydrocarbons in underexplored areas.

The South China Sea is also a key location for global LNG trade. Even a small military conflict within this region would cause a major blow to the global LNG market, considering EIA estimates that over half of global LNG trade passed through the South China Sea in 2011.[56] Around 56 percent of this volume landed in Japan, 24 percent in South Korea, and 19 percent in Mainland China and Taiwan. Qatar, Malaysia, Indonesia, and Australia supplied almost three-fourths of all LNG exports to the region.[57] China, Japan, Korea, Taiwan, and all other regional players that are highly dependent on LNG as an energy source have a mutual interest in preserving the free movement of this important commodity through the South China Sea.

Although LNG is generally an efficient form in which to import gas from multiple suppliers, for China, the existing tensions in the South China Sea make pipeline imports potentially more secure. This explains in part Beijing's intent to increase piped gas imports from Central Asia, Myanmar, and potentially Russia. Such imports circumvent the problem of sending LNG through the South China Sea, specifically the Strait of Malacca, which the US Navy can blockade. As part of its diversification and security of supply efforts, Beijing has proposed the "Silk Road Economic Belt" strategy of cooperation between China and Central Asian states. Another incentive may be related to the fact that, while traditional LNG suppliers like Australia have been highly reliable and have never interfered with supply of LNG to any country, new LNG suppliers such as the United States may pose geopolitical problems for China given complex relations between the two countries, including tensions in the South China Sea.[58]

Overall, China, Vietnam, and the Philippines have shown the most resolve to defend what they see as their territory in the region. From December 2013 to June 2015, China conducted massive projects in the

South China Sea, including land reclamation, upgrading island facilities, and the construction of infrastructure, which both angered and alarmed its neighbors.[59] Vietnam increased its defense spending by 128 percent in 2005–2014 and by nearly 10 percent in 2014 alone.[60] The Philippines also upgraded maritime forces with additional aircrafts, destroyers, submarines, and other military equipment, and in 2014 signed an Enhanced Defense Cooperation Agreement with the United States, which among other things could allow the Americans access to the country's naval bases for the first time since the Subic Bay naval base was closed in 1992.[61] In 2016, a code of conduct outlining the responsibilities of all parties in the South China Sea was negotiated between the Association of Southeast Asian Nations (ASEAN) and China that could prove to be a short-term way to prevent maritime clashes within the region, though many questions remain regarding its effectiveness. More importantly, China and United States—both of which share interest in the region—have agreed that differences should be resolved peacefully through diplomacy.[62]

While outright conflict may be avoided in the near term, the possibility should not be excluded. In 2016, the Permanent Court of Arbitration in The Hague ruled in favor of the Philippines, which argued against China's claims in the South China Sea. While Beijing has called for a peaceful resolution to the conflict, the Chinese military—the People's Liberation Army—is pushing President Xi Jinping for an armed response to the ruling against the United States and its Asian allies. As Liang Fang, professor at China's National Defense University, wrote, "The Chinese military will step up and fight hard and China will never submit to any country on matters of sovereignty."[63] Military drills will continue in the region—China and Russia have agreed to hold routine naval exercises starting September 2016—and although these drills are not to be directed at any country, they will attract US naval vessels to the South China Sea and thereby increase the risks of military miscalculation and possibly outright conflict.[64]

The South China Sea conflict continues to influence both political and energy relations among the Asian powers. It has even spilled over into another gas resource dispute in the East China Sea, where there are already tensions over the Japanese-controlled, uninhabited Senkaku Islands (Diaoyu Islands). In addition, Japan is upset at China's development of the Chunxiao offshore gas field, especially in view of the fact that in 2008 both agreed to jointly explore oil and gas in the disputed area of the East China Sea without either country claiming sole sovereignty over these resources. While Chunxiao is located in an undisputed area on China's side of the sea, Japan has argued that Chunxiao may be connected to reservoirs on Japan's side and thus has sought a share in the natural gas field.[65] Like the South China Sea conflict, disputes in the East China Sea over natural resources may create instability in the region and thereby threaten the interconnectedness of and the trade between different gas markets of the region.

Meeting Gas Demand through LNG Alone: Japan, South Korea, and India

China, Japan, South Korea, and India, the dominant players in gas trade in Asia, are expected to rely more on gas as an energy source.[66] Although these four countries are already or are destined to become some of the world's largest LNG consumers, Japan, South Korea, and India differ from China in two significant ways: they are relatively undiversified in their gas supplies, and for now they can only import LNG rather than a combination of LNG and piped gas. Diversification here also takes on a different meaning in comparison to the European context. Unlike some European states, Japan, South Korea, and India are not dependent on states they perceive as hostile or with whom they have problematic political or commercial relationships. Furthermore, the nature of LNG imports allows a gas buyer to purchase gas from multiple suppliers as opposed to being committed to pipeline imports from a single long-term gas provider. Yet in comparison to China, these three Asian demand centers do not encompass the same power of demand because they are not diversified to the same extent that China is. In other words, because they lack gas import pipeline infrastructure or domestic production, Japan, South Korea, and India do not have China's flexibility in how they obtain their supplies, and this, in turn, affects how these three nations integrate into the new global gas market. While the issue of energy insecurity tends to be addressed mainly in the Chinese context, it is important to understand the energy dilemmas faced by Japan, South Korea, and India, which are just as dependent on imports as China.

Japan

Long the most economically developed of the Asian powers, Japan has been a significant consumer of gas, reaching around 113 BCM in 2015, down from a peak of 118 BCM in 2014.[67] In contrast to China, Japan's domestic gas production (an annual average 5 BCM from 2003 to 2012) and gas reserves of 20.7 BCM are minimal, and thus the island country relies predominately on LNG imports.[68] Japan has been one of the earliest and since the 1970s by far the world's largest importer of LNG. In 2015, Japan imported 118 BCM of LNG from various countries, primarily Australia (25.7 BCM), Qatar (20.2 BCM), Malaysia (21.5 BCM), Russia (10.5 BCM), and Indonesia (8.9 BCM).[69] With no access to piped gas and with relatively high demand, Japan will continue to rely heavily on LNG imports.[70]

The Fukushima nuclear disaster of 2011 demonstrated the acuteness of Japan's dependence on LNG imports and also signaled the rise of global LNG trade. LNG vessels were redirected from Europe to Asia to compensate for Japan's surging gas demand in light of interrupted nuclear production. By 2014, however, gas demand was falling, and by the end of the year, few LNG tankers were redirected to this region as domestically produced nuclear power slowly came back on line, the share of renewable power increased, and the Japanese economy slowed down.[71] Overall, though, Japan has had difficulty getting its nuclear facilities operational to previous levels, and 2016 saw closure of one of only two operating nuclear plants in the country. Meanwhile, Japan has increasingly been investing in coal plants and solar power in order to meet its energy needs, and in its attempts to decrease its high dependence on

LNG imports. Still, LNG looks to remain a crucial part of Japan's energy mix going forward.

Japan has long eyed American LNG, and this relationship is likely to grow in the long term. Despite lacking a free trade agreement (FTA) with the United States, which typically makes the approval process for American energy exports to such countries more bureaucratic, Japan has been purchasing US LNG from Alaska's Kenai terminal since 1969 and is set to import gas from newly constructed US export projects. Of the four LNG export projects approved by the US Department of Energy (DOE) as of January 2015, Japanese companies contracted from two directly and one indirectly.[72] For Japan, US LNG serves as an insurance policy and gives the country leverage in negotiations with their other suppliers. If the new projects go forward, US LNG exports to Japan could reach 16.8 BCM per year by 2019 and account for almost 14 percent of Japan's LNG needs.[73] Thus Japan will be well supplied with LNG in the medium term, especially during periods of gas gluts. Yet in the event that the global overflow of LNG from the United States, Australia, Russia, and Qatar, to name a few, ends either with time, through conflict, or because of a natural disaster, Japan's inability to access gas through pipelines means that it is more insecure than China.

Because of the strategic importance of the South China Sea for LNG imports, Japan is also maneuvering for control over it. Given Tokyo's efforts to forge ties with Vietnam, the Philippines, and other Asian-Pacific countries, and given the potential conflicts with China in the East China Sea, Japan's growing presence in the South China Sea might exacerbate the unsteady situation in the Asian-Pacific region. Military confrontation seems unlikely, given that LNG imports for both Japan and China pass through the South and East China Seas, yet cooperation is also doubtful.

Strained relations between the two countries go back to the still salient military conflicts of the 1900s, and as of 2016 President Xi of China and Prime Minister Shinzō Abe of Japan have not engaged in any bilateral state visits since they both took office in 2012. Nonetheless, if China's demand drops lower than previously predicted and Russian gas makes its way to China, there could be more motivation to trade LNG between Japan and China despite the frosty and mistrustful relationship between the two governments. Overall, though Japan may be the world's largest LNG consumer, regional tensions highlight that the country's gas supply is still vulnerable to the swing of geopolitics.

South Korea

South Korea's domestic energy predicament echoes that of Japan. The Asian Tiger's consumption of natural gas has steadily increased over the last ten years due to its rapidly developing economy, reaching a peak of 52.5 BCM in 2013 and then declining to about 43.7 BCM in 2015.[74] With minimal gas reserves of under 6 BCM, tiny annual domestic production (322 million cubic meters in 2014),[75] and no international oil or natural gas pipelines, the country has had to rely exclusively on tanker shipments of LNG and crude oil to meet its demand.[76] LNG comes mostly from Qatar and Indonesia, among a mix of other countries.[77] South Korea's historic concerns about potential shortages of gas have disappeared with US gas slowly impacting the global gas market.

At the same time, weakened domestic demand due to the restart of several nuclear plants and unseasonably mild winters in 2013 and 2014 have tempered South Korea's recent LNG appetite.[78] Nonetheless, South Korea is the only major LNG importer to have signed an FTA with the United States, which means there will be fewer hurdles in importing or reexporting American LNG. While no US LNG cargoes from Sabine Pass have been exported to Korea as of the end of 2016, long-term contracts have been signed. In the future, in times of lower demand, Korean companies could resell US LNG cargoes to other markets. For instance, state-owned Korea Gas Corporation cut 20 percent of its planned import volumes from Sabine Pass in January 2014, long ahead of deliveries that are to start in 2017, and plans to sell this excess volume to the French gas company Total.[79]

There is no shortage of demand for US gas in Asia, which is something South Korea and Japan may need to consider in the long term.

Taiwan, for instance, began importing American gas on spot contracts from the Kenai LNG terminal in May 2015, the very first time in decades.[80] Moreover, French gas group GDF Suez will also deliver 800,000 tons of LNG to Taiwan annually for twenty years from Louisiana's Cameron LNG liquefaction plant starting in 2018, marking the first time that gas produced in United States will be delivered to Taiwan via long-term contracts.[81] In order to have greater bargaining power, South Korea is aiming to establish a regional bunkering hub—or a port where LNG is transferred onto a ship that utilizes the gas as fuel—which would increase the use of gas as a marine fuel, thus leading to increased LNG imports.[82]

South Korea has not escaped regional rivalries. It, too, is caught up in the East China Sea and other historical disputes, but Seoul seems to have been more open to cooperation. While China's involvement in the Korean War in the 1950s formerly cast a shadow over their relations, South Korea and China have been showing signs of rapprochement because each needs the other's diplomatic support in view of their shared concerns over North Korean and Japanese remilitarization. Compared with Xi's frosty relations with Abe, those he has established with South Korean President Park are very good.[83] As Park summarized, "China is Korea's largest trading partner, and China has a huge role to play in upholding peace and stability on the Korean Peninsula."[84] Indeed, trade between the two countries continues to increase, and in 2014 the two governments published a joint declaration to "expand energy cooperation."[85] Soon after, a starting point for gas cooperation between the enterprises of two countries came when China's state-owned Hanas New Energy Group signed a deal with South Korea's public SK Group, setting out rather broadly to "cooperate on the upstream, midstream and downstream sectors and also to explore future Asian LNG projects."[86] In the case of China and South Korea, the importance of energy diversification and the search for resource security are having a tempering effect on their decision-making processes and thus on regional political tensions.

India

India—the world's third largest economy and one whose growth outpaced China's in 2015—is a country to watch long term as the global gas sector changes. In 2016, the IEA estimated that India's gas demand is set to grow almost 6 percent annually between 2015 and 2021, and that overall energy demand will reach that of the United States by 2040.[87] Yet many analysts are still divided on the role of gas in the country's developing energy mix. Some do not expect that gas will displace either coal or oil by 2035,[88] while others see great potential in India with gas demand growing by as much as 155 percent by 2035.[89] Although India and China share a similar population size and GDP growth rate, China's primary energy consumption was more than four times that of India in 2015.[90] This is in no small part due to the fact that India's level of development still falls short of China's and its GDP per capita is less than half that of China. While both countries rely to a significant extent on coal and together drive the global coal trade, unlike China, India's coal consumption almost doubled over the past ten years to 407.2 MTOE in 2015. Because India has traditionally depended on coal for electricity generation, its natural gas consumption was merely 50.6 BCM in 2015, marking a record low of the previous five years.[91] The need to raise natural gas consumption over coal use has become particularly critical as India begins to face the same sort of pollution crises as China, with high smog levels posing risks to human health and impeding day-to-day activities, as seen from school closures in 2016.[92]

Historically India's gas needs have been so limited that before its consumption took off in 2004, the country met its demand through domestic production, which in 2015 stood at just over 29 BCM—down from a high of almost 50 BCM in 2010—while total proved reserves in 2015 stood at 1.5 TCM. In 2015, Indian natural gas imports, which came in the form of LNG primarily from Qatar (13.5 BCM) and the remainder from Nigeria, Yemen, and other countries, were under a modest 22 BCM, but that still makes India the fourth largest LNG importer in the world.[93] Despite its slow shift towards gas, India has enormous projected energy demand and an interest in finding cleaner and more efficient ways of satisfying such demand. A game changer for India may be the discovery of natural gas hydrates in the Bay of Bengal in the Indian Ocean in July of 2016. The findings by India's Oil Ministry, the US Geological Survey, and teams of scientists from Japan are the first of their kind in the region and, if producible, have the potential to significantly alter India's gas reality.[94]

India, like China, is expected to become a huge gas market. Yet India's overall position in the gas markets is quite different from that of China. Because of its lack of access to gas-importing pipelines and low domestic production, India's power of demand will remain weaker at least in the medium term. Indeed, its increasing demand could put India in Europe's position of dependence (albeit not on states India perceives as potentially hostile), rather than in China's position of strength in the gas markets. India's lack of leverage in the gas markets is evident in its relative weakness in negotiating attractive contract and pricing terms for LNG. Until January 2016, when Qatar agreed to lower the price of LNG to reflect prices on the spot market, India paid high gas prices of around $13 per one million British thermal units under its twenty-five-year contract (which expires in 2028). Likewise, only in July 2016 did the state-owned gas giant Gas Authority of India Limited (GAIL) express its interest in renegotiating its twenty-year, oil-indexed LNG purchase agreement with Gazprom, which was signed in 2012. Paying such high prices for gas has pushed GAIL to reduce or delay oil-indexed imports and find other less expensive sources of LNG.[95] Some of these volumes could be imported on the spot market, which will favor the premium Indian market and result in spare cargoes heading there instead of Europe. Other cargoes could come from the United States. India already received reexports of US gas from 2010 to 2012, and by August 2016 it received three cargoes as well as one reexport of American LNG from Cheniere's Sabine Pass project, the first Asian country to do so.[96]

India is trying to follow the footsteps of its East Asian neighbors by investing abroad in the LNG sector. Accordingly, Petronet LNG Limited—an Indian energy company formed by the Government of India for the purpose of importing LNG—expressed interest in financing the construction of a second LNG terminal in the Kutubdia Islands of Bangladesh and a 1 million tons per annum floating terminal in Sri Lanka, as well as buying a stake in the Yamal LNG project in Russia. India has also tried to enter the LNG shipbuilding business. For example, GAIL has sought nine LNG vessels to transport US LNG, with the caveat that one out of every three vessels was to be built at an Indian shipyard. This would by default bring the country prized shipbuilding know-how—niche expertise most ship engineering companies are not enthusiastic to share. However, LNG imports alone would not provide India with the same security of supply as China.[97]

India's dependence on LNG and lack of piped gas imports can partially be attributed to India's highly uneven gas demand, which is further exacerbated by poor domestic pipeline infrastructure. Industrial hubs Gujarat and Maharashtra and the agricultural state Uttar Pradesh together consume more than 65 percent of gas in India, while a large number of states have no access to gas at all.[98] The gap between natural gas supply and demand in India is likely to increase as the GDP continues to rise rapidly and as more of the underserved regions and the country's population begin to demand access to additional forms of energy.

The Indian government has been considering possible solutions to the country's gas demand and seeking (albeit for now unsuccessfully) piped gas imports. In the 2000s India lost out to China for access to Myanmar gas, even though India's Oil and Natural Gas Corporation (ONGC) and GAIL invested in the Sino-Myanmar gas pipeline along with companies from China, Myanmar, and South Korea.[99] More recently, Prime Minister Narendra Modi focused on establishing relations with the three gas-rich Central Asian nations of Uzbekistan, Kazakhstan, and Turkmenistan that already supply China.[100] In the medium term, however, the planned Turkmenistan-Afghanistan-Pakistan-India

(TAPI) gas pipeline seems like a pipe dream for three reasons: India still lacks the necessary infrastructure to realize this plan, pipeline infrastructure through war-torn Afghanistan would face security issues, and cooperation with rival Pakistan may at best prove difficult.[101] Moreover, India would have to compete with China and possibly even Europe for Turkmen gas. Other speculative projects include the Iran-Pakistan-India pipeline, which due to former international sanctions against Iran, the troubled areas of Pakistan that the pipeline would have to pass through, and the tense Pakistan-India relationship has so far prevented this project from getting off the ground. Likewise, the long-discussed India-Bangladesh pipeline has also failed to get beyond discussion stage for both commercial and political reasons.

As India's gas demand grows, it is important to consider whether it will face competition for resources from China or whether the two will cooperate. Possible Indian-Chinese competitive tensions could prove dangerous for the geopolitics of gas in Asia and for global stability in general considering China and India account for two-fifths of the world's total population and the majority of future global energy demand.[102] For now, growing diplomatic and economic ties between China and India seem to be laying a good foundation for potential cooperation in the energy sector as both countries seek to secure stable supplies of natural resources. Joint bidding by Chinese and Indian companies has enabled them to acquire various energy assets in Iran, Colombia, Sudan, Syria, and Peru, among others.[103] In 2012, India's ONGC and China's CNPC signed a memorandum of understanding extending existing upstream cooperation in Burma, Sudan, and Syria and planning further joint bids on foreign oil and gas fields.[104] Outside of the completed Sino-Myanmar gas pipeline, most of these projects are still in the agreement and memorandum stages. However, if any of them is successfully developed, it will set an important precedent for Indian-Chinese gas exploration and infrastructural projects and become another step forward for China in its diversification path, and a preliminary step forward for India to strengthen its position in the gas markets.

Russian Gas Prospects in China, Japan, and South Korea

Since the end of the Cold War, while the prospect of large-scale Russian gas exports in the form of LNG and especially pipeline to Asia and to China in particular has been highly anticipated, progress was disappointing in the 1990s and the 2000s. However, in the 2010s, changes in the global gas sector, political and commercial tensions in Russia's traditional market of Europe, where more LNG has been arriving and contract prices have been taking a dip due to greater market liquidity, and Western sanctions constricting revenue since 2014 have all put Russia under more pressure to seek new markets. Thus, Russia has reinvigorated its efforts to "pivot to Asia" in the hope of supplying the world's largest gas market by pipeline and LNG. Gaining the option to sell both to Europe and Asia would place Gazprom and Russia "at the heart of the global gas industry" in the words of gas experts James Henderson and Simon Pirani.[105]

Nonetheless, Russia's plans for massive LNG exports to Asia have not yet materialized. Of the four LNG projects on which Russia's exports to Asia depends, only the Sakhalin II plant is functioning and supplying Japanese and Korean energy utilities and other buyers in China, India, Thailand, and Taiwan.[106] Yamal LNG is in the works to supply China, while Vladivostok LNG and the scheme to expand Sakhalin II are delayed.[107] Moreover, Russia's declared pivot to Asia has turned into a pivot to China. Here, it is increasingly clear that Beijing, not Moscow, will dictate the terms of the emerging Sino-Russian gas relationship. China leverages its large market and ability not to depend on any single exporting country to drive a hard bargain with its suppliers. Beijing also shows a willingness to finance energy projects that other countries or companies cannot. For example, in 2013 the CNPC took a 20 percent stake in Russian Novatek's Yamal LNG project and will purchase at least 3 million tons per year of LNG. Without the support of Chinese banks and their loans during the period of Western sanctions, Russia would not have been able to finance this megaproject.

Russian pipeline gas exports to China remain the central but elusive goal of Sino-Russian energy cooperation. By 2014, in possibly the most awaited Eurasian gas deal for decades, Moscow and Beijing signed a colossal thirty-year pipeline gas deal worth a reported $400 billion. The deal appeared to signal a new axis of energy power relations and raised concerns about the geopolitical implications of Sino-Russian rapprochement, especially in light of the fact that the deal was struck soon after the deterioration of Western-Russian relations following Moscow's annexation of Crimea. Yet just a year later, many were already questioning whether this deal will ever be fulfilled, given the lackluster track record of Sino-Russian energy relations, especially when it comes to gas.

Chinese-Russian efforts at energy cooperation, which date back to the 1950s, have been the most dominant but also underwhelming aspect of the modern relationship between Moscow and Beijing.[108] In October 1949, the Soviet Union was the first country to establish diplomatic relations with the newly founded People's Republic of China,[109] and the two countries established their first joint-venture oil company in Xinjiang the following year. Despite the freeze in relations during the Cold War, China and Russia renewed their energy cooperation in the 1990s and created the Sino-Russia Oil and Gas Cooperation Leading Group in December 1994. It took until 2009 for this group to realize its first project: the East Siberia–Pacific Ocean (ESPO) oil pipeline from Angarsk, Russia, to Daqing, China, which was completed only after China agreed to provide a $25 billion loan to Rosneft and Transneft for the construction.[110]

Compared to the successful construction of the oil pipeline, the implementation of Sino-Russian gas pipeline projects have faced more challenges.[111] While China's Development and Planning Commission approved the import of natural gas from eastern Siberia in the 1990s and while the Russian and Chinese prime ministers signed off on studies for two natural gas pipelines in 1999, Gazprom and the CNPC signed the Agreement of Strategic Cooperation only in 2004 due to a variety of disagreements, such as the route of gas exports.[112] Two pipelines were considered: Altai and the Power of Siberia. Russia preferred and prioritized the 1,700-mile-long Altai pipeline, or the western route, from its already-developed western Siberian gas fields through to Xinjiang in China because of the shorter length of the pipeline and cheaper construction costs. Moreover, because Russia would be supplying China from the same western Siberian fields that supply Europe, Moscow would be able to shift between two customers and potentially to play them off one another. Beijing was hardly enthusiastic about Gazprom's "swing supplier" strategy; furthermore, Central Asia was already supplying gas to the regions of China that Altai would service. Instead, China preferred the 2,500-mile Power of Siberia pipeline via the eastern route. This route, originating from undeveloped resources in eastern Siberia, had the potential to meet gas demand in China's northeast industrial region and coastal cities, replacing coal and helping diminish severe air pollution. Moreover, these regions had no alternative gas supply options other than LNG imports.[113]

The 2000s failed to see progress despite endless negotiations, agreements, and involvement of political leadership at the highest level. In 2006, during President Vladimir Putin's visit to China, he and General Secretary Hu Jintao signed an agreement to build two pipelines from Russia to China; this was followed by a financial cooperation agreement signed by Gazprom CEO Alexei Miller and CNPC CEO Chen Geng.[114] But the two countries were still at odds over which pipeline should be given priority.[115] In 2010, Gazprom and the CNPC agreed on annual gas exports for thirty years, with 30 BCM designated for Altai and 38 BCM for Power of Siberia,[116] But while construction on Altai had started in Russia in 2006, in 2013, it was paused before significant progress was made in order to prioritize the construction of the Power of Siberia line.

Due mainly to a disagreement over price, the gas deal was also only formally finalized in 2014, when Miller and the chairman of the CNPC, Zhou Jiping, struck a compromise on price and signed the purchase and sales contract for Russian gas supply via the Beijing-preferred Power of Siberia route. The ability of China to stick to its own terms for over ten years demonstrates how China's access to diverse sources of gas supplies puts it in a strong bargaining position. President Putin himself inaugurated the construction of the Power of Siberia pipeline in Yakutsk in September 2014 and the construction of its Chinese section commenced in June 2015.[117] At the ceremony marking the start of construction, Zhang Gaoli, first-ranked vice premier of China, stated:

> East Route gas pipeline is the largest China-Russia cooperative project; it is also one of the world's largest energy projects and a symbol of the high level of the comprehensive strategic cooperative partnership between the two countries. The implementation of this project . . . is conducive to diversifying energy strategy and guaranteeing the energy security of both countries.[118]

Soon after the two sides inked a Heads of Agreement—a nonbinding agreement that sets out the key terms of the prospective agreement[119]—for pipeline gas supply from Russia via the western Altai route.[120] The first delivery through the eastern Power of Siberia line was scheduled for 2019, while deliveries from the western Altai route were to start in late 2015. In 2016, Gazprom signaled that the Power of Siberia gas deliveries to China would start as planned, though in June 2016 Gazprom cut expenditure for the project by 10 percent for the year. Likewise, in late 2015 the western Altai route stalled due to further disagreements over price and seemingly was postponed indefinitely.[121]

There are still many questions regarding how and to what extent the Sino-Russian gas relationship will materialize. Many of these questions go back decades and reflect the difficulty both sides have in reaching a compromise. However, in the past the main sources of disagreement between Beijing and Moscow were price, Russia's past refusal to grant the Chinese equity in fields or pipeline projects, China's alternative import options, and a lack of mutual trust between the two countries.[122]

Since 2014, falling energy prices and the global gas glut have further strengthened Beijing's bargaining position even as Moscow has been more willing to compromise. In particular, low oil prices have devastated Russia's capability to develop its side of the gas pipeline project, estimated to cost around $55 billion. Furthermore, Gazprom's gas contract with the CNPC is linked to a basket of oil product benchmarks, making conventional Russian gas exports to China under the conditions of very low global oil prices unprofitable.[123] Some experts suggest that decisions and negotiations on major Russian-Chinese gas projects could face considerable obstacles because of a combination of the fall in oil prices, the collapse of the ruble, and China's crackdown on corruption, which includes investigations into the country's top energy firms and the detention of several senior former and current officials.[124]

China's position of strength in the unfolding Sino-Russian energy relations is also evident in the unprecedented and increasing willingness of Moscow to grant Chinese companies equity in oil and gas exploration and production projects in both eastern Russia and in third countries, especially in projects operated by Russian state-owned oil company Rosneft. Some experts have even speculated that given Rosneft's successful oil exports to China, Rosneft rather than Gazprom may eventually supply gas to Asian markets.[125] In 2006, Sinopec and Rosneft established a joint venture, with 49 percent and 51 percent shares respectively, to manage Udmurtneft oil production in Russia's Middle Urals.[126] In 2013, the CNPC and Rosneft agreed to jointly develop oil and gas fields in eastern Siberia, with 49 percent and 51 percent shares in the project respectively, and then in 2014, a subsidiary of the CNPC gained access to a major oil field in Russia by acquiring a 10 percent stake in the subsidiary ZAO Vankorneft, which is developing a field that could produce 1 million barrels of oil per day by 2020.[127] Chinese involvement in these projects is crucial at a time when oil prices are low, sanctions are in full swing, and Russia is trying to maintain its position as a supplier to Europe while at the same time expanding eastward.[128]

Russia's desire to expand eastward has once made it look to South Korean and Japanese markets. South Korea has been eager for Russian piped gas since the 1980s and 1990s and has also expressed interest in participating in Sino-Russian gas projects. However, tensions between North Korea and South Korea hinder an onshore pipeline through the Korean peninsula, while the costs of an offshore pipeline from Russia directly to South Korea are prohibitive. As a result, discussions with Russia over such projects have been unsuccessful.[129]

Russia is also the closest gas-exporting country from which Japan could theoretically import pipeline gas. Already, Russia's Sakhalin II plant satisfies around 10 percent of Japan's LNG demand,[130] and in 2015, Rosneft CEO Igor Sechin stated that Russia has enough gas resources to satisfy all of its gas demand. Not only have stakes in yet another proposed but delayed LNG project—Far East LNG—been offered, but access to Russian shipyards is also on the table for Japanese investors.[131] Construction of a 1,500-kilometer, $6.7 billion undersea pipeline connecting Sakhalin Island and Hokkaido was even suggested, promising to decrease the cost of gas transportation to Japan by half. Yet constructing underwater infrastructure in the seismic-active region is problematic. Moreover, this pipeline ties seller and buyer together and binds Japan to continuously import Russian gas, even if it could import LNG from elsewhere.[132] Given its alliance with the United States and ongoing political and territorial tensions over Sakhalin and Kuril islands (Northern Territories), which were annexed by Russia at the end of the Second World War, Japan may or may not go through with this pipeline project, which has been discussed since the 1990s as part of a diversification strategy.

Sino-Russian gas cooperation is more promising and much more important strategically than that between Russia and Japan. When implemented, the Power of Siberia pipeline would have a significant impact not just on China but also on the global market as a whole. Additional pipeline supply would diversify China to the extent that it could alter the country's LNG demand, which would push additional gas volumes into the global market and further alter the nature of LNG trade.[133] On the other hand, if Russia were to send only LNG to Asia rather than large-scale piped exports, Moscow and Gazprom would fail to secure a long-term relationship and a guaranteed significant share in the world's largest growing gas market.[134]

The Sino-Russian gas relationship may eventually pave the way for greater economic interdependence between Russia and China and potentially bring the two countries into a broader alliance. Indeed, the 2014 gas deal was perceived as Moscow's effort to pivot east as Western governments imposed sanctions and pursued diplomatic initiatives to isolate Russia for its actions in Ukraine. The deal enabled Moscow to demonstrate that it could find partners, new markets, and revenues, while China could point to another indicator of its growing economic and political clout on the world stage.[135] Still, some are skeptical of this "alliance." Since the late 2000s, Bobo Lo, a scholar of Sino-Russian relations, has maintained that Moscow and Beijing share nothing more than an "axis of convenience," and in 2016 he argued that Russia is increasingly the weaker partner that needs China more than the other way around.[136] Indeed, the *Economist* compared the parties involved in the 2014 gas deal to "a rabbit and a boa constrictor."[137]

At the same time, Beijing and Moscow have historically built relations on their shared interest in facing the Western world, particularly the United States, together. Although mutual mistrust exists (including over control of the Pacific), neither would want to see its neighbor weakened while the United States reigns strong. Piped gas supplies from Russia would also serve as a buffer for any potential conflict in the South or East China Seas that could cause LNG supply disruptions. But while Beijing will continue pushing for Russian gas imports, it will at the same time strive to limit dependence on any single external energy source, especially Russia, which is not perceived as wholly reliable supplier.[138] Finally, the Chinese have demonstrated that they will deal with energy suppliers such as Russia only on their own terms.[139]

Global Gas, the Asian Price Differential, and the Potential of US LNG

Asia has traditionally been the key LNG import market. The world's top LNG buyers in 2015 were Japan (118 BCM), South Korea (43.7 BCM), China (26.2 BCM), India (21.7 BCM), and Taiwan (18.7 BCM), which jointly accounted for 71 percent of global LNG demand. Most of these imports come from traditional LNG exporters, namely Qatar or Malaysia.[140] The Asian countries have also traditionally paid among the highest prices in the world for LNG imports due to oil-linked contracts and a lack of alternative energy sources. That is why in the new global market, Asian importing states seek to diversify their supplies beyond their traditional suppliers in the Middle East and Southeast Asia by striking deals with emerging market players.

Although Asian importers have expressed interest in LNG from Russia, Australia, Canada, Mozambique, and Tanzania, much focus lies on US LNG. Traditionally, outside of Japan, Asian countries had very limited LNG trade flows with the United States. The United States does not have FTAs with most Asian countries (except for South Korea and Singapore), and US energy exports to non-FTA countries have historically been constrained, although new exceptions for gas trading may break this barrier to gas exports to Asia. The domestic political backlash against trade deals in United States during the 2016 presidential elections will probably postpone the Trans-Pacific Partnership (TPP) with Australia, Brunei, Canada, Chile, Japan, Malaysia, Mexico, New Zealand, Peru, Singapore, and Vietnam, which would have facilitated LNG exports to the growing Asian market. As of March 2016, the DOE approved thirteen companies for permits to export LNG to non-FTA countries with many more under review or pending; such a commercial hurdle would be eliminated if the TPP were to be ratified.[141] In case the TPP or other trade agreements are not ratified, US lawmakers have already been pushing alternative measures for the removal of restrictions on direct exports from US projects that currently do not have non-FTA approval to major gas buyers such as India.

In the context of the rising global gas market, US LNG has already fundamentally impacted Asian gas pricing and contracting schemes. Historically most Asian customers were locked into long-term, oil-indexed contracts linked to the Japan Custom-Cleared Crude index (JCC), also known as the "Japanese Crude Cocktail," calculated as the average price of customs-cleared crude oil imports into Japan, with LNG prices roughly 15 percent of the crude oil prices.[142] Asian spot gas prices have generally traded between JCC and National Balancing Point (NBP) prices. Long-term prices are generally indexed to crude oil but with a three-month lag. Yet US LNG became a game changer in Asia even before commercial LNG deliveries began for two reasons: in American

LNG pricing and contract structure, prices are linked to the Henry Hub, which reflects the supply-demand dynamic in gas prices, and there are no destination clauses, allowing buyers to more freely trade cargoes and thus increase liquidity in global gas trade. These two facets of US contract design have influenced Asia to begin making these sorts of changes in the Pacific region. American gas has the potential to help Asian buyers slowly move away from heavily oil-indexed contracts, gain more destination flexibility, and develop more leverage in their negotiations with traditional major suppliers such as Qatar and the potential new supplier of Russia. Even though it is not expected that oil-indexed contracts will disappear altogether, the percentage to which gas prices are linked to oil will decrease, the number of contracts that factor in spot or hub pricing will increase, and greater flexibility will become more standard. This would be a huge accomplishment for this gas trade region.[143]

Whether or not these transformations will actually occur, changing conditions in the global gas market are pushing for increased spot and short-term deals as well as the development of trading centers in Asia. In an example of the latter, in July 2015, China launched the Shanghai Oil and Gas Exchange, where traders can trade both pipeline gas and LNG, while the Singaporean Stock Exchange launched the Singapore SGX LNG Index Group (SLInG) to better reflect regional spot prices. Shanghai, Tokyo, and Singapore are trying to mirror the Henry Hub model and are vying to become the main price-setting hub for the Pacific region, with Singapore being the frontrunner in the race. Although these markets are working toward increasing the physical infrastructure for storage, regasification, and trade that is necessary for a hub and the establishment of trading platforms, Asia does not yet have a liquid and transparent enough exchange to function as the Henry Hub does. Before early 2016, there was little competitive gas supply and no significant domestic gas production within East Asian countries; technical issues such as the lack of LNG quality standardization and unloading and immediately reloading desired quantities also still have to be addressed. According to the Oxford Institute for Energy Studies, it may be ten to fifteen years before a hub with third-party access to pipelines and LNG terminals, bilateral trades, price disclosure, standardized trading contracts, brokered trading, nonphysical players, and futures exchanges will develop.[144]

Additionally, political tensions may hinder regional energy cooperation and prevent an Asian gas hub from being developed. As Holly Morrow, an expert on Asian energy politics, notes, "There are a lot of reasons why a gas hub does not exist in Asia the way it does in North America and Europe (namely geography and lack of market liquidity), but there is also the reality that none of these countries would be comfortable relying on the others for a strategic commodity. Japan is certainly not going to rely on a hub based in China for its LNG, and vice versa."[145] Thus, Asia may opt to follow Europe's lead and pursue multiple coexisting hubs for the region, each with its own index.[146] Even if Asia does work to develop a regional hub or hubs, until then, the US LNG trade model and exports will be the primary drivers of change for the region.

More than any other factor, the volumes of American LNG to be exported to Asia will depend on price. The difficulty in determining if the American LNG "price is right" for Asia stems from the difficulty of determining future prices of American Henry Hub prices versus those of Asian spot prices and the cost of shipping. For example, as far as shipping is concerned, the Tokyo Gas Company is already in talks with European LNG buyers to swap its US cargoes for cargoes being sent from exporting countries closer to Asia in an effort to reduce shipping time and costs.[147] Yet even without sizable US LNG volumes physically in Asia, Asian buyers have been demanding more attractive terms from their suppliers. US LNG will help increase liquidity and competition in the global gas market by increasing interregional trade and lowering prices, which will have the effect of "empowering" consumers.[148] In that regard, Asian consumer nations have been making headway. At the Asia-Pacific Economic Cooperation (APEC) Energy Ministerial of September 2015, for example, APEC countries were "encouraged to create favorable conditions for trade and investment to support the LNG market in the APEC region."[149] Furthermore, Japanese companies Kansai Electric Power and Tokyo Electric Power both signed contracts for US

LNG indexed to the Henry Hub, a first for Asia, while Chubu Electric Power signed a contract with Shell that is free of a destination clause. Similarly, China has been renegotiating its contract with Shell to import LNG at a more attractive and market-based price. Even India has been negotiating with Qatar to lower the price of its long-term LNG imports.[150] The Asian gas market has changed significantly in this new environment.

The shift in LNG trade and pricing structures, although generally a positive aspect of the new global gas market, has also prompted speculation regarding the prospects of large-scale American LNG exports to Asian countries. The dampened optimism surrounding American LNG exports to Asia demonstrates how LNG exports depend on a multitude of factors. The crude oil price collapse from $110 per barrel in the second half of 2014 to around $40 per barrel in mid-2015, and lower LNG demand in South Korea and China due to the economic downturn, together led to a sharp decline in both spot and long-term LNG prices. Additional factors include short-term LNG supply outstripping demand and a strong US dollar that makes buying American goods, including LNG, more expensive for other countries. Some energy experts have pointed to the fact that the energy price downturn and depressed Asian spot markets make US gas uncompetitive with cheap, locally produced coal and gradually cheaper renewable energy technologies.[151] Former IEA executive director Maria van der Hoeven observed in 2015 that "the belief that Asia will take whatever quantity of gas at whatever price is no longer a given. The experience of the past two years has opened the gas industry's eyes to a harsh reality: in a world of very cheap coal and falling costs for renewables, it was difficult for gas to compete."[152] The market conditions of 2016 and the forecasts for 2017 echo Maria van der Hoeven's statement.[153] Still, the industry is cautiously optimistic about the future of Chinese energy demand.

American LNG will continue to make its way to China even if the volumes will not be immediately as large as previously anticipated. The price differential between Russian gas and potential American LNG might give China leverage for future gas price negotiations with Russia, while a deal with Russia could push down the price China is willing to pay for American LNG. China successfully negotiated a lower gas price from Russia in 2014 due to China's access to cheaper Turkmen gas and sanctions on Russia by the EU and the United States.[154] In the end, China's gas strategy is focused on diversifying supplies and seeking energy self-sufficiency rather than dependence on a single external supplier. Thus Beijing is unlikely to focus its demand exclusively on Russian or Central Asian piped gas or American LNG supplies. After all, as former British prime minister Lord Palmerston said in the late nineteenth century, a country has no permanent allies or enemies, but only permanent interests.

Conclusion

In the long term, Asian gas demand will have a strong impact on the global gas markets. The main priority of the Asian countries, particularly China and India, remains security of supply. To achieve this aim, in the years to come, China will exemplify the politics of gas demand: Beijing will continue to pursue a careful strategy to diversify its imports among different producer countries, between piped gas and LNG, between long-term contracts and short-term trades, and overall between imports and domestic production, all conducted from a position of strength. The rise of a global gas market that favors the buyers rather than the sellers, and the market conditions of the mid-2010s, of low price and high supply, make it easier for China to implement its agenda. The country's geographic position, which enables access both to piped gas from Central Asia, Myanmar, and Russia and to coastal LNG imports from Qatar, Australia, as well as Russia and the United States, is another source of leverage. Even though China's gas demand slowed somewhat due to the slowing economy in the mid-2010s, it will remain significant due to Beijing's determination to accelerate the shift from a coal-driven to a clean (or cleaner) energy economy. The country's $3 trillion in reserves to stimulate its energy transformation sets it apart from most other countries. India on the other hand, while a growing demand center for gas will hold less leverage in the gas markets due to its lack of pipeline infrastructure, limited domestic production, and exclusive reliance on LNG imports.

The effects of the changing global gas market are already increasingly evident in Asia: increased trade flows, an abundance of gas, an evening out of gas prices across the regions, and greater flexibility in traditional contract schemes. These new trends in pricing and contract models will, according to Howard Rogers, director of the Oxford Institute for Energy Studies, "challenge the sclerotic orthodoxy of the Asian LNG business."[155] Low Asian gas prices in 2015 and 2016 have already given Asian consumers more leverage and choices. With the possibility of large-scale US gas exports arriving in 2017, Asian markets might no longer have to worry about potential gas shortages. Even if American gas exports to Asia turn out to be more modest than previously expected, US domestic production and the gas glut have freed up LNG from other sources for Asian markets. As a result, the need for Russian gas is reduced, and Moscow's influence within Northeast Asia is undermined. Nonetheless, Russia will remain a significant gas producer and exporter, and gas will be the Kremlin's main calling card in Asia as it seeks to reinvigorate or establish new alliances.

Asia's growing gas trade will have both market and geopolitical consequences. Competition between resources in national energy mixes—namely, between gas and coal in China and India or between gas and nuclear energy in Japan and South Korea—will impact local and regional policies as well as global climate change goals. In a region that has been marked by historical rivalries and bilateral tensions, where multilateral institutions play a limited role and security of energy supply has been the main preoccupation of national governments preventing interstate energy cooperation, an increase in gas trade can potentially foster mutually beneficial cooperation. LNG trade between China, South Korea, Japan, and beyond could over time create reliable regional gas partners, increase interconnectivity of the different Asian gas markets, and be a force in normalizing regional relations. Because of the size of the Asian gas market, such changes would add momentum to the emergence of a global gas market.

Conclusion

The golden age of gas is upon us, and a more robust global gas market is on the horizon. Having weathered a storm of low gas and oil prices since 2014, the US shale gas boom is here to stay, even if the new price conditions have dampened the appetite for additional investment and reduced the immediate urgency of a gas strategy for many import-dependent countries. In Washington, policymakers still debate the merits of US liquefied natural gas (LNG) exports and whether these can impact geopolitics and improve the energy security of America's allies. On the Eurasian continent, juggernaut China continues long-standing talks with Russia to meet its vast energy needs and attempts to strike energy deals that some fear could move Beijing and Moscow closer together. The United States, however, holds the potential for an equally powerful counteroffensive in Europe, Asia, Latin America, and beyond. In 2016, Cheniere's Sabine Pass launched its first LNG cargo deliveries to Brazil, India, the United Arab Emirates, Argentina, Portugal, Kuwait, Chile, Spain, China, Jordan, the Dominican Republic, and Mexico, thereby demonstrating US potential to become a significant LNG exporter across the globe in the years to come. Meanwhile in Europe, Asia, and other parts of the world, countries have plans in the works for numerous LNG import terminals to meet their gas demand, diversification objectives, and environmental concerns—but will they these plans be implemented if low gas prices persist? Markets can be fickle, and unexpected economic and technological developments can be game changers. We have seen how easily planned LNG import projects could disappear from the drawing board when the United States reversed its 2000s gas import plans in favor of exports following the shale boom.

Whether the United States and its allies fully grasp the new opportunities of the growing global gas market and the shale boom largely depends on their understanding of the new dynamics within the gas sector, the geopolitical context, and climate change considerations, as well as on their national and foreign policy priorities. The new US administration following the 2016 presidential and congressional elections will impact the energy and foreign policy priorities of the world's leading gas producer and consumer and its leading power. The Republican administration led by President Donald J. Trump and reinforced by a Republican Congress may prioritize natural gas and coal production and reduce environmental regulation of the natural gas industry. The Democratic opposition in Congress may seek greater regulation, particularly of fracking, and seek to boost green energy and renewables. However, at this point it, is increasingly unlikely that either the Republicans or the Democrats will try to significantly constrain the export of natural resources out of the country. Despite some political uncertainty, all the elements are in place for policymakers to leverage America's newfound gas and help shape a new world of gas diplomacy and trade, one that is already emerging from a changing gas market.

The Global Gas Sector Transformed

The global gas sector has experienced a significant shift, as gas moved from being a local and regional resource to a global commodity traded in an increasingly liquid market. A mix of market factors and technological breakthroughs has been driving this transformation. Long-term government policies and various public and government political and economic priorities have also been in play. The American shale boom is a prime example. Driven primarily by entrepreneurship and supported by capital markets and technological innovations, the shale breakthrough drew on government supported research programs of the 1970–1990s to map shale formations as well as to test hydraulic fracturing and horizontal drilling techniques. Some of these efforts were in no small part politically driven as the country found itself increasingly dependent on imported fossil fuels from producers in the Middle East on whom the United States felt it could not fully rely following the 1970s oil crisis. China's early shale production has been even more driven by Beijing's political will in face of the country's vast energy needs and the rising public pressure for cleaner energy in the face of overwhelming pollution. Chinese national energy companies continue to make tremendous investments to produce meager volumes of shale gas—a strategy that, to date, could not be feasible based purely on commercial considerations. If China eventually mirrors North America's success in the shale boom, gas markets will receive another boost and another shake-up of the status quo.

Outside of shale gas production, growing LNG trade has been another crucial factor in transforming the gas markets, serving to facilitate the transport and commerce of this previously localized resource. LNG trade started as a commercial operation in the United States, North Africa, and the Middle East. The future of American LNG exports will also be driven primarily by commercial considerations, financing opportunities, and infrastructure availability. Politics can also potentially play a role by removing some of the roadblocks, especially in countries and regions where there is a strategic need for gas import diversification. Indeed, politics was in the background of the early growth of LNG trade, which took place largely between allied countries. For instance, the first ever LNG shipment was from the United States to the United Kingdom. Likewise Japan became the world's greatest LNG importer in the 1970s and 1980s, importing primarily from its ally, the United States, which was not exporting LNG anywhere else. Similarly, the recent European entrants to the LNG markets, such as Lithuania and Poland, have been motivated in equal measure by political and commercial considerations. Decades spent languishing at the mercy of the monopolist Gazprom, which seemed also to peg prices to political considerations, together with high energy prices of the late 2000s and early 2010s, encouraged these countries to seek LNG import infrastructure.

The trending buildup of new pipelines and other gas infrastructure in Europe and Asia likewise shows both forces—markets and politics—at play. The demand for gas among some Western European countries, like Germany, along with Russia's commercially and politically motivated desire to eliminate an uncooperative Ukraine from its former role in gas transit, has driven the construction of pipelines like Nord Stream and plans for Nord Stream II. The on-again, off-again Turk Stream pipeline project reflects the volatile political relationship between Moscow and Ankara. On the other hand, Gazprom's heavy hand as a supplier to import-dependent states has inspired the European Union (EU) to seek alternative sources and pipeline routes, such as the Southern Gas Corridor project, to bring Caspian gas to Europe. Moreover, Moscow's resurgence, demonstrated by campaigns in Ukraine and Syria, has raised alarm more broadly. The risk of unstable Russian gas flows via Ukraine has sped up the EU's implementation of pipeline interconnectors and efforts to reverse the flow of pipelines from several different directions to supply the adversely affected European countries.

In the Caucasus and Central Asia, strategic new gas infrastructure has emerged since the 2000s. A new pipeline now supplies Georgia with Azeri rather than Russian gas. New pipelines from Central Asia help supply China's rising demand for gas, but they have also served to bring the two regions closer together and, in the case of the Central Asian countries, farther out of Moscow's orbit. Still, pipeline infrastructure in China, Southeast Asia, and India continues to lag behind the plans of these countries and certainly behind those of Europe or North America.

The changes in the gas sector driven by shale, LNG, and infrastructure have in turn shifted trade flows and put pressure on gas pricing and traditional contract schemes. These new market forces, which contribute to the rise of a global gas market, also have significant political implications. The shifting trade flows create new supplier countries, such as the United States, and establish new trade patterns, new relationships, and mutual dependencies between new sets of countries. New commercial relationships can build the foundation for closer political cooperation and even spark new alliances between new sets of states. The deepening ties of mutual dependence for gas imports and exports between China and Central Asia are a notable case of this phenomenon. The start of US LNG exports to Europe, Latin America, Asia, and the Middle East in 2016 will likewise create new partnerships in the years to come. Meanwhile, a new abundance of gas and rising liquidity in the markets have put pressure on gas prices and their historic indexation to the price of oil. In the process, Cheniere's LNG exports are increasing the prevalence of linking gas prices to the America's Henry Hub levels. These market pressures have also been challenging long-term contracts in favor of spot trades, and these new practices are upending the basis of long-term relationships between established sets of importing and exporting countries, which constituted the traditional modus operandi in the geopolitics of gas.

The New Geopolitics of Gas

What does the rising global gas market mean for the geopolitics of gas? It is a new era, in which gas markets rather than politics are having a bigger role. Thus, new and diverse gas suppliers are playing according to the revised rules of increased competition and greater market pressure rather than following the former playbook that allowed monopolists to supply captive or near-captive markets. Long-term gas supply relationships still matter, but opportunities for spot trading and establishing mutually beneficial short-term relationships are abundant. Large-scale infrastructure, with sizable investment requirements and long-term commitments, also still plays a significant role, but new technology such as floating LNG, compressed natural gas, and other innovations, are increasing the flexibility of buyers. It is a market with more options not only because there are more buyers and sellers, but also because there are more transport routes—created, for example, by the expansion of the Suez Canal in 2015 and of the Panama Canal to accommodate the largest LNG tankers, such as the Q-Max, in 2016. That summer, the first LNG shipment from Sabine Pass navigated the Panama Canal on its way to China. As a result of these various changes, gas importing states will be able to choose among a number of suppliers and different forms of gas imports such as LNG or piped gas, and they will be able to adjust the volumes of their imports based on seasonal demand, economic growth or contractions, and possibly even political considerations. Gas importers could likewise reexport gas to other countries either to offload unneeded volumes or to assist allied states in times of gas shortages or other crises.

The rise of the global gas market and the growing interdependency of the energy markets that it represents has been met with different appraisals and a dose of caution. Some have argued that growing interconnectedness and interdependency of the global energy markets will bring increased volatility and uncertainty. Political upheavals, terrorist attacks, energy production or import decisions in one part of the world impact consumers, producers, and policymakers in other parts of the world.[1] Other energy scholars also envision more instability but for different reasons. Working from an assumption that interdependency between supplier and importer states creates stable supply relations, they argue that the breakdown of long-term gas supply relationships and the growing reliance on LNG trade will threaten interdependence and thus stability. According to this line of thinking, the rise in cross-border natural gas trade could create more opportunity for politics to impact energy relations.[2]

The assessments of this book are different and more optimistic, predicting that growing global gas trade will create opportunities for increased stability and less politicization of gas supplies. First, I believe that stability from interdependency is a false promise because it is more likely to occur in trade theory rather than in real trade. The asymmetry of dependence constantly tips from one side to another in gas trade relations, thus challenging the notion of stable relations per se. Second, in the older era of gas trade, when importers depended on a few gas suppliers and a few or a single set of import infrastructures, gas was a less fungible commodity, and liquidity and optionality were almost nonexistent for many importers. As a result, gas relations were often established for the long term between two sets of states and were much more strategic and politically tinged than those of oil. The interconnectedness and interdependence created by the new global gas markets are not between a set of states as before—the importer and the exporter; rather, the new interconnectedness and interdependence are between the importing or exporting state on the one hand and the global gas market on the other. Thus, while this interconnectedness can have elements of instability, such instability is more diffused. Dependence and relationships are more diversified than they were historically, when regional gas markets and states were in vulnerable and unbalanced dependencies on one another for imports or exports.

Despite the ascendency of markets over politics and of optionality over dependence, the new geopolitics of gas will not escape politics altogether. In this new era, the politics of supply—or the commercial and political leverage that gas-producing states hold over the importing states—will be more constrained. This is largely because the conditions of the rising global gas trade, such as interconnection and liquidity, reflect a buyer's market more than a seller's market. As a result, importing countries that take advantage of the new opportunities provided by the markets will have greater capacity to diversify their imports or limit imports from specific gas exporting states. In the long-term, countries will even have greater flexibility to pursue sanctions against gas exporting states, in contrast to the EU's inability in 2014 to include energy resources in the sanctions against Russia because alternatives to meet Europe's energy needs were still lacking. Overall, this new reality will make it easier for countries to pursue the politics of demand rather than fall into the politics of dependence.

The United States

Looking country by country or region by region provides the best perspective for assessing the new geopolitics of gas. The United States is on the brink of emergence as an energy superpower with its growing potential for substantial LNG exports. What tone will America set for the new geopolitical energy order? The old debate in Washington about whether the country's energy abundance (especially oil rather than LNG) should be hoarded at home to ensure sustained energy independence or be open instead for export across the globe seems to be dying down, but it could be a sleeping dragon. There are divisions of opinion on the optimum approach among different political parties, branches of government, and different agencies such as the White House, the Senate, the Department of Energy (DOE), and the State Department. The change in administration following the election of Donald Trump could reawaken these disagreements. Already ahead of the elections in 2016, a group of US senators wrote a letter to the DOE requesting it to slow its approval of large volumes of LNG over concerns that such exports can increase domestic natural gas prices and impact competitiveness of US industries, among other concerns.[3] Assuming that the support for domestic energy production is not challenged, and assuming that the United States broadly sticks to the principles of free trade, which were indeed questioned by Trump and the American public during the electoral campaign, the country will likely emerge as a significant exporter of LNG and possibly even oil. Even if the United States were to limit exports of its natural resources, it will meet its domestic gas demand with domestic production, which will free up natural gas previously destined for US imports in the international markets. This in itself will have positive consequences for America's energy-dependent allies and constrain its energy-producing rivals.

The State Department and former secretary of state John Kerry had suggested in 2014 that the United States should not and would not use energy as a weapon or as a means of coercion, and it opposes such use of energy by other countries.[4] This stance must be taken with a grain of salt. The United States, as the world's leading power, is highly active on the international stage and has pursued its interests, leveraged its assets, and not shielded away from leadership in nearly every part of the world. Thus, the United States would likely use its newfound energy power to support allies, contain foes, and reign in rivals. America may come to exemplify the new politics of gas supply in the decades to come. Its historical use of sanctions against perceived rival or rogue states such as Cuba, Iran, the Soviet Union, and most recently Russia also suggests that energy sanctions will continue to be part of its toolkit. Possible methods could include blocking its own exports to some countries or together with its allies enacting bans against imports from others, the latter of which will be made easier by the abundance of gas and variety of delivery routes in global gas markets. At the same time this gas freedom will somewhat constrain the United States' energy leverage because the sanctioned gas producers would likewise potentially have more options to find buyers around the world. On balance, American leadership and its preference for free energy trade will support the rise of an independent and robust global gas market. American technologies, investments, and know-how could also spread the shale revolution beyond North America and further reinforce the abundance, commoditization, and depoliticization of gas.

Russia

Russia's role in the new geopolitical gas order is likely to decline. This is not to say that Russia will supply fewer countries or export lower volumes of gas. Instead, its strategic importance as the sole or primary supplier for a number of European and Eurasian countries will continue to be challenged. Moreover, Moscow's monopoly on gas transit infrastructure in the former Soviet republics has already started to crumble, as can be seen from new pipelines in the Caucasus and in Central Asia. Russia's weakening position as a supplier of Europe is also evident in Gazprom's efforts to turn eastward to supply China and other Asian countries. Here as well, though, it will not be Moscow but rather Beijing that will dictate the terms of the gas relationship. The abundance of gas on global gas markets and the changing terms of gas trade, such as the decline of long-term contracts and destination clauses that have been Gazprom's modus operandi, will weaken the position of both Gazprom and Russian gas. How Russia plays its politics of supply will continue to change.

The implications of reduced energy dependence on Russia go beyond eliminating strong-arm tactics that Moscow wields against energy vulnerable states. It can also cure some of the residual problems of Russian gas supplies such as behind-the-scenes domestic political influence, corruption, and cycle of rents that exist in post-Soviet *and* EU states. However, these changes should not be taken for granted. As a power that relies on energy exports for half of its state budget, Moscow will work hard to maintain existing gas relationships and to maintain and attract new European and Eurasian countries as customers, as can be seen from its efforts to push through the Nord Stream II and the Turk Stream pipelines. Along with income, these gas relationships will serve a revisionist Moscow as a source of influence and power. In the foreseeable future, Russia's role in the gas markets will be diminished and more constrained, but certainly not eliminated.

Europe

It is on the European continent and in the post-Soviet states where the status quo of gas will be most transformed. The EU's half-century-long importation of Russian gas, fueled by periods of falling domestic production and rising demand, could soon lose its strategic importance. Here, Russia's diminished power of supply will be most evident if the EU succeeds with the majority of its internal pipeline interconnection plans and supply diversification efforts. Moreover, Brussels will gain most if it finally manages to negotiate with one voice and forge an effective energy union. At the same time the EU, more than any other group of states, is determined to reduce its reliance on fossil fuels and increase energy efficiency and use of renewables. Thus, the overall importance of gas in EU states is likely to diminish in the medium to long term. Still, this new geopolitical gas order rests on the shaky assumption that the EU, which has been showing signs of strain in the mid-2010s, will remain unified. Economic stagnation, pressures on the Eurozone (as in the case of Greece), the challenge to liberal democratic values (as demonstrated in Hungary and elsewhere), and the difficulty of maintaining a singular policy response to Russia's military resurgence since 2014 and the Syrian refugee crisis of 2015 could be revealing of cracks in the foundation of the EU.

On Europe's frontier, Ukraine and Belarus will continue to play a role as gas transit states and as swing states in the new geopolitical gas order. However, with LNG and reverse flows, Europe's gas supplies will be less tied to a small set of pipelines such as the ones via Ukraine and Belarus. As a result, their strategic importance as transit states for Europe and for Russia (if it succeeds in building additional export infrastructure) will decline. At the same time, Ukraine has the potential to escape some constraints that stem from its current, vulnerable position of energy dependence on Russia with simultaneous pressures from the politics of transit. The impetus to change in Ukraine has come from its territorial and military conflict with Russia since 2014 more so than from the change in the global gas sector. Regime change in Kyiv since late 2013 and new domestic and foreign policy priorities make possible reforms that could reduce corruption tied to gas transit and end the inefficient gas subsidies that have been the target of the International Monetary Fund and other lenders. Ukraine could also boost its own domestic energy production and in the long term secure additional gas import routes, such as LNG terminals, to complement reverse flow supplies from EU states. However, Ukraine's domestic politics remain in a precarious position, and these plans will be no easy task to implement for a country where corruption remains entrenched, where Russian gas interests have disproportionate resources and wealth, and where a Kremlin-stoked conflict continues to threaten the integrity of the state.

Belarus, on the other hand, remains much more isolated politically, economically, and geographically from the transforming global gas market. As a landlocked country run by "the last dictator in Europe," and with a largely state-planned economy that is reliant on heavily subsidized gas from its closest ally of Russia, Belarus is unlikely either to diversify its gas imports or to seek to depoliticize its gas supply relationship. Nevertheless, as with Ukraine, the Belarusian regime or relations with Moscow may not be static forever. Any relational change would necessitate a change in the country's politics of gas and could also be aided by new opportunities presented in a transformed global gas sector.

The Caucasus and Central Asia

The gas-rich states of the Caucasus and Central Asia have, for the most part, already mitigated their position as isolated gas producers. This activity has been driven in part by the markets, specifically the high prices of the 2000s and the external forces of demand. Most importantly, these countries have already escaped their dependence on Russian pipelines for exports. They are building new pipelines to supply Europe, in the case of Azerbaijan, and have been supplying China since 2009, in the case of Turkmenistan, Uzbekistan, and Kazakhstan. Yet outside the gas producing states, the results in this region have been mixed when it comes to leveraging the opportunities presented by the transforming global gas markets or improving their energy security. Only Georgia has diversified its gas supplies and emerged as an important transit state for Europe's Southern Gas Corridor project. Meanwhile, Armenia, Kyrgyzstan, and Tajikistan remain highly vulnerable in their gas supplies or remain dependent on Russia, among other domestic political and economic woes in these countries. Going forward, their isolated geography (only Georgia has access to the sea) and borders with the volatile states of Iran and Afghanistan will limit their access to new gas trading partners and thus will constrain their full participation in new global gas markets.

Asia

Asia is possibly the most interesting and important part of the puzzle in the transforming global gas sector and the resulting new geopolitical order of gas in Eurasia. The old geopolitics of gas was exemplified by the relationship between Europe and Russia, but the declining demand for gas of the former and the declining strategic significance of supply by the latter gives way for new players to take center stage. Asia's vast gas demand, though somewhat slowing in relation to earlier expectations, will make it the most important player in the politics of demand going forward. The Asia-Pacific's share of global energy demand is expected to rise to 47 percent by 2035, which is more than twice as much as the second largest energy-consuming region of Europe, Russia and the post-Soviet space, with 18 percent of global demand.[5] Who will supply this gas demand will determine much of the dynamics of the new geopolitics of gas. For now this role is still largely open. In the buyer's market and under the new conditions of the global gas market the demand will most likely be met by a variety of suppliers. American LNG exports have a significant role to play here, but there will be competition from Russian and Central Asian piped gas and from traditional LNG suppliers in this region such as Australia, Qatar, and Indonesia, and the potential for considerable local gas production, including shale exploration in countries such as China. Diversification and domestic production will be the centerpiece of Beijing's gas strategy to leverage its vast market and appetite for gas and parlay it into a position of strength rather than vulnerability vis-à-vis potential suppliers. In Asia, as in Europe, the new geopolitics of gas will more likely be driven by the politics of demand than the politics of supply.

Although this book has focused on the countries of Eurasia and on the United States, countries such as Australia, Qatar, and Indonesia will remain important players in gas supply and, along with Canada and Argentina, may also experience a boost in shale gas production. Likewise, Iran may eventually reemerge as an important gas exporter. Its proximity to European markets may in the long term help increase the EU's diversification and be another factor in diminishing Russia's strategic role as a supplier. Finally, competition for gas in the last frontier of the Arctic and in other waters such as the South China Sea will create tensions among global and regional players in the new geopolitics of gas.

Policy Implications

For the United States and its allies, there are a number of immediate and long-term policies that can be pursued to support the emergence of a global gas market and to leverage its full geopolitical potential. While the US gas sector will predominantly be driven by business and markets, that does not mean it does not warrant support from policymakers. Here, the success of US shale production can be further ensured by the support of relevant state commissions and federal regulatory bodies for the buildup of necessary pipeline infrastructure connecting gas-producing regions with LNG export terminals. If gas in producing regions, including the Bakken formation in Montana and North Dakota (and parts of Canada), can be successfully marketed abroad, it could go a long way toward sustaining the US shale boom and eliminating environmentally and economically harmful practices such as gas flaring. US LNG exports and export terminals will likewise depend on the continued timely permitting and licensing by the Federal Energy Regulation Commission, the US Maritime Administration, the Coast Guard, and the DOE.

Passing new trade legislation could be a major game changer for exports and bring a whole host of new importing countries under the free trade agreement (FTA) umbrella, most notably Japan, thus facilitating energy exports beyond the current twenty countries that have signed an FTA with the United States. However, the popular backlash against trade agreements in the 2016 US presidential elections suggests that agreements like the Trans-Pacific Partnership with the Pacific Rim countries and the Transatlantic Trade Investment Partnership with EU countries are unlikely to be ratified in the near to medium term. Nonetheless, US legislation increasingly reflects the new reality of energy abundance. The effects are being felt beyond gas. Most tellingly, the US government repealed a decades'-long ban on crude oil exports in December 2015.

Energy security is increasingly on the agenda of the North Atlantic Treaty Organization (NATO) and has attained higher visibility in

Washington and the capitals of other allied nations. Still, NATO is only in the process of formulating an approach to energy. So far that approach has been concerned mainly with how disruption of energy supplies could impact NATO countries and NATO operations. This is the focus of NATO's Energy Security Center of Excellence in Vilnius, Lithuania, which opened in 2012, but the center has yet to take on a larger role of considering new opportunities and constraints for European gas or energy security. In particular, the center could serve as a medium for gas policy coordination among NATO member states or serve as a hub that could provide both technical and policy advice concerning energy security to interested parties. The United States as a budding energy superpower, and the only significant energy producer among NATO members outside of Norway, should take a leading role in formulating NATO's energy agenda, and together with Norway could have a role to play in assisting their allies in cases of an energy cut-off from hostile powers. However, given President Trump's expressed skepticism toward NATO during the electoral campaign, it is uncertain if the United States will play a greater role in the alliance during his administration. Nonetheless, greater energy cooperation among NATO members is also necessary, considering Russia's ongoing efforts to woo NATO member states into an energy bloc via new pipeline projects. Turkey and the states of Southeast Europe have been the primary targets, as seen from the on-again, off-again Turk Stream pipeline project. Regardless of how the volatile Ankara-Moscow relations will unfold, Turkey is strategically important to the energy balance of Europe. It holds the keys to potential LNG flows in the Black Sea to countries like Ukraine, though Ankara currently objects to LNG tankers traversing the Bosphorus Strait due to safety concerns and congestion.

US agencies and gas industry could also have a role to play in sharing their expertise or financing feasibility studies for LNG terminals, gas storage caverns, and pipeline infrastructure in the energy-vulnerable states of Europe and its ally countries in Asia and beyond. This sort of cooperation is far from unprecedented. In fact, the initial feasibility study for Lithuania's new LNG terminal was financed by the United

States. Moreover, the United States could share its experience and technologies to promote investment in shale gas exploration among allies that are interested in tapping their domestic resources and decreasing dependence on imports. However, it is important not to have unrealistic expectations of Washington's role because its government institutions, unlike those of the EU, do not have funding for hard assets and have little sway over the private energy companies that drive the American energy sector. In contrast, Brussels could be in a better position to provide financial assistance in getting steel in the ground for infrastructural projects. With its Third Energy Directive, the EU has funding available for pipeline interconnectors, LNG import facilities, and electricity distribution networks. The EU has been making progress, but not quickly enough, in order to reassess its energy priorities, take advantage of new opportunities in the gas markets, and mitigate the risks posed by a newly resurgent Russia. As a result, in the years to come, all eyes will be on the United States to rise to the challenge and take the lead in the new geopolitics of natural gas.

Notes

Introduction

1. *Are We Entering a Golden Age of Gas? World Energy Outlook 2011,* special report (Paris: International Energy Agency, 2011), http://www.worldenergyoutlook.org/media/weowebsite/2011/WEO2011_GoldenAgeofGasReport.pdf; Sharad Apte and Julian Critchlow, *Are We on the Edge of a Truly Global Gas Market?* (Bangkok: Bain & Company, 2011), http://www.bain.com/Images/BAIN_BRIEF_Are_we_on_the_edge_of_a_truly_global_gas_market.pdf; Jason Czerwiec, "Breaking Russia's Natural-Gas Chokehold," nationalreview.com, April 2, 2015, http://www.nationalreview.com/article/416312/breaking-russias-natural-gas-chokehold-jason-czerwiec; Matt Badiali, "Are Low Oil Prices Killing the U.S. Shale Boom?," The Growth Stock Wire, June 26, 2015, http://growthstockwire.com/4099/are-low-oil-prices-killing-the-u-s-shale-boom-.

2. Brenda Shaffer, *Energy Politics* (Philadelphia: University of Pennsylvania Press, 2011), 1; Daniel Yergin, "Energy Security and Markets," in *Energy and Security,* ed. Jan H. Kalicki and David L Goldwyn (Washington, DC: Woodrow Wilson Center Press, 2013), 69–71.

3. Shaffer, *Energy Politics,* 36.

4. John Deutsch and James R. Schlesinger, *National Security Consequences of U.S. Oil Dependency,* Independent Task Force report (Washington, DC: Council on Foreign Relations, 2006), xi, http://www.cfr.org/oil/national-security-consequences-us-oil-dependency/p11683; US Energy Information Administration (henceforth EIA), *Annual Energy Outlook 2008*

with Projections to 2030, June 2008, 78, http://www.eia.gov/forecasts/archive /aeo08/pdf/0383(2008).pdf; EIA, *Annual Energy Outlook 2009 with Projections to 2030*, March 2009, 78, http://www.eia.gov/forecasts/archive/aeo09 /pdf/0383(2009).pdf.

5. Jason Bordoff and Trevor Houser, "American Gas to the Rescue?" (New York: Columbia University Center on Global Energy Policy, September 2014), http://energypolicy.columbia.edu/sites/default/files/energy /CGEP_American%20Gas%20to%20the%20Rescue%3F.pdf.

6. EIA, *Annual Energy Outlook 2016 with Projections to 2040,* August 2016, http://www.eia.gov/forecasts/aeo/pdf/0383(2016).pdf.

7. EIA, "How Much Shale Gas Is Produced in the United States?," Frequently Asked Questions, June 14, 2016, http://www.eia.gov/tools/faqs /faq.cfm?id=907&t=8; British Petroleum (BP), "Energy Charting Tool," n.d., http://www.bp.com/en/global/corporate/about-bp/energy-economics/energy -charting-tool.html; EIA, "International Energy Statistics," n.d., http:// www.eia.gov/cfapps/ipdbproject/iedindex3.cfm?tid=3&pid=26&aid=1&cid =regions&syid=1980&eyid=2014&unit=BCF.

8. EIA, "International Energy Statistics"; EIA, "Shale in the United States," October 22, 2015, http://www.eia.gov/energy_in_brief/article/shale _in_the_united_states.cfm; International Gas Union, *World LNG Report—2015 Edition* (Oslo, Norway: International Gas Union, June 2015), 6, 30, http:// www.igu.org/sites/default/files/node-page-field_file/IGU-World%20 LNG%20Report-2015%20Edition.pdf.

9. Michael Levi, *The Power Surge: Energy, Opportunity, and the Battle for America's Future* (Oxford: Oxford University Press, 2013); Yergin, "Energy Security and Markets," 74.

10. Damien Gaul, *U.S. Natural Gas Imports and Exports: Issues and Trends 2005*, EIA, Office of Oil and Gas, February 2007, 1, http://www.eia .gov/naturalgas/importsexports/annual/archives/2007/ngimpexp05.pdf; Kiran Dhillon, "Why Are U.S. Oil Imports Falling?," *Time,* April 17, 2014, http://time.com/67163/why-are-u-s-oil-imports-falling/; EIA, "How Much Petroleum Does the United States Import and from Where?," September 14, 2015, http://www.eia.gov/tools/faqs/faq.cfm?id=727&t=6; "Frequently Asked Questions: How Much Oil Consumed by the United States Comes from Foreign Countries?," EIA, March 8, 2016, http://www.eia.gov/tools/faqs/faq .cfm?id=32&t=6; EIA, "U.S. Natural Gas Imports & Exports 2015," May 31,

2016, https://www.eia.gov/naturalgas/importsexports/annual/; EIA, "Natural Gas Consumption by End Use (Million Cubic Feet)," July 29, 2016, https://www.eia.gov/dnav/ng/ng_cons_sum_dcu_nus_a.htm.

11. Shaffer, *Energy Politics,* 15.

12. David G. Victor, "The Promise of Gas," in *Energy and Security,* ed. Jan H. Kalicki and David L Goldwyn (Washington, DC: Woodrow Wilson Center Press, 2013), 88.

13. John Deutsch, "The Good News about Gas: The Natural Gas Revolution and Its Consequences," *Foreign Affairs,* February 2011, https://www.foreignaffairs.com/articles/2011-01-01/good-news-about-gas; Jonathan Stern, *Natural Gas in Asia: The Challenges of Growth in China, India, Japan, and Korea,* 2nd ed. (Oxford: Oxford University Press, 2008), 381–384; Carlos Pascual, "The New Geopolitics of Energy" (New York: Columbia Center on Global Energy Policy, September 2015), http://energypolicy.columbia.edu/sites/default/files/energy/The%20New%20Geopolitics%20of%20Energy_September%202015.pdf; Bordoff and Houser, "American Gas to the Rescue?"; "Step on It," *Economist,* January 30, 2016, http://www.economist.com/news/finance-and-economics/21689644-it-will-take-time-fragmented-market-verge-going-global-step; James Henderson and Simon Pirani, eds., *The Russian Gas Matrix: How Markets Are Driving Change* (Oxford: Oxford University Press, 2014), 3, 10, 378–380.

14. Roger Owen and Sevket Pamuk, *A History of Middle East Economies in the Twentieth Century* (Cambridge, MA: Harvard University Press, 1999), 216.

15. Jan Daniel Yergin, *The Quest: Energy, Security and the Remaking of the Modern World* (New York: Penguin, 2011).

16. International Gas Union, *2016 World LNG Report* (Oslo, Norway: International Gas Union, June 2016), http://www.igu.org/publications/2016-world-lng-report.

17. Richard Devetak, Anthony Burke, and Jim George, eds., *An Introduction to International Relations* (Cambridge: Cambridge University Press, 2012), 492.

18. Saul Bernard Cohen, *Geopolitics of the World System* (Lanham, MD: Rowman & Littlefield, 2002), 11; Colin S. Gray and Geoffrey Sloan, *Geopolitics, Geography and Strategy* (London: Routledge, 2013); Zbigniew Brzezinski, *Game Plan: A Geostrategic Framework for the Conduct of the U.S.-Soviet Contest* (Boston: Atlantic Monthly Press, 1986).

19. Aleh Cherp and Jessica Jewell, "The Concept of Energy Security: Beyond the Four As," *Energy Policy* 75 (December 2014): 415–421.

20. Andreas Wenger, Robert W. Orttung, and Jeronim Perovic, eds., *Energy and the Transformation of International Relations: Toward a New Producer-Consumer Framework* (Oxford: Oxford University Press, 2009); Jonathan Stern, *The Future of Russian Gas and Gazprom* (Oxford: Oxford University Press, 2005); Thane Gustafson, *Wheel of Fortune: The Battle for Oil and Power in Russia* (Cambridge, MA: Harvard University Press, 2012); James Henderson and Simon Pirani, *The Russian Gas Matrix: How Markets Are Driving Change* (Oxford: Oxford University Press, 2014).

21. Shaffer, *Energy Politics;* Levi, *The Power Surge;* Yergin, *The Quest;* Jan H. Kalicki and David L Goldwyn, eds., *Energy & Security* (Washington, DC: Woodrow Wilson Center Press, 2013); Steve A. Yetiv, *Myths of the Oil Boom: American National Security in a Global Energy Market* (Oxford: Oxford University Press, 2015).

22. David G. Victor, Amy M. Jaffe, and Mark H. Hayes, eds., *Natural Gas and Geopolitics: From 1970 to 2040* (Cambridge: Cambridge University Press, 2008), 5.

23. David G. Victor, "The Gas Promise" in *Energy & Security,* ed. Jan H. Kalicki and David L Goldwyn (Washington, DC: Woodrow Wilson Center Press, 2013), 92.

24. EIA, "International Energy Statistics," n.d., http://www.eia.gov/beta/international/data/browser/index.cfm#.

25. Fareed Zakaria, *The Post-American World* (New York: W. W. Norton, 2008); Stefan Fröhlich, *The New Geopolitics of Transatlantic Relations: Coordinated Responses to Common Dangers* (Baltimore, MD: Johns Hopkins University Press, 2012).

26. Charles Clover, "Dreams of the Eurasian Heartland: The Reemergence of Geopolitics," *Foreign Affairs* 78, no. 2 (March–April 1999), https://www.foreignaffairs.com/articles/asia/1999-03-01/dreams-eurasian-heartland-reemergence-geopolitics; Mark Boyle, *Human Geography: A Concise Introduction* (Chichester, UK: Wiley-Blackwell), 108.

27. Brzezinski, *Game Plan*, 22.

28. Kadri Liik, "Introduction: Russia's Pivot to (Eur)asia," in *Russia's "Pivot" to Eurasia,* ed. Kadri Liik (London: European Council on Foreign Relations, May 2014), 6, http://www.ecfr.eu/page/-/ECFR103_RUSSIA

_COLLECTION_290514_AW.pdf; Dmitry Shlapentokh, ed., *Russia between East and West: Scholarly Debates on Eurasianism* (Leiden: Brill, 2007), 148.

29. Nayef R. F. Al-Rodhan, *Neo-Statecraft and Meta-Geopolitics: Reconciliation of Power, Interests and Justice in the 21st Century* (Berlin: LIT Verlag, 2009), 114.

30. Liik, "Introduction," 8.

31. David Baldwin, *Economic Statecraft* (Princeton, NJ: Princeton University Press, 1985); Joseph S. Nye Jr. and Robert O. Keohane, *Power and Interdependence,* 4th ed. (Boston: Longman, 2011).

32. Shaffer, *Energy Politics,* 4, 161.

33. Ibid., 4.

34. Pascual, *The New Geopolitics of Energy,* 10–11.

35. Asjylyn Loder, "Oil Bust of 1986 Reminds U.S. Drillers of Price War Risks," Bloomberg, November 26, 2014, http://www.bloomberg.com/news/articles/2014-11-26/oil-bust-of-1986-reminds-u-s-drillers-of-price-war-risks; Andrew Topf, "Did the Saudis and the US Collude in Dropping Oil Prices?," *Oil Price,* December 23, 2014, http://oilprice.com/Energy/Oil-Prices/Did-The-Saudis-And-The-US-Collude-In-Dropping-Oil-Prices.html.

36. Pascual, *The New Geopolitics of Energy,* 10–11.

37. Interview with Guy Caruso, Washington, DC, January 2016.

38. "A Better Mix: Shale Gas Will Improve Global Security of Energy Supplies," *Economist,* July 12, 2012, http://www.economist.com/node-/21558455.

39. Shaffer, *Energy Politics,* 4, 25.

40. Margarita M. Balmaceda, *The Politics of Energy Dependency: Ukraine, Belarus, and Lithuania between Domestic Oligarchs and Russian Pressure* (Toronto: University of Toronto Press, 2013).

41. Katja Yafimava, *The Transit Dimension of EU Energy Security: Russian Gas Transit across Ukraine, Belarus, and Moldova* (Oxford: Oxford Institute for Energy Studies, 2011), 5, 319–320.

1. The Changing Global Gas Sector

1. Howard V. Rogers, "The Impact of a Globalising Market on Future European Gas Supply and Pricing: The Importance of Asian Demand and North American Supply," OIES Paper NG 59, Oxford Institute for Energy Studies, January 2012, iii, http://www.oxfordenergy.org/wpcms/wp-content

/uploads/2012/01/NG_59.pdf; US Energy Information Administration (henceforth EIA), "Global Natural Gas Markets Overview: A Report Prepared by Leidos, Inc., Under Contract to EIA," Working Paper Series, August 2014, overview. 1, chap. 1, 1, http://www.eia.gov/workingpapers/pdf/global_gas.pdf.

2. Definition by the UK Office of Gas and Electricity Markets, "What Is Liquidity and How Do You Measure That?," *GasTerra,* January 30, 2015, http://www.gasterra.nl/en/news/what-is-liquidity-and-how-do-you-measure-that.

3. John Deutsch, "The Good News about Gas; The Natural Gas Revolution and Its Consequences," *Foreign Affairs,* February 2011, https://www.foreignaffairs.com/articles/2011-01-01/good-news-about-gas; Jonathan Stern, *Natural Gas in Asia: The Challenges of Growth in China, India, Japan, and Korea,* 2nd ed. (Oxford: Oxford University Press, 2008), 381–384; Carlos Pascual, "The New Geopolitics of Energy" (New York: Columbia Center on Global Energy Policy, September 2015), http://energypolicy.columbia.edu/sites/default/files/energy/The%20New%20Geopolitics%20of%20Energy_September%202015.pdf; Jason Bordoff and Trevor Houser, "American Gas to the Rescue? The Impact of US LNG Exports on European Security and Russian Foreign Policy" (New York: Columbia University, September 2014), http://energypolicy.columbia.edu/sites/default/files/energy/CGEP_American%20Gas%20to%20the%20Rescue%3F.pdf; "Step on It," *Economist,* January 30, 2016, http://www.economist.com/news/finance-and-economics/21689644-it-will-take-time-fragmented-market-verge-going-global-step; David Stokes, Howard Rogers, and Olly Spinks, "US Exports Are Now a Reality," Timera Energy, January 11, 2016, http://www.timera-energy.com/us-exports-are-now-a-reality/; Tsuyoshi Inajima, "SGX Seeks to Break LNG's Link to Oil with Singapore SLIng," Bloomberg, January 25, 2016, http://www.bloomberg.com/news/articles/2016-01-25/sgx-seeks-to-break-lng-s-price-link-to-oil-with-singapore-sling; James Henderson and Simon Pirani, eds., *The Russian Gas Matrix: How Markets Are Driving Change* (Oxford: Oxford University Press, 2014), 3, 10, 378–380.

4. Deutsch, "The Good News about Gas."

5. Constanza Jacazio, Senior Analyst at the International Energy Agency, presentation and interview, Center for Strategic and International Studies, Washington, DC, June 28, 2016.

6. Tim Boersma, Charles K. Ebinger, and Heather L. Greenley, "An Assessment of U.S. Natural Gas Exports," Natural Gas Issue Brief (Washington, DC: Brookings, July 2015), 14–15, http://www.brookings.edu/~/media/research/files/papers/2015/07/us-natural-gas-exports/lng_markets.pdf.

7. Jeanette Lee, "Why LNG Doesn't Trade like Oil," Alaska Natural Gas Transportation Projects, US Arctic Research Commission, September 10, 2014, http://www.arcticgas.gov/why-lng-does-not-trade-like-oil; EIA, "Global Natural Gas Markets Overview," 1; Jonathan Stern and Howard Rogers, "The Dynamics of a Liberalised European Gas Market: Key Determinants of Hub Prices, and Roles and Risks of Major Players" (Oxford: Oxford Institute for Energy Studies, December 2014).

8. EIA, "Natural Gas Explained," April 11, 2014, http://www.eia.gov/energyexplained/index.cfm?page=natural_gas_home; EIA, "Oil: Crude and Petroleum Products Explained," May 29, 2015, http://www.eia.gov/energyexplained/index.cfm?page=oil_home#tab2; Robin Webster, "What's the Difference between Natural Gas, Liquid Natural Gas, Shale Gas, Shale Oil and Methane? An Oil and Gas Glossary," *Carbon Brief,* August 23, 2013, http://www.carbonbrief.org/blog/2013/08/oil-and-gas-glossary/; Don Hofstrand, "The Role of Natural Gas in Tomorrow's Energy Economy," Agricultural Marketing Resource Center, March 2010, http://www.agmrc.org/renewable_energy/energy/the-role-of-natural-gas-in-tomorrows-energy-economy/; "Separating Oil, Natural Gas and Water," *Adventures in Energy,* n.d., http://www.adventuresinenergy.org/Exploration-and-Production/Separating-Oil-Natural-Gas-and-Water.html; EIA, "Definitions, Sources and Explanatory Notes," n.d., https://www.eia.gov/dnav/ng/TblDefs/ng_enr_nprod_tbldef2.asp.

9. Robert W. Kolb, *The Natural Gas Revolution: At the Pivot of the World's Energy Future* (Upper Saddle River, NJ: Pearson, 2014), http://ptgmedia.pearsoncmg.com/images/9780133353518/samplepages/0133353516.pdf; "Oil and Gas in Everyday Life," *Voice of Australia's Oil and Gas Industry,* n.d., http://www.appea.com.au/oil-gas-explained/benefits/oil-and-gas-in-everyday-life/; American Petroleum Institute, "Natural Gas and Its Uses," n.d., http://www.api.org/oil-and-natural-gas-overview/exploration-and-production/natural-gas/natural-gas-uses; Scott Disavino, "Natural Gas Likely Overtook Coal as Top U.S. Power Source in 2015," Reuters, January 20, 2016, http://www.reuters.com/article/us-usa-natgas-coal-idUSKCN0UY2LT.

10. Daniel Yergin, *The Prize: The Epic Quest for Oil, Money, and Power* (New York: Simon & Schuster, 1991), 177.

11. Ibid., 12–13, 14, 23–24, 26–27.

12. Ibid., 177.

13. EIA, "U.S. Field Production of Crude Oil," June 30, 2015, http://www.eia.gov/dnav/pet/hist/LeafHandler.ashx?n=PET&s=MCRFPUS1&f=A.

14. Tyler Priest, "The Dilemmas of Oil Empire," in "Oil in American History," special issue, *Journal of American History* 99, no. 1 (June 2012): 236, doi:10.1093/jahist/jas065.

15. Yergin, *The Prize,* 429.

16. Saeid Mokhatab, William A. Poe, and James G. Speight, eds., *Handbook of Natural Gas Transmission and Processing,* 2nd ed. (Waltham, MA: Elsevier Science, 2012), 1–2; Lee H. Solomon, "Natural Gas," *Encyclopedia Britannica,* April 15, 2014, http://www.britannica.com/science/natural-gas; Yergin, *The Prize,* 379, 428, 430, 668; Vaclav Smil, *Energy Transitions: History, Requirements, Prospects* (Santa Barbara, CA: Praeger, 2010), 37–38.

17. "The Russian Federation," The Brookings Foreign Policy Studies, Energy Security Series, Brookings Institution, October 2006, 5, http://www.brookings.edu/~/media/research/files/reports/2006/10/russia/2006russia.pdf; Paul Stevens, "The History of Gas," Polinares, EU Policy on Natural Resources, September 2010, 13, http://www.polinares.eu/docs/d1-1/polinares_wp1_history_gas.pdf; Jonathan Steele, *Soviet Power: The Kremlin's Foreign Policy—Brezhnev to Chernenko* (New York: Simon & Schuster, 1984), 204; US Senate, *Energy Supplies in Eurasia and Implications for U.S. Energy Security,* 2005, http://www.hudson.org/content/researchattachments/attachment/454/senateenergysuppliesineurasia-september05.pdf.

18. Yergin, *The Prize,* 668; Jonathan Stern, "Natural Gas in Europe—The Importance of Russia," Oxford Institute for Energy Studies, 2011, 1, http://www.centrex.at/en/files/study_stern_e.pdf; Stevens, "The History of Gas," 11–12; Carol A. Dahl, *International Energy Markets: Understanding Pricing, Policies, and Profits,* 2nd ed. (Tulsa, OK: Pennwell Books, 2015), 239.

19. Priest, "The Dilemmas of Oil Empire," 237.

20. Roger Owen and Sevket Pamuk, *A History of Middle East Economies in the Twentieth Century* (Cambridge, MA: Harvard University Press, 1999), 212, 216; Axel M. Wietfeld, "Understanding Middle East Gas Exporting Behavior," *Energy Journal* 32, no. 2 (2011): 205, 209–210, http://bwl.univie.ac

.at/fileadmin/user_upload/lehrstuhl_ind_en_uw/lehre/ws1112/SE_Int._Energy_Mgmt_1/GasMidlleEast.pdf.

21. Yergin, *The Prize,* 59.

22. "Early Oil Transportation: A Brief History," 1–4, in *Oil150,* n.d., http://www.oil150.com/files/early-oil-transportation-a-brief-history.pdf; N. A. Krylov, A. A. Bokseman, and E. R. Stavrovsky, *Oil Industry of the Former Soviet Union—Reserves, Extraction and Transportation* (Amsterdam, Netherlands: CRC Press, 1998), 187.

23. Solomon, "Natural Gas"; Mokhatab, Poe, and Speight, *Handbook of Natural Gas Transmission and Processing,* 2; Yergin, *The Prize,* 379, 428, 430; "LNG Shipping at 50," A Commemorative Society of International Gas Tanker and Terminal Operators / International Group of Liquefied Natural Gas Importers Publication (Enfield, UK: Riviera Maritime Media, 2014), 11, 13, http://www.giignl.org/sites/default/files/PUBLIC_AREA/Publications/lng-shipping-at-50compressed.pdf.

24. Yergin, *The Prize,* 586.

25. Ibid., 586, 590–591, 607; British Petroleum (henceforth BP), "Energy Charting Tool," n.d., http://www.bp.com/en/global/corporate/about-bp/energy-economics/energy-charting-tool.html.

26. Sarah O. Ladislaw, Maren Leed, and Molly A. Walton, *New Energy, New Geopolitics; Background Report 1: Energy Impacts* (Washington, DC: Center for Strategic and International Studies, May 2014), 4–5; Joel Darmstadter, "Reflections on the Oil Shock of 40 Years Ago," Resources for the Future, 2014, http://www.rff.org/Publications/Resources/Pages/186-Reflections-on-the-Oil-Shock-of-40-Years-Ago.aspx; Thomas Helbling, "On the Rise," International Monetary Fund, March 2013, http://www.imf.org/external/pubs/ft/fandd/2013/03/helbling.htm; Michael L. Ross, "How the 1973 Oil Embargo Saved the Planet," *Foreign Affairs,* October 15, 2013, https://www.foreignaffairs.com/articles/north-america/2013-10-15/how-1973-oil-embargo-saved-planet; Loren King, Ted Nordhaus, and Michael Shellenberger, "Lessons from the Shale Revolution; A Report on the Conference Proceedings" (Oakland, CA: Breakthrough Institute, April 2015), 13, http://thebreakthrough.org/images/pdfs/Lessons_from_the_Shale_Revolution.pdf.

27. John Deutsch and James R. Schlesinger, *National Security Consequences on U.S. Oil Dependency,* Independent Task Force Report (New York:

Council on Foreign Relations, 2006), xi, http://www.cfr.org/oil/national-security-consequences-us-oil-dependency/p11683.

28. Ibid., 9–10, 54.

29. Interview with Robert Anderson, US Department of Energy (henceforth DOE), Washington, DC, December 29, 2015; T. R. Malthus, *An Essay on the Principle of Population* (Oxford: Oxford World Classics, 2008).

30. Dieter Helm, "Peak Oil and Energy Policy—a Critique," *Oxford Review of Economic Policy* 27, no. 1 (November 1, 2011): 69–74, doi:10.1093/oxrep/grr003; G. Maggio and G. Cacciola, "When Will Oil, Natural Gas, and Coal Peak?," *Fuel* 98 (August 2012): 111–123; Colin J. Campbell, "Understanding Peak Oil," Association for the Study of Peak Oil & Gas International, n.d., http://www.peakoil.net/about-peak-oil; R. W. Allmendinger, "Peak Oil?," Cornell University Energy Studies in the College of Engineering, 2007, http://www.geo.cornell.edu/eas/energy/the_challenges/peak_oil.html.

31. "The Nobel Peace Prize 2007," Nobelprize.org, n.d., http://www.nobelprize.org/nobel_prizes/peace/laureates/2007/.

32. Nick Mabey, Liz Gallagher, and Tom Burke, "Judging the COP21 Outcome and What's Next for Climate Action," Third Generation Environmentalism, December 12, 2015, http://www.e3g.org/library/judging-cop21-outcome-and-whats-next-for-climate-action; Urs Luterbacher and Detlef F. Sprinz, eds., *International Relations and Global Climate Change* (Cambridge, MA: MIT Press, 2011), 23–24, 26–27, http://graduateinstitute.ch/files/live/sites/iheid/files/sites/admininst/shared/doc-professors/luterbacher%20chapter%202%20102.pdf; "Kyoto Protocol," United Nations Framework Convention on Climate Change, 2014, http://unfccc.int/kyoto_protocol/items/2830.php; American Institute of Physics, "The Discovery of Global Warming," February 2015, https://www.aip.org/history/climate/timeline.htm; "The Nobel Peace Prize 2007"; "Cancun Climate Change Conference, November 2010," United Nations Framework Convention on Climate Change, http://unfccc.int/meetings/cancun_nov_2010/meeting/6266.php; "Draft Decision—CF.17; Establishment of an Ad Hoc Working Group on the Durban Platform for Enhanced Action," United Nations Framework Convention on Climate Change, November 2011, http://unfccc.int/files/meetings/durban_nov_2011/decisions/application/pdf/cop17_durbanplatform.pdf; "Doha Climate Change Conference, November 2012," United Nations

Framework Convention on Climate Change, http://unfccc.int/meetings /doha_nov_2012/meeting/6815/php/view/decisions.php; Mark Golden, "Stanford-Led Study Assesses the Environmental Costs and Benefits of Fracking," Stanford News Service, September 12, 2014, http://news.stanford .edu/pr/2014/pr-fracking-costs-benefits-091214.html; US Environmental Protection Agency, "Fact Sheet: Clean Power Plan Overview; Cutting Carbon Pollution from Power Plants," n.d., https://www.epa.gov /cleanpowerplan/fact-sheet-clean-power-plan-overview.

33. BP, "BP Statistical Review of World Energy June 2016," 4, 20, http://www.bp.com/content/dam/bp/pdf/energy-economics/statistical -review-2016/bp-statistical-review-of-world-energy-2016-full-report.pdf; International Energy Agency (henceforth IEA), *Medium-Term Market Report 2016; Market Analysis and Forecasts to 2021—Executive Summary,* Paris, 2016, http://www.iea.org/Textbase/npsum/MTGMR2016SUM.pdf; International Gas Union, *2016 World LNG Report* (Oslo, Norway: International Gas Union, June 2016), 5, 10, http://www.igu.org/publications/2016-world-lng -report; Chou Hui Hong, "LNG Supply Seen Tight after 2020 as Oil Fall Slows Projects," Bloomberg, February 5, 2015, http://www.bloomberg.com /news/articles/2015-02-05/lng-supply-seen-falling-after-2020-as-oil-plunge -delays-projects; Kerri Maddock and Peter Lambert, "Despite Low Oil Prices and a Weak Medium-Term Outlook, LNG Has a Bright Future," *McKinsey Energy Insights,* June 2015, https://www.mckinseyenergyinsights .com/insights/positive-outlook-for-lng.aspx.

34. "Energy Charting Tool"; BP, "BP Energy Outlook 2030," January 2013, 8–11, http://www.bp.com/content/dam/bp/pdf/Energy-economics/Energy -Outlook/BP_Energy_Outlook_Booklet_2013.pdf; EIA, "Demand: Non-OECD," n.d., http://www.eia.gov/finance/markets/demand-nonoecd.cfm; Spencer Dale, BP, "Energy in 2014: After a Calm Comes the Storm," June 10, 2015, 7, http://www.bp.com/content/dam/bp/pdf/Energy-economics /statistical-review-2015/bp-statistical-review-of-world-energy-2015-spencer -dale-presentation.pdf.

35. BP, "BP Statistical Review of World Energy June 2016," 20, 22.

36. EIA, "Conventional Oil and Natural Gas Production," n.d., http:// www.eia.gov/tools/glossary/index.cfm?id=C#conv_oil_nat_gas_prod; EIA, "Unconventional Oil and Natural Gas Production," n.d., http://www .eia.gov/tools/glossary/index.cfm?id=u.

37. "Conventional Oil and Natural Gas Production"; "Unconventional Oil and Natural Gas Production"; IEA, "Unconventional Oil & Gas Production," Energy Technology Network, Energy Technology Systems Analysis Programme, May 2010, 1–2, http://www.iea-etsap.org/web/E-TechDS/PDF/P02-Uncon%20oil&gas-GS-gct.pdf.

38. DOE, Office of Fossil Fuel, "Natural Gas from Shale: Questions and Answers; What Is Shale Gas?," April 2013, http://energy.gov/sites/prod/files/2013/04/f0/complete_brochure.pdf; DOE, Office of Fossil Fuel, "Natural Gas from Shale: Questions and Answers; How Is Shale Gas Produced?," April 2013, http://energy.gov/sites/prod/files/2013/04/f0/how_is_shale_gas_produced.pdf; Energy Mining Advisory Partnership, "Impact of US LNG Exports on Europe" London, n.d., 1, http://energyminingadvisorypartnership.github.io/impact_of_us_lng_on_exports_on_europe.pdf; Lynn Helms, "Horizontal Drilling," *North Dakota Department of Mineral Resources Newsletter* 35, no. 1 (2008): 1–3; Geological Society of America, "Hydraulic Fracturing Defined," n.d., http://www.geosociety.org/criticalissues/hydraulicFracturing/defined.asp.

39. Hamid Al-Megren, ed., *Advances in Natural Gas Technology* (Rijeka, Croatia: InTech, 2012), chap. 1, http://cdn.intechopen.com/pdfs/35285/InTech-Shale_gas_development_in_the_united_states.pdf; Barry Brady et al., "Cracking Rock: Progress in Fracture Treatment Design," *Oilfield Review,* October 1992, 4–5.

40. Alex Trembath, Michael Shellenberger, Jesse Jenkins, and Ted Nordhaus, "Where the Shale Gas Revolution Came From; Government's Role in the Development of Hydraulic Fracturing in Shale" (Oakland, CA: Breakthrough Institute Energy and Climate Program, May 2012); National Energy Technology Laboratory, "Shale Gas: A Short History from NETL," May 25, 2011, 1, http://cce.cornell.edu/EnergyClimateChange/NaturalGasDev/Documents/PDFs/Shale%20Gas-%20a%20short%20history%20from%20NETL.pdf; Zhongmin Wang and Alan Krupnick, "A Retrospective Review of Shale Gas Development in the United States: What Led to the Boom?" (Washington, DC: Resources for the Future, April 2013), 3, http://www.rff.org/RFF/documents/RFF-DP-13-12.pdf; Al-Megren, *Advances in Natural Gas Technology,* 8–9, 12.

41. Jon Gertner, "George Mitchell, Father of Fracking," *New York Times Magazine,* December 21, 2013, http://www.nytimes.com/news/the-lives-they

-lived/2013/12/21/george-mitchell/; Al-Megren, *Advances in Natural Gas Technology,* 12–13, 17, 20, 23; "The Father of Fracking," *Economist,* August 3, 2013, http://www.economist.com/news/business/21582482-few-businesspeople-have-done-much-change-world-george-mitchell-father; Wang and Krupnick, "A Retrospective Review of Shale Gas Development in the United States," 3; Paul Stevens, "The 'Shale Gas Revolution': Developments and Changes" (London: Chatham House, August 2012), 2, http://www.chathamhouse.org/sites/files/chathamhouse/public/Research/Energy%2C%20Environment%20and%20Development/bp0812_stevens.pdf; EIA, "How Much Shale Gas Is Produced in the United States?," Frequently Asked Questions, June 14, 2016, http://www.eia.gov/tools/faqs/faq.cfm?id=907&t=8; EIA, *Annual Energy Outlook 2016 with Projections to 2040,* August 2016, http://www.eia.gov/forecasts/aeo/pdf/0383(2016).pdf.

42. Peter E. Paraschos, "American Shale Revolution: Key Aspects & Implications for Japan," *Japan Spotlight,* August 2014, 26–27; Laura Alfaro, Richard H. K. Vietor, and Hilary White, "The U.S. Shale Revolution: Global Rebalancing?" (Cambridge, MA: Harvard Business School, March 13, 2014), 1–2, 5, https://www.alumni.hbs.edu/Documents/reunions/Vietor_US%20Shale%20Case.pdf; "The Once and Future U.S. Shale Gas Revolution," *Knowledge @ Wharton,* August 29, 2012, http://knowledge.wharton.upenn.edu/article/the-once-and-future-u-s-shale-gas-revolution/; Leonardo Maugeri, "The Shale Oil Boom: A U.S. Phenomenon," Geopolitics of Energy Project (Cambridge, MA: Harvard Kennedy School Belfer Center for Science and International Affairs, June 2013), 2, http://belfercenter.ksg.harvard.edu/files/The%20US%20Shale%20Oil%20Boom%20Web.pdf.

43. Faouzi Aloulou, "Four Countries Added to Global Shale Oil and Natural Gas Resource Assessment," EIA, December 14, 2015, http://www.eia.gov/todayinenergy/detail.cfm?id=24132; EIA, "Argentina and China Lead Shale Development outside North America in First-Half 2015," June 26, 2015, http://www.eia.gov/todayinenergy/detail.cfm?id=21832; John Krohn and Grant Nuelle, "Marcellus, Utica Provide 85% of U.S. Shale Gas Production Growth Since Start of 2012," EIA, July 28, 2015, http://www.eia.gov/todayinenergy/detail.cfm?id=22252&src=email; EIA, "Shale Oil and Shale Gas Resources Are Globally Abundant," June 10, 2013, http://www.eia.gov/todayinenergy/detail.cfm?id=11611.

44. Adam Sieminski, "Implications of the U.S. Shale Revolution," US-Canada Energy Summit, Chicago, October 17, 2014, 10, http://www.eia.gov/pressroom/presentations/sieminski_10172014.pdf; "Argentina and China Lead Shale Development outside North America in First-Half 2015."

45. Faouzi Aloulou, "Shale Gas and Tight Oil Are Commercially Produced in Just Four Countries," EIA, February 13, 2015, http://www.eia.gov/todayinenergy/detail.cfm?id=19991; "Argentina and China Lead Shale Development outside North America in First-Half 2015"; Al-Megren, *Advances in Natural Gas Technology,* 23; Adam Sieminski, "Outlook for U.S. Shale Oil and Gas," Argus Americas Crude Summit, Houston, January 22, 2014, 19, http://www.eia.gov/pressroom/presentations/sieminski_01222014.pdf; Ladislaw, Leed, and Walton, "New Energy, New Geopolitics," 17; Wang and Krupnick, "A Retrospective Review of Shale Gas Development in the United States," 31; Maugeri, "The Shale Oil Boom," 1, 3.

46. IEA, "Are We Entering a Golden Age of Gas? World Energy Outlook 2011," Paris: 2011, 7, http://www.worldenergyoutlook.org/media/weowebsite/2011/WEO2011_GoldenAgeofGasReport.pdf.

47. Tim Boersma, Charles K. Ebinger, and Heather L. Greenley, "An Assessment of U.S. Natural Gas Exports," Natural Gas Issue Brief (Washington, DC: Brookings, July 2015), 2, http://www.brookings.edu/~/media/research/files/papers/2015/07/us-natural-gas-exports/lng_markets.pdf; Alfaro, Vietor, and White, "The U.S. Shale Revolution," 17; "Technically Recoverable Shale Oil and Shale Gas Resources: An Assessment of 137 Shale Formations in 41 Countries Outside the United States," DOE, Independent Statistics and Analysis, June 2013, 10, http://www.eia.gov/analysis/studies/worldshalegas/pdf/fullreport.pdf.

48. Melissa C. Lott, "Methane Hydrates—Bigger than Shale Gas, 'Game Over' for the Environment?," *Scientific American,* March 19, 2013, http://blogs.scientificamerican.com/plugged-in/methane-hydrates-bigger-than-shale-gas-game-over-for-the-environment/; DOE, Office of Fossil Energy, "Methane Hydrate," n.d., http://energy.gov/fe/science-innovation/oil-gas-research/methane-hydrate.

49. EIA, "The Global Liquefied Natural Gas Market: Status & Outlook," December 2003, 3, http://www.eia.gov/oiaf/analysispaper/global/pdf/eia_0637.pdf; California Energy Commission, "Frequently Asked Questions about LNG," 2015, http://www.energy.ca.gov/lng/faq.html.

50. International Gas Union, *World LNG Report—2015 Edition* (Oslo, Norway: International Gas Union, June 2015), 6–11, http://www.igu.org/sites/default/files/node-page-field_file/IGU-World%20LNG%20Report-2015%20Edition.pdf; International Gas Union, *2016 World LNG Report,* 4, 6–9, 45.

51. James T. Jensen, "The Development of a Global LNG Market: Is It Likely? If So When?," Oxford Institute for Energy Studies, 2004, 8, http://www.jai-energy.com/pubs/Oxfordbook.pdf.

52. DOE, Office of Fossil Energy, "Liquefied Natural Gas: Understanding the Basic Facts," August 2005, 4, http://energy.gov/sites/prod/files/2013/04/f0/LNG_primerupd.pdf; International Gas Union, *World LNG Report—2015 Edition,* 7; Jensen, "The Development of a Global LNG Market," 8–10.

53. "The LNG Supply Chain," *Gas in Focus,* n.d., http://www.gasinfocus.com/en/focus/the-lng-supply-chain/; DOE, "Liquefied Natural Gas," 8; Klynveld Peat Marwick Goerdeler (KPMG) Global Energy Institute, "Unlocking the Supply Chain for LNG Project Success," March 2015, 7, https://assets.kpmg.com/content/dam/kpmg/pdf/2015/03/unlocking-supply-chain-LNG-project-success.pdf; Matt Smith, "Commentary: 10 Challenges Faced by the Global LNG Market," *Fuel Fix,* August 27, 2015, http://fuelfix.com/blog/2015/08/27/10-challenges-faced-by-the-global-lng-market/.

54. DOE, "Liquefied Natural Gas," 12; "LNG Shipping at 50," 16; "How Do FPSOs Work?," Rigzone, n.d., http://www.rigzone.com/training/insight.asp?insight_id=299; Victoria Zaretskaya, "Floating LNG Regasification Is Used to Meet Rising Natural Gas Demand in Smaller Markets," EIA, April 27, 2015, http://www.eia.gov/todayinenergy/detail.cfm?id=20972; Shell Global, "Floating Liquefied Natural Gas (FLNG)," n.d., http://www.shell.com/global/future-energy/natural-gas/flng.html; International Gas Union, *World LNG Report—2015 Edition,* 6, 12, 22, 24, 37, 40; International Gas Union, *2016 World LNG Report,* 5, 34, 45, 51.

55. "Small Scale LNG Map," *Gas Infrastructure Europe,* May 2015, http://www.gie.eu/index.php/maps-data/gle-sslng-map; International Gas Union, *World LNG Report—2015 Edition,* 6; BOC, "Leading Edge Technology," n.d., http://www.boc-gas.com.au/en/industries/power_and_energy/lng/leading-edge-technology.html; Alternative Fuels Data Center, "Natural Gas Fuel Basics," n.d., http://www.afdc.energy.gov/fuels/natural_gas_basics.html; Consumer Energy Center, California Energy Commission, "Compressed

Natural Gas (CNG) as a Transportation Fuel," n.d., http://www.consumerenergycenter.org/transportation/afvs/cng.html.

56. Trans Adriatic Pipeline, "Southern Gas Corridor," n.d., http://www.tap-ag.com/the-pipeline/the-big-picture/southern-gas-corridor.

57. "Slovakia Opens Reverse-Flow Pipeline to Carry Gas to Ukraine," *Deutsche Welle,* September 2, 2014, http://www.dw.com/en/slovakia-opens-reverse-flow-pipeline-to-carry-gas-to-ukraine/a-17895333; Tom Knox, "Bulletin Issued about Reverse-Flow Oil Pipeline Failures," *Columbus Business First,* November 18, 2014, http://www.bizjournals.com/columbus/news/2014/11/18/bulletin-issued-about-reverse-flow-oil-pipeline.html; John Krohn and Katie Teller, "New Pipeline Projects Increase Northeast Natural Gas Takeaway Capacity," EIA, January 28, 2016, http://www.eia.gov/today inenergy/detail.cfm?id=24732.

58. Paolo Sorbello, "Europe and Turkmenistan Make Nice," *Diplomat,* May 17, 2015, http://thediplomat.com/2015/05/europe-and-turkmenistan-make-nice/.

59. International Gas Union, *2016 World LNG Report,* 30–32; "Cheniere Takes Control of Sabine Pass LNG Train 1," Argus Media, May 31, 2016, http://www.argusmedia.com/news/article/?id=1249430; John Krohn, Nicholas Skarzynski, and Katie Teller, "Growth in Domestic Natural Gas Production Leads to Development of LNG Export Terminals," EIA, March 4, 2016, http://www.eia.gov/todayinenergy/detail.cfm?id=25232.

60. Ladislaw, Leed, and Walton, "New Energy, New Geopolitics," 24; Abache Abreu, "Growing LNG Liquidity to Boost Spot, Short-Term Trade: BP," S&P Global Platts, September 9, 2015, http://www.platts.com/latest-news/natural-gas/singapore/growing-lng-liquidity-to-boost-spot-short-term-26204710.

61. Ladislaw, Leed, and Walton, "New Energy, New Geopolitics," 24–25; Government of the Netherlands, "Groningen Gas Extraction Further Reduced to 30 Billion Cubic Meters in 2015," June 23, 2015, https://www.government.nl/latest/news/2015/06/23/groningen-gas-extraction-further-reduced-to-30-billion-cubic-metres-in-2015; Will Dalrymple, "Blue Gold," Machinery.co.uk, September 1, 2015, http://www.machinery.co.uk/machinery-features/natural-gas-future-market-prospects/88927/; Stephen Tindale, "Europe Needs More Gas, So Should Develop Shale Gas," Interfax Global Energy, August 20, 2015, http://interfaxenergy.com/analytics/article/442

/europe-needs-more-gas-so-should-develop-shale-gas; Jason Bordoff and Trevor Houser, "American Gas to the Rescue? The Impact of US LNG Exports on European Security and Russian Foreign Policy" (New York: Columbia University, September 2014), 3, http://energypolicy.columbia.edu/sites/default/files/energy/CGEP_American%20Gas%20to%20the%20Rescue%3F.pdf.

62. Anthony J. Melling, "Natural Gas Pricing and Its Future: Europe as the Battleground" (Washington, DC: Carnegie Endowment for International Peace, 2010), 7, 20, 134–135, http://carnegieendowment.org/files/gas_pricing_europe.pdf; Jonathan Stern and Howard Rogers, "The Transition to Hub-Based Gas Pricing in Continental Europe," Oxford Institute for Energy Studies, March 2011, 2, http://www.oxfordenergy.org/wpcms/wp-content/uploads/2011/03/NG49.pdf; Jonathan Stern and Howard V. Rogers, "The Dynamics of a Liberalised European Gas Market: Key Determinants of Hub Prices, and Roles and Risks of Major Players," OIES Paper NG 94, Oxford Institute for Energy Studies, December 2014, 2, http://www.oxfordenergy.org/wpcms/wp-content/uploads/2014/12/NG-94.pdf; Eleonora Waektare, "Territorial Restrictions and Profit Sharing Mechanisms in the Gas Sector: The Algerian Case," *Competition Policy Newsletter* 3 (March 2007): 19, http://ec.europa.eu/competition/publications/cpn/2007_3_19.pdf; Andrei A. Konoplianik, "Russian Gas to Europe: From Long-Term Contracts, On-Border Trade & Destination Clauses to . . . ?," Third EU Energy Law and Regulation Workshop, Florence, September 23, 2004, 8, http://konoplyanik.ru/speeches/12-E-Florence-23-24.09.pdf.

63. Katja Yafimava, "Outlook for the Long Term Contracts in a Globalizing Market (focus on Europe)," United Nations Economic Commission for Europe, Fifth Gas Centre Industry Forum, Geneva, January 19, 2014, 3, https://www.unece.org/fileadmin/DAM/energy/se/pp/geg/gif5_19Jan2015/s1_1_Yafimava.pdf; Melling, "Natural Gas Pricing and Its Future," 21–22; Silvia Colombo, Mohamed El Harrak, and Nicolo Sartori, eds., *The Future of Natural Gas; Markets and Geopolitics* (Hof van Twente, Netherlands: Lenthe / European Energy Review and Istituto Affari Internazionali, 2016), 43, http://www.iai.it/sites/default/files/iai-ocp_gas.pdf; Howard V. Rogers and Jonathan Stern, "Challenges to JCC Pricing in Asian LNG Markets," OIES Paper NG 81, Oxford Institute for Energy Studies, February 2014, 17–18, https://www.oxfordenergy.org/wpcms/wp-content/uploads/2014/02/NG-81.pdf.

64. Stern and Rogers, "The Transition to Hub-Based Gas Pricing in Continental Europe," 2; Patrick Heather, "Continental European Gas Hubs: Are They Fit for Purpose?," OIES Paper NG 63, Oxford Institute for Energy Studies, June 2012, 5, http://www.oxfordenergy.org/wpcms/wp-content /uploads/2012/06/NG-63.pdf; EIA, "Spot Market (Natural Gas)," n.d., http://www.eia.gov/tools/glossary/index.cfm?id=S; OECD-IEA, "The Asian Quest for LNG in a Globalising Market," Partner Country Series, Paris, 2014, 20, http://www.iea.org/publications/freepublications/publication/PartnerCountrySeriesTheAsianQuestforLNGinaGlobalisingMarket.pdf; Luca Franza, "Long-Term Gas Import Contracts in Europe; The Evolution in Pricing Mechanisms" (The Hague: Clingendael International Energy Programme, 2014), 22, http://www.clingendaelenergy.com/inc/upload/files /Ciep_paper_2014-08_web_1.pdf; Melling, "Natural Gas Pricing and Its Future," 7, 10, 23; European Commission, *Quarterly Report on European Gas Markets, Brussels,* Directorate-General for Energy, third quarter 2015, 17, 20, https://ec.europa.eu/energy/sites/ener/files/documents/quarterly _report_on_european_gas_markets_q3_2015.pdf; Colombo, El Harrak, and Sartori, *The Future of Natural Gas,* 45.

65. OECD-IEA, "The Asian Quest for LNG in a Globalising Market," 17–18, 21; "Global Natural Gas Markets Overview," 5; Colombo, El Harrak, and Sartori, *The Future of Natural Gas,* 43.

66. In older gas contracts, part of the pricing formula included 14 percent of the cost of Brent, or 14 percent the cost of a barrel of oil. So if oil cost $100 a barrel, then gas would cost 14 percent of 100, or $14 plus some additional costs. Now the percentage indexation has dropped to 11–13 percent range times the price of a barrel of oil.

67. Melling, "Natural Gas Pricing and Its Future," 10; Franza, "Long-Term Gas Import Contracts in Europe," 11; Stern and Rogers, "The Dynamics of a Liberalised European Gas Market," 3.

68. "Spot Market (Natural Gas)."

69. Lee, "Why LNG Doesn't Trade like Oil"; Michael Stoppard, "Low Oil Prices and LNG," *IHS Energy CERAWeek 2015,* April 22, 2015, 1, http:// ceraweek.com/2015/wp-content/uploads/2015/04/WSJ2015-04-22final.pdf; Howard V. Rogers, "The Impact of Lower Gas and Oil Prices on Global Gas and LNG Markets," OIES Paper NG 99, Oxford Institute for Energy Studies, July 2015, 34–35, http://www.oxfordenergy.org/wpcms/wp-content/uploads

/2015/07/NG-99.pdf; Howard Rogers, "Transition from JCC Pricing in Asian LNG Markets," Timera Energy, February 24, 2014, http://www.timera-energy.com/transition-away-from-jcc-pricing-in-asian-lng-markets/; Oleg Vukmanovic and Sarah McFarlane, "Update 2—Trading Houses Dominate in Egypt's $2.2 Bln LNG Tender," Reuters, January 27, 2015, http://www.reuters.com/article/2015/01/27/egypt-lng-tender-idUSL6N0V63EY20150127; "The Next Leg Down in Commodity Prices," Timera Energy, August 31, 2015, http://www.timera-energy.com/the-next-leg-down-in-commodity-prices/; Colombo, El Harrak, and Sartori, *The Future of Natural Gas,* 46, 54.

70. EIA, "Global Natural Gas Markets Overview," 5–6, http://www.eia.gov/workingpapers/pdf/global_gas.pdf; Franza, "Long-Term Gas Import Contracts in Europe," 20, 32; Energy Mining Advisory Partnership, "Impact of US LNG Exports on Europe," 3; Dale Nijoka, Barry Munro, and Foster Mellen, "Competing for LNG Demand: The Pricing Structure Debate" (Toronto, Canada: Ernst & Young, 2014), 5, http://www.ey.com/Publication/vwLUAssets/Competing-for-LNG-demand/$FILE/Competing-for-LNG-demand-pricing-structure-debate.pdf.

71. Clifford Krauss, "Oil Prices: What's behind the Plunge? Simple Economics," *New York Times,* September 1, 2015, http://www.nytimes.com/interactive/2015/business/energy-environment/oil-prices.html?_r=0; "The Next Leg Down in Commodity Prices," Timera Energy, August 31, 2015, http://www.timera-energy.com/the-next-leg-down-in-commodity-prices/; Delia Morris, "With No Supply Response in Sight, WTI Crashes through $27," Rigzone, January 20, 2016, http://www.rigzone.com/news/oil_gas/a/142585/With_No_Supply_Response_in_Sight_WTI_Crashes_Through_27; "Changing Market Realities in Light of Continuing Low Gas Prices," Natural Gas Europe, March 14, 2016, http://www.naturalgaseurope.com/changing-market-realities-in-light-of-continuing-low-gas-prices-28618.

72. Lester R. Brown, "Eco-Economy: Building an Economy for the Earth," Earth Policy Institute, n.d., http://www.earthpolicy.org/mobile/books/eco/eech5_ss6?phpMyAdmin=1d6bec1fea35111307d869d19bcd2ce7; IEA, "Are We Entering a Golden Age of Gas?," 7; Scott Jell, "Electricity from Natural Gas Surpasses Coal for First Time, but Just for One Month," EIA, July 31, 2015, http://www.eia.gov/todayinenergy/detail.cfm?id=22312&src=email; Damian Carrington, "Fracking Boom Will Not Tackle Global Warming, Analysis Warns," *Guardian,* October 15, 2014, http://www

.theguardian.com/environment/2014/oct/15/gas-boom-from-unrestrained-fracking-linked-to-emissions-rise; Ladislaw, Leed, and Walton, "New Energy, New Geopolitics," 25; Owen Comstock, "Natural Gas-Fired Electricity Generation Expected to Reach Record Level in 2016," EIA, July 2016, http://www.eia.gov/todayinenergy/detail.cfm?id=27072.

73. Christina Nunez, "Switch to Natural Gas Won't Reduce Carbon Emissions Much, Study Finds," *National Geographic,* September 25, 2014, http://news.nationalgeographic.com/news/energy/2014/09/140924-natural-gas-impact-on-emissions/.

74. Carrington, "Fracking Boom Will Not Tackle Global Warming"; Thomas K. Grose, "As U.S. Cleans Its Energy Mix, It Ships Coal Problems Abroad," *National Geographic,* March 17, 2013, http://news.nationalgeographic.com/news/energy/2013/03/130315-us-coal-exports/.

75. Alfaro, Vietor, and White, "The U.S. Shale Revolution," 16. Carrington, "Fracking Boom Will Not Tackle Global Warming"; Trevor Houser and Shashank Mohan, *Fueling Up: The Economic Implications of America's Oil and Gas Boom* (Washington, DC: Peterson Institute for International Economics, 2014).

76. Nunez, "Switch to Natural Gas Won't Reduce Carbon Emissions Much"; Ladislaw, Leed, and Walton, "New Energy, New Geopolitics," 27.

77. Michael A. Levi, Implications of the 2015 Paris Climate Conference, interview by Irina A. Faskianos, conference call, December 14, 2015, http://www.cfr.org/climate-change/implications-2015-paris-climate-conference/p37370; Joe Nocera, "Shale Gas and Climate Change," *New York Times,* July 14, 2015, http://www.nytimes.com/2015/07/14/opinion/joe-nocera-shale-gas-and-climate-change.html?emc=edit_th_20150714&nl=todaysheadlines&nlid=67917937&_r=2.

78. Colombo, El Harrak, and Sartori, *The Future of Natural Gas,* 25, 27, 30, 39; Marjolein Helder, "Renewable Energy Is Not Enough: It Needs to Be Sustainable," World Economic Forum, September 2, 2015, https://www.weforum.org/agenda/2015/09/renewable-energy-is-not-enough-it-needs-to-be-sustainable/; Josh Freed, Matt Bennett, and Matt Goldberg, "The Climate Challenge: Can Renewables Really Do It Alone?," *Third Way,* December 16, 2015, http://www.thirdway.org/report/the-climate-challenge-can-renewables-really-do-it-alone.

79. Linda Dong, "What Goes in and out of Hydraulic Fracturing," Dangersoffracking.com, n.d., http://www.dangersoffracking.com/;

Joe Hoffman, "Potential Health and Environmental Effects of Hydrofracking in the Willston Basin, Montana," Science Education Resource Center at Carleton College, n.d., http://serc.carleton.edu/NAGTWorkshops/health/case_studies/hydrofracking_w.html; Natural Resources Defense Council, "Unchecked Fracking Threatens Health, Water Supplies," n.d., http://www.nrdc.org/energy/gasdrilling/; Tom Zeller Jr., "Yes, Fracking Can Be Directly Linked to Earthquakes," *Forbes,* January 6, 2015, http://www.forbes.com/sites/tomzeller/2015/01/06/yes-fracking-can-be-directly-linked-to-earthquakes/#5b2ffe2e49dd; "Earthquakes," *California Frack Facts*, n.d., http://www.cafrackfacts.org/impacts/seismology/; Roger Real Drouin, "On Fracking Front, a Push to Reduce Leaks of Methane," *Yale Environment 360,* April 7, 2014, http://e360.yale.edu/feature/on_fracking_front_a_push_to_reduce_leaks_of_methane/2754/; Michael Levi, "Fracking and the Climate Debate," *Democracy Journal,* Summer 2015, http://democracyjournal.org/magazine/37/fracking-and-the-climate-debate/.

80. "Natural Gas and the Environment," NaturalGas.org, September 20, 2013, http://naturalgas.org/environment/naturalgas/; Patrick J. Kiger, "Green Fracking? 5 Technologies for Cleaner Shale Energy," *National Geographic,* March 21, 2014, http://news.nationalgeographic.com/news/energy/2014/03/140319-5-technologies-for-greener-fracking/; "Fracking FAQs," *Gasland the Movie,* n.d., http://www.gaslandthemovie.com/whats-fracking/faq/technology-safety; Mario Parker and Mark Drajem, "Obama's Methane Limits Seen Wiping Out the Marginal Driller," Bloomberg, August 18, 2015, http://www.bloomberg.com/news/articles/2015-08-18/obama-proposes-deep-cut-in-methane-leaks-from-oil-gas-operators; Mark Jones, "Summary of Countries That Have Taken Action against Fracking May 2015," *Protect Our Limestone Coast,* May 21, 2015, http://www.protectlimestonecoast.org.au/summary-of-countries-that-have-taken-action-against-fracking-may-2015/.

81. Colombo, El Harrak, and Sartori, *The Future of Natural Gas,* 51–54.

82. Leonardo Maugeri, "The Coming Global Gas-Market Bust," *National Interest,* February 24, 2015, http://nationalinterest.org/feature/the-coming-global-gas-market-bust-12316.

2. The Politics and Commerce of American LNG Exports

1. US Energy Information Administration (henceforth EIA), "U.S. Natural Gas Prices," October 2016, http://www.eia.gov/dnav/ng/ng_pri

_sum_dcu_nus_m.htm; “Appendix B: Natural Gas,” in *QER Report: Energy Transmission, Storage, and Distribution Infrastructure,* Washington, DC, April 2015, NG-12, http://energy.gov/sites/prod/files/2015/09/f26/QER_AppendixB_NaturalGas.pdf.

2. Michael A. Levi, *Power Surge: Energy, Opportunity, and the Battle for America’s Future* (New York: Oxford University Press, 2013), 23–24.

3. EIA, “EIA FAQs: How Much Shale Gas Is Produced in the United States?,” 2015, http://www.eia.gov/tools/faqs/faq.cfm?id=907&t=8.

4. EIA, “US Natural Production reaches record high in 2015,” April 15, 2016, http://www.eia.gov/todayinenergy/detail.cfm?id=25832&src=email.

5. EIA, “Shale in the United States,” in *EIA Energy in Brief,* 2016, http://www.eia.gov/energy_in_brief/article/shale_in_the_united_states.cfm; EIA, “Natural Gas Gross Withdrawals and Production: Dry Production,” October 31, 2016, http://www.eia.gov/dnav/ng/ng_prod_sum_a_EPG0_FPD_mmcf_m.htm.

6. Natural Gas Supply Association (henceforth NGSA), “Top 40 Producers US Natural Gas Production,” 2015, http://www.ngsa.org/download/analysis_studies/Top%2040%202014%204th%20quarter.pdf.

7. EIA, “Technically Recoverable Shale Oil and Shale Gas Resources: An Assesment of 137 Shale Formations in 34 Countries Outside of the US,” in *EIA Analysis and Projections,* 2013, http://www.eia.gov/analysis/studies/worldshalegas/.

8. Jason Bordoff and Trevor Houser, “American Gas to the Rescue?” (New York: Columbia University Center on Global Energy Policy, September 2014), http://energypolicy.columbia.edu/sites/default/files/energy/CGEP_American%20Gas%20to%20the%20Rescue%3F.pdf.

9. EIA, “Technically Recoverable Shale Oil and Shale Gas Resources.”

10. Jordan W. Brock, “IE Questions: What Are Unproved Reserves and Why Should You Care?,” *Inside Energy,* December 2014, http://insideenergy.org/2014/12/22/ie-questions-what-are-unproved-reserves-and-why-should-you-care/.

11. EIA, “International Energy Statistics: Proved Reserves of Natural Gas,” 2015, http://www.eia.gov/cfapps/ipdbproject/IEDIndex3.cfm?tid=3&pid=3&aid=6.

12. Ibid.

13. Ibid.; EIA, Shale Gas Reserves, unpublished raw data, December 4, 2014, http://www.eia.gov/dnav/ng/ng_enr_shalegas_dcu_nus_a.htm.

14. EIA, "Shale Gas (Billion Cubic Feet)," Data Series, November 19, 2015, http://www.eia.gov/dnav/ng/ng_enr_shalegas_a_EPG0_R5301_Bcf_a.htm; EIA, "Estimated Dry Natural Gas Contained in Total Natural Gas Proved Reserves (Billion Cubic Feet)," Data Series, November 19, 2015, http://www.eia.gov/dnav/ng/ng_enr_dry_a_EPG0_r11_bcf_a.htm.

15. "About Marcellus Shale," *Marcellus Shale Electronic Field Guide,* Penn State College of Agricultural Sciences, n.d., http://marcellusfieldguide.org/index.php/about/.

16. Warren Wilczewski and William Brown, "Hydrocarbon Gas Liquids Production and Related Industrial Development, in *Annual Energy Outlook 2016,* EIA, July 6, 2016, https://www.eia.gov/forecasts/aeo/hgl.cfm.

17. EIA, *Drilling Productivity Report for Key Tight Oil and Shale Gas Regions,* 2015, http://www.eia.gov/petroleum/drilling/pdf/dpr-full.pdf.

18. "Information on the Barnett Shale," *Naturalgasintel.com*, 2015, http://www.naturalgasintel.com/barnettinfo.

19. EIA, "Drilling Productivity Report," November 2016, https://www.eia.gov/petroleum/drilling/pdf/bakken.pdf.

20. "Henry Hub Gas Price Assessment," S&P Global Platts, 2015, http://www.platts.com/price-assessments/natural-gas/henry-hub.

21. Scott Disavino, "Henry Hub, King of US Natural Gas Trade, Losing Crown to Marcellus." Reuters, September 25, 2014, http://www.reuters.com/article/2014/09/25/us-natgas-henryhub-marcellus-analysis-idUSKCN0HK17E20140925.

22. Ibid.

23. Freeport LNG, "LNG Export Project Video," 2015, http://www.freeportlng.com/LNG_Export_Video.asp.; Rosalind Krasny, "Update 2—U.S. FERC Approves Dominion's Cove Point LNG Export Facility," Reuters, September 30, 2014, http://www.reuters.com/article/2014/09/30/usa-energy-dominion-resourc-idUSL2N0RV00J20140930.

24. Timothy Cama and Cristina Marcos. "House Passes Bill to Speed Up Liquefied Natural Gas Exports," *The Hill,* January 28, 2015, http://thehill.com/blogs/floor-action/house/230990-house-passes-bill-to-speed-up-liquefied-natural-gas-exports.http://thehill.com/blogs/floor-action/house/230990-house-passes-bill-to-speed-up-liquefied-natural-gas-exports; Amy Harder, "Democrats Increasingly Backing Oil and Gas Industry," *Wall Street Journal,* August 11, 2014, http://www.wsj.com/articles/democrats-increasingly-backing-oil-and-gas-industry-1407790617.

25. Bordoff and Houser, "American Gas to the Rescue?," 9, 21.

26. The eleven US terminals include one facility in Puerto Rico. Center for Liquefied Natural Gas, "US Import Terminals," n.d, http://www.lngfacts.org/lng-market/imports/; Michael Ratner, Paul W. Parfomak, Linda Luther, and Ian F. Furgesson, "US Natural Gas Exports: New Opportunities, Uncertain Outcomes," issue brief no. R42074 (Washington, DC: Congressional Research Service, 2015), 2, https://www.fas.org/sgp/crs/misc/R42074.pdf; International Gas Union, *2016 World LNG Report* (Oslo, Norway: International Gas Union, June 2016), 65–68, http://www.igu.org/publications/2016-world-lng-report.

27. Michael Ford, "Projections Show US Becoming a Net Exporter of Natural Gas," EIA, 2015, http://www.eia.gov/todayinenergy/detail.cfm?id=20992.

28. Department of Energy Presentation, The Oil and Gas Forum, Washington, DC, July 16, 2016; EIA, "US Natural Gas Imports by Country," raw data, September 30, 2015, http://www.eia.gov/dnav/ng/NG_MOVE_IMPC_S1_A.htm.

29. North East Gas Association, "The Role of LNG in the North East Natural Gas (and Energy) Market," n.d., http://www.northeastgas.org/about_lng.php.

30. Freeport LNG, "LNG Export Project Video."

31. EIA, "U.S. Natural Gas Exports and Re-Exports by Country," August 31, 2015, http://www.eia.gov/dnav/ng/ng_move_expc_s1_a.htm.

32. Ratner et al., "US Natural Gas Exports," 10.

33. "Kenai Peninsula Borough Mayor's Office: LNG Project," August 26, 2015, http://www.kpb.us/mayor/lng-project.http://www.kpb.us/mayor/lng-project.

34. Ibid.

35. Cheniere Energy Partners, "Liquefaction Facilities: Trains 1–6," September 26, 2016, http://www.cheniere.com/terminals/sabine-pass/trains-1-6/.

36. "Cheniere Takes Control of Sabine Pass LNG Train 1," Argus Media, May 31, 2016, http://www.argusmedia.com/news/article/?id=1249430.

37. "Cheniere Granted Permit to Start LNG Exports from Sabine Pass Train 2," *LNG World News*, October 13, 2016, http://www.lngworldnews.com/cheniere-granted-permit-to-start-lng-exports-from-sabine-pass-train-2/?utm_source=emark&utm_medium=email&utm_campaign=daily-update-lng-world-news-2016-10-14&uid=27940.

38. Interview with Naila Jamoussi of Cheniere, December 14, 2015.

39. Federal Energy Regulatory Commission (henceforth FERC), "North American LNG Import / Export Terminals Approved," 2015, http://www.ferc.gov/industries/gas/indus-act/lng/lng-approved.pdf.

40. Ibid.

41. EIA, "About US Natural Gas Pipelines—Transporting Natural Gas," 2009, http://www.eia.gov/pub/oil_gas/natural_gas/analysis_publications/ngpipeline/southwest.html#overview; FERC, "North American LNG Import / Export Terminals Approved."

42. David Ledesma, James Henderson, and Nyrie Palmer, "The Future of Australian LNG Exports: Will Domestic Challenges Limit the Future of LNG Export Capacity?," OIES Paper NG 90, Oxford Institute for Energy Studies, September 2014, 19, http://www.oxfordenergy.org/wpcms/wp-content/uploads/2014/09/NG-90.pdf.

43. FERC, "North American LNG Export Terminals Proposed," January 6, 2016, https://www.ferc.gov/industries/gas/indus-act/lng/lng-proposed-export.pdf.

44. Ibid., 34–35; Christopher Langdon, Paul R. Cassidy, and Gordon N. Nettleton, *Canadian LNG Projects; Key Development and Financing Issues* (Toronto: McCarthy Tetrault LLP, 2015), 3, https://www.mccarthy.ca/pubs/Canadian_LNG_Yearbook_2015_update.pdf.

45. Rebecca Penty and Bradley Olson, "Global Liquefied Gas Market Faces Potential Oversupply," Bloomberg, March 5, 2014, http://www.bloomberg.com/news/articles/2014-03-05/global-liquefied-gas-market-faces-potential-oversupply.

46. Freeport LNG, "LNG Export Project Video."

47. Interview with Warren Wilczewski, Washington, DC, August 10, 2015; ICF International, *Access Northeast Project—Reliability Benefits and Energy Cost Savings to New England*, 2015, 3, http://accessnortheastenergy.com/wp-content/uploads/2015/02/ICF-Report-on-Access-Northeast-Project1.pdf.

48. Mary Serreze, "Kinder Morgan Pulls the Plug on Northeast Energy Direct-Tennessee Gas Pipeline," *Mass Live*, April 20, 2016, http://www.masslive.com/news/index.ssf/2016/04/kinder_morgan_pulls_the_plug_o.html; Erin Ailworth, "New York Environmental Regulators Deny Permit for Constitution Pipeline," *Wall Street Journal*, April 22, 2016, http://

www.wsj.com/articles/new-york-environmental-regulators-deny-permit-for-constitution-pipeline-1461366759.

49. ExxonMobil, "Understanding Flares," n.d., http://www.exxonmobil.com/AP-English/about_who_profile_understanding_flares.aspx.

50. Ratner et al., "US Natural Gas Exports," 26.

51. EIA, "Natural Gas Gross Withdrawals and Production," 2016, http://www.eia.gov/dnav/ng/ng_prod_sum_a_EPG0_VGV_mmcf_a.htm; EIA, "North Dakota Flaring Targets Challenged by Rapid Production Growth," 2013, http://www.eia.gov/todayinenergy/detail.cfm?id=23752; Henning Gloystein, "EU Gas Imports from Russia Could Drop a Quarter by 2020," Reuters, April 9, 2014, http://www.reuters.com/article/2014/04/09/us-ukraine-crisis-gas-idUSBREA3818J20140409.

52. Gazprom, "Gas Marketing in Europe," 2013, http://www.gazprom.com/about/marketing/europe/.

53. Harder "Democrats Increasingly Backing Oil and Gas Industry."

54. "Senators Stabenow and Markey Lead Bipartisan Group of 22 Senators Urging President Obama to Consider the Impact of Large-Scale Natural Gas Exports on American Manufacturing, Families," Stabenow.senate.gov, May 9, 2014, http://www.stabenow.senate.gov/?p=press_release&id=1338.

55. Ratner et al., "US Natural Gas Exports," 15–16.

56. Ibid., 11.

57. Atlantic Council Task Force on the US Energy Boom and National Security, comp., *Empowering America: How Energy Abundance Can Strengthen US Global Leadership* (Washington DC: Atlantic Council, July 2015); Bordoff and Houser, "American Gas to the Rescue?"

58. Claudia Squeglia and Rafaello Matarazzo, "The US-EU Energy Trade Dilemma" (Rome: Istituto Affari Internazionali, September 2015), 8, http://www.iai.it/sites/default/files/iaiwp1528.pdf; interview with Warren Wilczewski of EIA, February 2, 2016.

59. Enerknol Research, "FERC Advances Oregon LNG Export Facility as Congress Moves to Fast-Track New Trade Legislation," *Breaking Energy* (blog), May 4, 2014, http://breakingenergy.com/2015/05/04/ferc-advances-oregon-lng-export-facility-as-congress-moves-to-fast-track-new-trade-legislation/.

60. Ibid.

61. Norman Bay, Natural Gas Roundtable Talk, Washington, DC, July 21, 2015.

62. Enerknol Research, "FERC Advances Oregon LNG Export Facility."

63. See list: Office of the United States Trade Representative, "Free Trade Agreements," n.d., https://ustr.gov/trade-agreements/free-trade-agreements.

64. Enerknol Research, "FERC Advances Oregon LNG Export Facility."

65. Office of Fossil Energy, "How to Obtain Authorization to Import and / or Export Natural Gas and LNG," n.d., http://energy.gov/fe/services/natural-gas-regulation/how-obtain-authorization-import-andor-export-natural-gas-and-lng; United States Mission to the European Union, "Remarks at the European Parliament's LNG Roundtable," 2015, http://useu.usmission.gov/rm-05292015.html.

66. Cama and Marcos, "House Passes Bill to Speed up Liquefied Natural Gas Exports."

67. Ambrose Evans-Pritchard, "US to Launch Blitz of Gas Exports, Eyes Global Energy Dominance," *Telegraph*, April 26, 2015, http://www.telegraph.co.uk/finance/newsbysector/energy/11563761/US-to-launch-blitz-of-gas-exports-eyes-global-energy-dominance.html.

68. "CEQ NEPA Guidance 'Another Hurdle' for U.S. LNG Exports, CLNG Says," *LNG World News,* August 3, 2016, http://www.lngworldnews.com/ceq-nepa-guidance-another-hurdle-for-u-s-lng-exports-clng-says/?utm_source=emark&utm_medium=email&utm_campaign=daily-update-lng-world-news-2016-08-04&uid=27940.

69. Office of Fossil Energy, DOE, "LNG Monthly Report—October 2016," http://energy.gov/sites/prod/files/2016/10/f33/LNG%202016_0.pdf; "Second U.S. LNG Cargo Delivered to Europe," *LNG World News,* July 25, 2016, http://www.lngworldnews.com/second-u-s-lng-cargo-delivered-to-europe/; "China Receives U.S. LNG Cargo," *LNG World News,* August 22, 2016, http://www.lngworldnews.com/china-receives-u-s-lng-cargo/.

70. Chesapeake Climate Action Network, "Known, Signed Liquefaction Tolling Agreements from LNG Facilities that Have Received U.S. Department of Energy Approval to Export Gas to Foreign Countries," 2014, http://chesapeakeclimate.org/wp/wp-content/uploads/2014/03/Known-LNG-export-contracts.pdf.

71. Leslie Palti-Guzman, "Gas under Pressure," *Foreign Affairs,* January 8, 2016, https://www.foreignaffairs.com/articles/united-states/2016-01-08/gas-under-pressure; Chesapeake Climate Action Network, "Known, Signed Liquefaction Tolling Agreements."

72. Fereidun Fesharaki, "The Global LNG Market Outlook: Too Many Sellers, Not Enough Buyers," presentation, Center for Strategic and International Studies, Washington, DC, September 25, 2015.

73. "Regasification of LNG: Strategic Access to Markets," *Total,* n.d., http://www.total.com/en/energies-expertise/oil-gas/trading-shipping/fields-expertise/transportation-storage/regasification-lng-strategic-access-markets.

74. Ibid.

75. Boris Ertl, Isa Mohammed, Charles Durr, David Coyle, and Stanley Huang, "New LNG Receiving Terminal Concepts," World Petroleum Congress, January 1, 2005; Wim van Wijngaarden, Hein Oomen, and Jos van Doorn, "Offshore LNG Terminals: Sunk or Floated?," Offshore Technology Conference, Houston, 2004, 1.

76. Saeid Mokhatab, David Wood, Jaleel Valappil, and John Y. Mak, *Handbook of Liquefied Natural Gas* (Kidlington, UK: Gulf Professional, 2014), 24.

77. Agnia Grigas, "Standing Up to Gazprom: What Ukraine Can Learn from Lithuania," *Atlantic Council,* June 23, 2014, http://www.atlanticcouncil.org/blogs/new-atlanticist/standing-up-to-gazprom-what-ukraine-can-learn-from-lithuania.

78. Michael D. Tusiani and Gordon Shearer, *LNG: A Nontechnical Guide* (Tulsa, OK: PennWell, 2007), 83.

79. "Höegh LNG Turns Down Lithuania's FSRU Advanced Sale Proposal," Natural Gas Europe, August 13, 2015, http://www.naturalgaseurope.com/norway-hoegh-turns-down-lithuanis-proposal-fsru-advance-sale-25006.

80. International Gas Union, *World LNG Report—2015 Edition* (Oslo, Norway: International Gas Union, June 2015), 6, 24, http://www.igu.org/sites/default/files/node-page-field_file/IGU-World%20LNG%20Report-2015%20Edition.pdf.

81. Keith Schaefer, "Floating LNG: The Revolution in Natural Gas," Oilandgas-investments.com, March 14, 2012, http://oilandgas-investments

.com/2012/natural-gas/floating-lng-the-revolution-in-natural-gas; Small Cap Traders, "Liquefied Natural Gas Limited," June 30, 2014, smallcap-traders.com/wp-content/uploads/2014/06/LNGLF-MM-BB-SCT.pdf; Jurgita Lapienytė, "Iki 2025 metų Lietuvai SGD terminalas kainuos 2,4 mlrd. Lt," *15min,* May 30, 2012, http://www.15min.lt/verslas/naujiena/finansai/suskystintu-gamtiniu-duju-terminalo-kaina-3-mlrd-lt-662-222627.

82. PWC Energy, "The Progression of an LNG Project: Canadian LNG Projects" (Ontario: Pricewaterhouse Coopers, 2014), 5, 12, http://www.pwc.com/en_GX/gx/mining/publications/assets/pwc-lng-progression-canada.pdf.

83. The project was planned to be operational in 2019 but lacked secured commitment as of 2016. Jasmina Kuzmanovic, "Croatia, Hungary, Ukraine to Create Adriatic Gas Corridor," Bloomberg, October 23, 2013, http://www.bloomberg.com/news/articles/2013-10-23/croatia-hungary-ukraine-to-create-adriatic-gas-corridor; "EU Gravitates toward LNG Expansion," Natural Gas Europe, May 28, 2015, http://www.naturalgaseurope.com/european-union-lng-expansion-23637; Bartłomiej Sawicki, "Third Time Lucky—Croatia to Start Building Terminal LNG on Island Krk Next Year," Visegradplus, May 6, 2015, http://visegradplus.org/third-time-lucky-croatia-to-start-building-terminal-lng-on-island-krk-next-year/.

84. Credit Suisse, "The Shale Revolution," December 2012, 31, derivative.credit-suisse.com/get.cfm?id=BD714796-A110-4FFB-BFD7-B9D8CB5B3C16.

85. Michaela Jellicoe and Michael S. Delgado, "Quantifying the Effects of Underground Natural Gas Storage on Nearby Residents," National Agricultural and Rural Development Policy Center (NARDeP), brief 24, June 2014.

86. EIA, "The Basics of Underground Natural Gas Storage," August 2004, http://www.eia.gov/pub/oil_gas/natural_gas/analysis_publications/storagebasics/storagebasics.html.

87. Barry Stevens, "Natural Gas Storage Is Vital for Future Industry Growth," *Oilprice,* June 11, 2012, http://oilprice.com/Energy/Natural-Gas/Natural-Gas-Storage-is-Vital-for-Future-Industry-Growth.html.

88. "Liquefied Natural Gas: Understanding the Basic Facts," August 2005, 3, http://energy.gov/sites/prod/files/2013/04/f0/LNG_primerupd.pdf.

89. "The European LNG Terminal Infrastructure 2015: Status and Outlook," *Gas Infrastructure Europe,* June 17, 2015, http://www.gie.eu/index.php/publications/gle/doc_download/24224-abstract-lng-map-investment

-database-2015-the-european-lng-terminal-infrastructure-2015-status-and-outlook; International Gas Union, *World LNG Report—2015 Edition,* 79–82.

90. Ibid.

91. Ibid.

92. Lietuvos Energetikos Institutas, "LNG Terminal Project in Lithuania," November 2012, http://www.lei.lt/energy-security-conference/index_files/Masiulis.pdf.

93. Not included among these is a terminal in Puerto Rico and two in Mexico, where one is built by a US company to serve as an import terminal for the California market.

94. International Gas Union, *World LNG Report—2015 Edition,* 79–82; Lynn Doan and Harry Weber, "Mitsui to Supply Colombia Its First Liquefied Natural Gas Cargo," Bloomberg, May 27, 2016, http://www.bloomberg.com/news/articles/2016-05-27/mitsui-to-supply-colombia-its-first-liquefied-natural-gas-cargo.

95. United States International Trade Commission, *Natural Gas Services: Recent Reforms in Selected Markets, Investigation No. 332-426* (Washington, DC, 2001).

96. Tim Boersma, Tatiana Mitrova, Geert Greving, and Anna Galkina, "Business as Usual: European Gas Market Functioning in Times of Turmoil and Increasing Import Dependence," Brookings Institution, policy brief 14-05, October 2014, 1–5.

97. "Energy Union and Climate," European Commission, n.d., https://ec.europa.eu/priorities/energy-union-and-climate_en; "Market Legislation," European Commission: Energy, n.d., https://ec.europa.eu/energy/en/topics/markets-and-consumers/market-legislation.

98. King Abdullah Petroleum Studies and Research Center, "Natural Gas: Entering the New Dark Age?," August 2015, kapsarc.org/en/Publications/KS-1519-WB17A-Natural%20Gas_%20Entering%20the%20New%20Dark%20Age.pdf.

99. Eric Yep, "Spot LNG Prices Hit Record in Asia," *Wall Street Journal,* February 14, 2014, http://www.wsj.com/articles/SB10001424052702304315004579382261761664926.

100. Edward Dodge, "Impact of Falling Oil Prices on LNG," *Breaking Energy,* January 21, 2015, breakingenergy.com/2015/01/21/impact-of-falling-oil-prices-on-lng/.

101. Cheniere Energy, "J. P. Morgan West Coast Energy Infrastructure / MLP 1x1 Forum," March 2016, 23, 25, http://phx.corporate-ir.net/phoenix.zhtml?c=101667&p=irol-presentations; Jack Farchy, "Global Gas Market Braced for Price War," *Financial Times,* February 3, 2016, https://next.ft.com/content/c9c44750-ca50-11e5-a8ef-ea66e967dd44; James Henderson, "Gazprom—Is 2016 the Year for a Change of Pricing Strategy in Europe?," Oxford Institute for Energy Studies, January 2016, 7, https://www.oxfordenergy.org/wpcms/wp-content/uploads/2016/01/Gazprom-Is-2016-the-Year-for-a-Change-of-Pricing-Strategy-in-Europe.pdf; Deloitte, "Oil and Gas Reality Check 2015; A Look at the Top Issues Facing the Oil and Gas Sector" (London: Deloitte, 2015), http://www2.deloitte.com/content/dam/Deloitte/global/Documents/Energy-and-Resources/gx-er-oil-and-gas-reality-check-2015.pdf.

102. Interview with Peter Ragauss, Houston, Texas, October 2015.

103. "Tolling Agreement," Risk.net, 2015, http://www.risk.net/energy-risk/glossary/2041727/tolling-agreement.

104. Silvia Colombo, Mohamed El Harrak, and Nicolo Sartori, eds., *The Future of Natural Gas: Markets and Geopolitics* (Hof van Twente, Netherlands: Lenthe / European Energy Review and Istituto Affari Internazionali, 2016), 54–57, 228, http://www.iai.it/sites/default/files/iai-ocp_gas.pdf; interview with Warren Wilczewski, Washington, DC, August 9, 2016.

105. Energy Mining Advisory Partnership, "Impact of US LNG exports on Europe," 2015, 4. http://energyminingadvisorypartnership.github.io/impact_of_us_lng_on_exports_on_europe.pdf.

106. Interview with Fred H. Hutchinson, Washington, DC, April 18, 2016.

107. Ibid.; "LITGAS Signs Deal to Buy LNG from Cheniere Marketing from 2016," Natural Gas Europe, March 2, 2015, http://www.naturalgaseurope.com/litgas-signs-deal-to-buy-lng-from-cheniere-marketing-from-2016-22453.

108. Clifford Krauss, "Drillers Answer Low Oil Prices with Cost-Saving Innovations," *New York Times,* May 11, 2015, http://www.nytimes.com/2015/05/12/business/energy-environment/drillers-answer-low-oil-prices-with-cost-saving-innovations.html?_r=0.

109. Ibid.; Russell Gold, "Drillers Unleash 'Super-Size' Natural Gas Output," *Wall Street Journal,* September 1, 2015, http://www.wsj.com/articles/drillers-get-super-size-natural-gas-output-1441127955.

110. Krauss, "Drillers Answer Low Oil Prices"; Starr Spencer, "Q2 Oil Results Show Strength of US 'Super-Shale' for the Price of a Slice," S&P Global Platts, The Barrel, August 21, 2015, http://blogs.platts.com/2015/08/21/q2-oil-us-super-shale-price-slice/; Dan Murtaugh, "Texas Isn't Scared of $30 Oil," Bloomberg, February 3, 2016, http://www.bloomberg.com/news/articles/2016-02-03/texas-toughness-in-oil-patch-shows-why-u-s-still-strong-at-30.

111. Spencer, "Q2 Oil Results Show Strength of US 'Super-Shale' for the Price of a Slice."

112. Krauss, "Drillers Answer Low Oil Prices."

113. Grant Nuelle, "EIA Expects Near-Term Decline in Natural Gas Production in Major Shale Regions," EIA, August 26, 2015, http://www.eia.gov/todayinenergy/detail.cfm?id=22672&src=email; "Happy Fracking, America," *American Interest,* September 2, 2015, http://www.the-american-interest.com/2015/09/02/happy-fracking-america/.

114. EIA, *Annual Energy Outlook 2016 with Projections to 2040,* August 2016, ES-5, ES-6, https://www.eia.gov/forecasts/aeo/pdf/0383(2016).pdf.

115. EIA, "Global Natural Gas Consumption Doubled from 1980 to 2010," April 12, 2012, http://www.eia.gov/todayinenergy/detail.cfm?id=5810.

116. Ibid.

117. Andrea Gilardoni, *The World Market for Natural Gas: Implications for Europe* (Berlin: Springer, 2008), 40.

118. British Petroleum, "BP Energy Outlook 2035," February 2016, 30, 65, https://www.bp.com/content/dam/bp/pdf/energy-economics/energy-outlook-2016/bp-energy-outlook-2016.pdf.

119. Rob Verdonck and Isis Almeida, "IEA Cuts Natural Gas Growth Outlook as Asian Demand Falters," Bloomberg, June 4, 2015, bloomberg.com/news/articles/2015-06-03/iea-cuts-gas-growth-outlook-a-third-year-as-asian-demand-falters.

120. International Energy Agency, "Medium-Term Gas Market Report 2015 Executive Summary: Market Analysis and Forecasts to 2020," 2015, 3–4, http://www.iea.org/Textbase/npsum/MTGMR2015SUM.pdf.

121. Jonathan Gaventa, Manon Dufour, and Dave Jones, "Europe's Gas Demand Is Falling: Doesn't Anybody Notice?," *Energy Post,* July 9, 2015, energypost.eu/europes-gas-demand-falling-doesnt-anybody-notice.

122. Ibid.

123. Matt Smith, "Commentary: 10 Challenges Faced by the Global LNG Market," FuelFix, August 27, 2015, http://fuelfix.com/blog/2015/08/27/10-challenges-faced-by-the-global-lng-market/.

124. Joe Fisher, "LNG First-Mover Algeria Seeking to Grow Exports," Natural Gas Intelligence, October 17, 2014, http://www.naturalgasintel.com/articles/100098-lng-first-mover-algeria-seeking-to-grow-exports; Naser Al-Tamimi, "Navigating Uncertainty: Qatar's Response to the Global Gas Boom," Brookings Doha Center, March 2015, http://www.brookings.edu/~/media/Events/2015/03/25-brookings-doha-energy/En-Tamimi-PDF.pdf?la=en; Subbu Bettadapura, "Malaysia LNG Outlook," Frost and Sullivan, October 25, 2011, http://www.slideshare.net/FrostandSullivan/malaysia-lng-outlook.

125. Anthony Fensom, "China: The Next Shale-Gas Superpower?," *National Interest*, October 9, 2014, http://nationalinterest.org/feature/china-the-next-shale-gas-superpower-11432.

126. Gaurav Agnihotri, "What the Iran Deal Could Mean for Natural Gas Markets," Oil Price.com, July 17, 2015, http://oilprice.com/Energy/Crude-Oil/What-The-Iran-Deal-Could-Mean-For-Natural-Gas-Markets.html.

127. "Fitch: Major Iranian Gas Exports Will Take at Least Five Years," *Fitch Ratings*, July 10, 2015, https://www.fitchratings.com/site/fitch-home/pressrelease?id=987711.

128. John Vidal, "Five G7 Nations Increased Their Coal Use over a Five-Year Period, Research Shows," *Guardian*, June 8, 2015, http://www.theguardian.com/environment/2015/jun/08/five-g7-nations-increased-their-coal-use-over-a-five-year-period-research-shows.

129. Citi Group, the Brookings Institution, Ernst and Young, the International Monetary Fund (IMF), the Congressional Research Service, and Columbia University Center on Global Energy Policy are among the many groups referenced in this chapter.

130. Daniel Yergin, *The Quest: Energy, Security and the Remaking of the Modern World* (New York: Penguin, 2012), 344.

131. Atlantic Council Task Force, *Empowering America*, 22

132. Tim Boersma, Charles K. Ebinger, and Heather L. Greenley, *An Assessment of U.S. Natural Gas Exports*, Natural Gas Issue Brief 4 (Washington, DC: Brookings Institution, 2015), 16 http://www.brookings.edu/~/media/research/files/papers/2015/07/us-natural-gas-exports/lng_markets.pdf.

133. Gal Luft, "U.S. Gas Exports: The Pipe Dream," Natural Gas Europe, August 3, 2015, http://www.naturalgaseurope.com/us-gas-exports-the-pipe-dream-lng-24879.

134. Atlantic Council Task Force, *Empowering America,* 20.

135. Edward L. Morse, Eric Lee, and Anthony Yuen, "Energy 2020 out of America: The Rapid Rise of the United States as a Global Energy Superpower," *Citi GPS: Global Perspectives and Solutions,* November 2014, 11–12, https://ir.citi.com/rBWYa6YM4Scr4LsmpuQB8DZubx61JU8NqGw5jsCxMMfxMRxBp4u4gwbWrQgfNlfhzMSpx1Jv3qA%3D.

136. Benjamin Hunt, Dirk Muir, and Martin Sommer, "The Potential Macroeconomic Impact of the Unconventional Oil and Gas Boom in the United States," IMF Working Paper 92, International Monetary Fund Research Department, 2015, http://www.imf.org/external/pubs/ft/wp/2015/wp1592.pdf.

137. Yergin, *The Quest,* 344.

138. Atlantic Council Task Force, *Empowering America,* 14.

139. Timothy Cama, "House Panel to Vote on Lifting Oil Export Ban," *The Hill,* September 8, 2015, http://thehill.com/policy/energy-environment/252924-house-panel-to-vote-on-lifting-oil-export-ban.

140. Jason Bordoff, "Why the US Should Not Want Energy Independence," *Wall Street Journal,* November 16, 2015, http://blogs.wsj.com/experts/2015/11/16/why-the-u-s-should-not-want-energy-independence/.

141. Elena Holodny, "Here's What Russian Intervention in Syria Means for Oil," Business Insider, October 11, 2015, http://www.businessinsider.com/what-russian-intervention-in-syria-means-for-oil-2015-10?IR=T; Gabriel Collins, "Russian Intervention in Syria Could Drive Crude Prices Deep into the $30s," OilPro, October 2015, http://oilpro.com/post/19057/russian-intervention-syria-could-drive-crude-prices-deep-into-30s; David Butter, "Russia's Syria Intervention Is Not All About Gas," Carnegie Endowment for International Peace, November 19, 2015, http://carnegieendowment.org/sada/?fa=62036; Robert F. Kennedy Jr., "Why the Arabs Don't Want Us in Syria," Politico, February 23, 2016, http://www.politico.eu/article/why-the-arabs-dont-want-us-in-syria-mideast-conflict-oil-intervention/.

142. Victoria Zaretskaya and Scott Bradley, "Natural Gas Prices in Asia Mainly Linked to Crude Oil, but Use of Spot Indexes Increases," EIA, September 29, 2015, https://www.eia.gov/todayinenergy/detail.cfm?id=23132.

143. Tim Daiss, "Singapore's LNG Trading Hub Ambitions Press Forward," *Forbes,* March 21, 2016, http://www.forbes.com/sites/timdaiss/2016/03/21/singapores-lng-trading-hub-ambitions-press-forward/#1ff268b273f8.

144. Oleg Vukmanovic, Nina Chestney, and Dmitry Zhdannikov, "Gazprom Prepares Cold Reception for U.S. Super-Cooled Gas," Reuters, June 22, 2016, http://www.reuters.com/article/europe-gas-gazprom-idUSL8N19D37B.

3. The Politics of Supply

1. US Energy Information Administration (henceforth EIA), "Russia," July 28, 2015, http://www.eia.gov/beta/international/analysis.cfm?iso=RUS.

2. Glenn E. Curtis, ed., *Russia: A Country Study* (Washington, DC: Government Printing Office for the Library of Congress, 1996), http://countrystudies.us/russia/61.htm.

3. James Henderson and Simon Pirani, eds., *The Russian Gas Matrix: How Markets Are Driving Change* (Oxford: Oxford Institute for Energy Studies, 2014), 2.

4. British Petroleum (henceforth BP), "BP Statistical Review of World Energy," 2005, https://www.bp.com/content/dam/bp/pdf/energy-economics/statistical-review-2015/bp-statistical-review-of-world-energy-2015-full-report.pdf.

5. Henderson and Pirani, *The Russian Gas Matrix,* 41, values for 2013 and 2014 provided by Jonathan Stern of the Oxford Institute for Energy Studies; Gazprom, "Production," n.d., http://www.gazprom.com/about/production/.

6. Henderson and Pirani, *The Russian Gas Matrix,* 108–109.

7. Organization of Petroleum Exporting Countries, "World Natural Gas Exports by Country, 1980–2007," http://www.opec.org/library/Annual%20Statistical%20Bulletin/interactive/2007/FileZ/XL/T55.XLS.; EIA, "Dry Natural Gas Production—2014," https://www.eia.gov/beta/international/.

8. BP, "BP Statistical Review of World Energy."

9. Jonathan Stern, "Soviet Natural Gas in the World Economy," in *Soviet Natural Resources in the World Economy,* ed. Robert G. Jensen, Theodore Shabad, and Arthur W. Wright (Chicago: University of Chicago Press, 1983), 373.

10. Gazprom Export, "Delivery Statistics," n.d., http://www.gazpromexport.ru/en/statistics/.

11. Bank of Russia, “Russian Federation: Natural Gas Exports, 2000–15,” June 25, 2015, http://www.cbr.ru/eng/statistics/credit_statistics/print.aspx?file=gas_e.htm.

12. Henderson and Pirani, *The Russian Gas Matrix,* 54.

13. Ibid., 41.

14. John Logh, “Russia’s Energy Diplomacy,” Chatham House briefing paper, May 2011, https://www.chathamhouse.org/sites/files/chathamhouse/19352_0511bp_lough.pdf.

15. Henderson and Pirani, *The Russian Gas Matrix,* 8.

16. David Sherfinski, “McCain: ‘Russia Is a Gas Station Masquerading as a Country,’ ” *Washington Times,* March 16, 2014, http://www.washingtontimes.com/news/2014/mar/16/mccain-russia-gas-station-masquerading-country/.

17. Henderson and Pirani, *The Russian Gas Matrix,* 6–8; Sarah O. Ladislaw, Maren Leed, and Molly A. Walton, “New Energy, New Geopolitics,” Center for Strategic and International Studies, June 2014, 5, http://csis.org/files/publication/140605_Ladislaw_NewEnergyNewGeopolitics_background2_Web.pdf.

18. Silvana Tordo, Brandon S. Tracy, and Noora Arfaa, “National Oil Companies and Value Creation,” World Bank Working Paper no. 218, 2011, 16, http://siteresources.worldbank.org/INTOGMC/Resources/9780821388310.pdf.

19. Daniel Yergin, *The Prize: The Epic Quest for Oil, Money, and Power* (New York: Simon & Schuster, 1991), 58–59.

20. Lauren Goodrich and Marc Lanthemann, “The Past, Present and Future of Russian Energy Strategy,” *Stratfor Global Intelligence,* February 12, 2013, https://www.stratfor.com/weekly/past-present-and-future-russian-energy-strategy.

21. Mark H. Hayes and David G. Victor, “Introduction to the Historical Case Studies: Research Questions, Methods, and Case Selection,” in *Natural Gas and Geopolitics: From 1970 to 2040,* ed. David G. Victor, Amy M. Jaffe, and Mark H. Hayes (Cambridge: Cambridge University Press, 2006), 44–47.

22. Leslie Dienes and Theodore Shabad, *The Soviet Energy System: Resource Use and Policies* (Washington, DC: V. H. Winston, 1979), 75; Andreas Heinrich and Heiko Pleines, eds., *Export Pipelines from the CIS Region* (Stuttgart: Ibidem Press, 2014), 29.

23. Dienes and Shabad, *The Soviet Energy System,* 75–76; Eustream, "History of Company: More than 40 Years of Safe and Reliable Gas Transmission Services," 2014, http://www.eustream.sk/en_company-eustream/en_history.

24. Stern, "Soviet Natural Gas in the World Economy," 373.

25. Dienes and Shabad, *The Soviet Energy System,* 76.

26. George Ginsburgs and Robert M. Slusser, *A Calendar of Soviet Treaties, 1958–1973* (Alphen aan den Rijn, Netherlands: Sijthoff & Noordhoff, 1981), 442.

27. Per Högselius, *Red Gas: Russia and the Origins of European Energy Dependence* (New York: Palgrave Macmillan, 2013), 101, 218, 227–228.

28. Yergin, *The Prize,* 336.

29. Willy Brandt, *A Peace Policy for Europe,* trans. Joel Carmichael (New York: Holt, Rinehart and Winston, 1969).

30. Ginsburgs and Slusser, *A Calendar of Soviet Treaties,* 523.

31. Stern, "Soviet Natural Gas in the World Economy," 373–374.

32. Ginsburgs and Slusser, *A Calendar of Soviet Treaties,* 525.

33. Gazprom, "40th Anniversary of Russian Gas Supplies to Italy," December 10, 2009, http://www.gazprom.com/about/history/events/italy40/; Hayes and Victor, "Introduction to the Historical Case Studies," 44–47.

34. Ibid.

35. David M. Adamson, "Soviet Gas and European Security," *Energy Policy* 13, no. 1 (February 1985): 15.

36. Dienes and Shabad, *The Soviet Energy System,* 87.

37. Arthur W. Wright, "Contrasts in Soviet and American Energy Policies," *Energy Policy* (March 1975): 38–39.

38. Heinrich and Pleines, *Export Pipelines from the CIS Region,* 34.

39. Dienes and Shabad, *The Soviet Energy System,* 87–94.

40. "Soviets, Suppliers Agrees on Pipeline, but Bankers Leery," *Financial Times of London,* June 29, 1981.

41. Clyde H. Farnsworth, "U.S. Pushes Soviet Gas Alternative," *New York Times,* July 29, 1981.

42. R. Copulos Milton, "Is the Soviet Gas Pipeline a Steel Noose?," Heritage Foundation, March 15, 1982, http://www.heritage.org/research/reports/1982/03/is-the-soviet-gas-pipeline-a-steel-noose.

43. Bruce W. Jentleson, *Pipeline Politics: The Complex Political Economy of East-West Energy Trade* (Ithaca, NY: Cornell University Press, 1986), 183.

44. Commission of the European Communities, Communication from the Commission to the Council Concerning Natural Nas, COM (81) 530 final, October 1, 1981, http://aei.pitt.edu/5050/1/5050.pdf.

45. Thane Gustafson, *Soviet Negotiating Strategy: The East-West Gas Pipeline Deal, 1980–1984* (Santa Monica, CA: Rand, 1985), 8–9; Richard M. Weintraub, "President Lifts Sanctions on Soviet Pipeline," *Washington Post,* November 12, 1982.

46. Högselius, *Red Gas,* 197.

47. Gazprom, "Chronicle," n.d., http://www.gazprom.com/about/history/chronicle/1989-1995/.

48. Konstantin Dushenko, "Mi choteli kak lutshe," 2012, http://www.dushenko.ru/quotation_date/121221/.

49. President of the Russian Federation, "Ukaz Prezidenta Rossiyskoi Federatsii ot 05.11.1992 #1333," November 5, 1992, http://www.kremlin.ru/acts/bank/2349.

50. Stephen Bierman, "Rosneft Surpasses Gazprom as Russia's Most Valuable Company," Bloomberg, April 11, 2016, http://www.bloomberg.com/news/articles/2016-04-14/iea-sees-oil-oversupply-almost-gone-in-second-half-on-shale-drop.

51. Gazprom, "Shares," May 7, 2015, http://www.gazprom.com/investors/stock/.

52. Scott B. MacDonald and Jonathan Lemco, *State Capitalism's Uncertain Future* (Santa Barbara, CA: ABC-Clio, 2015), 81–82.

53. Agnia Grigas, *The Politics of Energy and Memory between the Baltic States and Russia* (Farnham, Surrey, UK: Ashgate, 2013), 103.

54. "Putin's Placemen Pull the Strings," *Telegraph,* March 7, 2004, http://www.telegraph.co.uk/finance/2879197/Putins-placemen-pull-the-strings.html.

55. Grigas, *The Politics of Energy and Memory,* 103–104.

56. Jonathan Stern, *The Future of Russian Gas and Gazprom* (Oxford: Oxford University Press, 2005), 220–221.

57. Andreas Goldthau, "Gazprom Is Sitting between a Rock and a Hard Place," May 27, 2015, http://goldthau.com/2015/05/27/gazprom-is-sitting-between-a-rock-and-a-hard-place/; "There Is No Silver Bullet for Gazprom," Natural Gas Europe, May 11, 2015, http://www.naturalgaseurope.com/andreas-goldthau-there-is-no-silver-bullet-for-gazprom-23575.

58. Gazprom Export, "Yamal-Europe," n.d., http://www.gazpromexport.ru/en/projects/4/.

59. Henderson and Pirani, *The Russian Gas Matrix,* 209.

60. Andrey Vavilov, Galina Kovalishina, and Georgy Trofimov, "The New Export Routes and Gazprom's Strategic Opportunities in Europe," in *Gazprom: An Energy Giant and Its Challenges in Europe,* ed. Andrey Vavilov (New York: Palgrave Macmillan, 2015), 183.

61. Michèle Bodmer, Janet Anderson, Henrieke Neitzel, et. al., "Secure Energy for Europe: The Nord Stream Pipeline Project," April 2014, 15–16, http://www.nord-stream.com/media/documents/pdf/en/2014/04/secure-energy-for-europe-full-version.pdf.

62. Vavilov, Kovalishina, and Trofimov, "The New Export Routes,"185.

63. Directorate-General for Energy and Transport, "Trans-European Energy Networks," TEN-E Priority Projects (Brussels: European Commission, 2004), 24–25.

64. Hans Michael Kloth, "Indirect Hitler Comparison: Polish Minister Attacks Schröder and Merkel," Spiegel Online, May 1, 2006, http://www.spiegel.de/international/indirect-hitler-comparison-polish-minister-attacks-schroeder-and-merkel-a-413969.html.

65. Kirsten Westphal, "Germany and the EU-Russia Energy Dialogue," in *The EU-Russian Energy Dialogue: Europe's Future Energy Secuirty,* ed. Pami Aalto (Hampshire, UK: Ashgate, 2008), 109.

66. Bendik Solum Whist, "Nord Stream: Not Just a Pipeline," Fridtjof Nansen Institute, November 2008, http://www.fni.no/doc&pdf/FNI-R1508.pdf.

67. Agnia Grigas, "Opinion: Schroeder and Putin: An Unseemly Alliance," *Lithuania Tribune,* May 8, 2014, http://www.lithuaniatribune.com/67800/opinion-schroeder-and-putin-an-unseemly-alliance-201467800/.

68. Westphal, "Germany and the EU-Russia Energy Dialogue," 106.

69. Tony Paterson, "Merkel Fury after Gerhard Schroeder backs Putin on Ukraine," *Telegraph,* March 14, 2014, http://www.telegraph.co.uk/news/worldnews/europe/ukraine/10697986/Merkel-fury-after-Gerhard-Schroeder-backs-Putin-on-Ukraine.html.

70. "SPIEGEL Interview with Ex-Chancellor Gerhard Schroeder: "I'm Anything but an Opponent of America," Spiegel Online International, October 23, 2006, http://www.spiegel.de/international/spiegel/spiegel

-interview-with-ex-chancellor-gerhard-schroeder-i-m-anything-but-an-opponent-of-america-a-444069.html.

71. Directorate-General for Energy, “Quarterly Report Energy on European Gas Markets” (Brussels: European Commission, 2014), 7; Gazprom Export, “Nord Stream,” n.d., http://www.gazpromexport.ru/en/projects/5/.

72. Gazprom, “Gazprom, BASF, E.ON, ENGIE, OMV and Shell Sign Shareholders Agreement on Nord Stream II Project,” September 4, 2015, http://www.gazprom.com/press/news/2015/september/article245837/.

73. Dan Alexe, “EU Eastern Countries Reject Russia’s Nord Stream Plans,” *New Europe,* November 27, 2015, http://neurope.eu/article/eu-eastern-countries-reject-russias-nord-stream-plans/.

74. Gabriela Baczynska and Alissa de Carbonnel, “EU’s Juncker Says Doubts over Nord Stream 2 Pipeline Plan ‘Beyond Legal,’ ” Reuters, June 16, 2016, http://www.reuters.com/article/us-energy-nordstream-eu-juncker-exclusiv-idUSKCN0Z229I.

75. Konrad Szymanski, “Russia’s Gas Pipeline Threatens European Unity,” *Financial Times*, October 21, 2016, https://www.ft.com/content/25a17928-96c3-11e6-a1dc-bdf38d484582.

76. “Analysis: The South Stream Gas Pipeline,” *Friedl News,* June 25, 2014, http://www.friedlnews.com/article/analysis-the-south-stream-gas-pipeline.

77. Georgi Gotev, “Putin: We Haven’t Given Up the South Stream Project,” EurActiv.com, February 18, 2015, http://www.euractiv.com/sections/energy/putin-we-havent-given-south-stream-project-312224.

78. Agnia Grigas and Nora Fisher Onar, “From Russia with Love: The Moscow-Ankara Energy Affair,” *Forbes,* May 13, 2015, http://www.forbes.com/sites/realspin/2015/05/13/from-russia-with-love-the-moscow-ankara-energy-affair/.

79. “Black Sea Partners Toast Gas Deal,” *BBC News,* November 17, 2005, http://news.bbc.co.uk/2/hi/europe/4445158.stm.

80. Jonathan Stern, Simon Pirani, and Katja Yafimova, “Does the Cancellation of South Stream Signal a Fundamental Reorientation of Russian Gas Export Policy?,” Oxford Institute for Energy Studies, January 2015, 7, http://www.oxfordenergy.org/wpcms/wp-content/uploads/2015/01/Does-cancellation-of-South-Stream-signal-a-fundamental-reorientation-of-Russian-gas-export-policy-GPC-5.pdf.

81. Edward C. Chow and Zachary D. Cuyler, "New Russian Gas Export Projects—From Pipe Dreams to Pipelines," Center for Strategic and International Studies, July 22, 2015, http://csis.org/publication/new-russian-gas-export-projects-pipe-dreams-pipelines.

82. Agnia Grigas, "Deteriorating Relations between Turkey and Russia Will Have Broader Consequences," *Forbes,* November 25, 2015, http://www.forbes.com/sites/realspin/2015/11/25/deteriorating-relations-between-russia-and-turkey-will-have-broader-consequences/.

83. Nora Fisher Onar, "What the Failed Coup Means for Turkey's Foreign Policy," Transatlantic Academy, July 18, 2016, http://www.transatlanticacademy.org/node/948.

84. Henderson and Pirani, *The Russian Gas Matrix,* 23.

85. Keith C. Smith, "Defuse Russia's Energy Weapon," Center for Strategic and International Studies, January 17, 2006, http://csis.org/press/csis-in-the-news/defuse-russias-energy-weapon; Keith C. Smith, "Russia and European Energy Security," Center for Strategic and International Studies, October 2008, http://csis.org/files/media/csis/pubs/081024_smith_russiaeuroenergy_web.pdf.

86. Anita Orban, *Power, Energy, and the New Russian Imperialism* (Westport, CT: Praeger Security International, 2008).

87. Marshall I. Goldman, *Petrostate: Putin, Power, and the New Russia* (Oxford: Oxford University Press, 2008), 139, 152.

88. Janusz Bugajski, *Expanding Eurasia: Russia's European Ambitions* (Washington, DC: Center for Strategic and International Studies, 2008), 73.

89. James Sherr, *Hard Diplomacy and Soft Coercion* (London: Chatham House, 2013).

90. Margarita M. Balmaceda, *Energy Dependency, Politics and Corruption in the Former Soviet Union: Russia's Power, Oligarchs' Profits and Ukraine's Missing Energy Policy, 1995–2006* (New York: Routledge, 2007), 143.

91. Margarita M. Balmaceda, *The Politics of Energy Dependency: Ukraine, Belarus, and Lithuania between Domestic Oligarchs and Russian Pressure* (Toronto: University of Toronto Press, 2013), 286–287.

92. Edward Lucas, "Edward Lucas: Russia Is Winning," *Delfi,* September 4, 2014, http://en.delfi.lt/opinion/edward-lucas-russia-is-winning.d?id=65745272; Bohdan Tsioupine, "The Old New Cold War," *Ukrainian*

Week, April 3, 2014, http://ukrainianweek.com/Politics/106680; Edward Lucas, *The New Cold War* (New York: Palgrave Macmillan, 2014).

93. Cited in Jeffrey Mankoff, *Russian Foreign Policy—The Return of Great Power Politics* (Blue Ridge Summit, PA: Rowman & Littlefield, 2009), 35.

94. Roman Kupchinsky, "Russian Energy Strategy—The Domestic Political Factor," Jamestown Foundation, October 8, 2009, http://www.jamestown.org/single/?no_cache=1&tx_ttnews%5Btt_news%5D=35598#.Viz-pdKrTDc.

95. Jonathan Stern, "The Russian-Ukrainian Gas Crisis of January 2006," Oxford Institute for Energy Studies, January 16, 2006, http://www.oxfordenergy.org/wpcms/wp-content/uploads/2011/01/Jan2006-RussiaUkraineGasCrisis-JonathanStern.pdf; Simon Pirani, Jonathan Stern, and Katja Yafimova, "The Russo-Ukrainian Gas Dispute of January 2009: A Comprehensive Assessment," Oxford Institute for Energy Studies, February 2009, http://www.oxfordenergy.org/wpcms/wp-content/uploads/2010/11/NG27-TheRussoUkrainianGasDisputeofJanuary2009AComprehensiveAssessment-JonathanSternSimonPiraniKatjaYafimava-2009.pdf.

96. Andreas Goldthau, "Russia's Energy Weapon Is a Fiction," *Europe's World,* February 1, 2008, http://europesworld.org/2008/02/01/russias-energy-weapon-is-a-fiction/#.ViPUsdIrLDc.

97. Tim Boersma, "The End of the Russian Energy Weapon (That Arguably Was Never There)," Brookings, March 5, 2015, http://www.brookings.edu/blogs/order-from-chaos/posts/2015/03/05-end-of-russian-energy-weapon-boersma.

98. "Gazprom's Grip: Russia's Leverage over Europe," Radio Free Europe / Radio Liberty, November 7, 2015, http://www.rferl.org/content infographics/gazprom-russia-gas-leverage-europe/25441983.html.

99. Jonathan Stern, "Gazprom: A Long March to Market-Based Pricing in Europe?," *Forum: A Quarterly Journal for Debating Energy Issues,* Issue 101, August 2014, Oxford Institute for Energy Studies, http://www.oxfordenergy.org/wpcms/wp-content/uploads/2015/09/OEF-101.pdf.

100. Vavilov, Kovalishina, and Trofimov, "The New Export Routes," 180.

101. Franziska Holz, Hella Engerer, Claudia Kemfert, Philipp M. Richter, and Christian von Hirschhausen, "European Natural Gas Infrastructure: The Role of Gazprom in European Natural Gas Supplies" (Berlin: Deutsches Institut für Wirtschaftsforschung, May 2014), http://www.diw.de/documents/publikationen/73/diw_01.c.465334.de/diwkompakt_2014-081.pdf.

102. Gazprom, "Yamal-Europe," n.d., http://www.gazprom.com/about/production/projects/pipelines/yamal-evropa/.

103. Stern, *The Future of Russian Gas and Gazprom,* 119.

104. "PGNiG Takes Control of Strategic Pipeline," Radio Poland, July 17, 2015, http://www.thenews.pl/1/12/Artykul/214097,PGNiG-takes-control-of-strategic-pipeline.

105. Gazprom, "Gazprom and EuRoPol GAZ to Cooperate under Yamal—Europe-2 Gas Pipeline Project," April 5, 2013, http://www.gazprom.com/press/news/2013/april/article159672/.

106. Jan Cienski, "Poland to Gazprom: What Pipeline?," *Financial Times,* April 5, 2013, http://blogs.ft.com/beyond-brics/2013/04/05/poland-to-gazprom-what-pipeline/.

107. Vladimir Socor, "Polish Government Sheds Light on Gazprom-EuroPolGaz MOU," *Eurasia Daily Monitor* 10, no. 8 (April 29, 2013), http://www.jamestown.org/single/?tx_ttnews%5Btt_news%5D=40808&no_cache=1#.VdScleEVjug.

108. Gazprom Export, "Storage," n.d., http://www.gazpromexport.ru/en/projects/storage/.

109. Gazprom, "RAG and WINGAS Commission Phase 2 of Haidach Underground Gas Storage Facility," May 19, 2011, http://www.gazprom.com/press/news/2011/may/article112523/.

110. Gazprom, "Gazprom and MND Group Embark on New UGS Facility Construction in Czech Republic," March 20, 2013, http://www.gazprom.com/press/news/2013/march/article158442/.

111. Alexei Lossan, "Gazprom Recognizes EU Restrictions for the First Time," *Russia beyond the Headlines*, March 29, 2016, http://rbth.com/business/2016/03/29/gazprom-recognizes-eu-restrictions-for-the-first-time_579955; "Natgas Tariffs Will Inevitably Climb after Latvijas Gaze Is Unbundled—Itera Latvija CEO," Latvian News Service, August 22, 2016, http://www.latviannewsservice.lv/en/topic/1905/news/51705527/.

112. Holz et al., "European Natural Gas Infrastructure," 21.

113. Sandu-Daniel Kopp, *Politics, Markets and EU Gas Supply Security: Case Studies of the UK and Germany* (Wiesbaden: Springer, 2015), 193.

114. "BASF and Gazprom Sign Asset Swap Agreement," Wintershall, December 23, 2013, http://www.wintershall.com/en/press-news/detail/news/basf-and-gazprom-sign-asset-swap-agreement.html.

115. Christopher Alessi and Sarah Sloat, "BASF, Gazprom Call Off Asset Swap amid Political Tensions," *Wall Street Journal,* December 18, 2014, http://www.wsj.com/articles/basf-gazprom-call-off-asset-swap-due-to-current-political-environment-1418928527.

116. Kopp, *Politics, Markets and EU Gas Supply Security,* 192.

117. Westphal, "Germany and the EU-Russia Energy Dialogue," 102.

118. Stylianos A. Sotiriou, *Russian Energy Strategy in the European Union, the Former Soviet Union Region, and China* (London: Lexington Books, 2015), 136.

119. Gerrit Wiesmann and Isabel Gorst, "Eon Offloads Its 3.5% Stake in Gazprom," *Financial Times,* December 1, 2010, http://www.ft.com/cms/s/0/aa932fe8-fd52-11df-b83c-00144feab49a.html.

120. Gazprom, "Alexey Miller Appoints Burckhard Bergmann as His Advisor," June 30, 2011, http://www.gazprom.com/press/news/2011/june/article114559/.

121. Westphal, "Germany and the EU-Russia Energy Dialogue," 102.

122. Grigas, *The Politics of Energy and Memory,* 93–94.

123. Sotiriou, *Russian Energy Strategy,* 150.

124. Gazprom Export, "Italy," n.d., http://www.gazpromexport.ru/en/partners/italy/.

125. James Henderson and Alastair Ferguson, *International Partnership in Russia: Conclusions from the Oil and Gas Industry* (New York: Palgrave Macmillan, 2014), 52.

126. Kathrin Hille and Guy Chazan, "Russian Groups Buy Eni's Stake in Severenergia," *Financial Times,* November 20, 2013, http://www.ft.com/intl/cms/s/0/ba3f4caa-5200-11e3-8c42-00144feabdc0.html.

127. Sotiriou, *Russian Energy Strategy,* 152.

128. Guy Chazan, "ENI in Spot Market Gas Deal with Gazprom," *Financial Times,* May 23, 2014, http://www.ft.com/cms/s/0/3b79b0e4-e284-11e3-a829-00144feabdc0.html#axzz3hjPnFu5k.

129. Sylvie Cornot-Gandolphe and Ralf Dickel, "Flexibility in Natural Gas: Supply and Demand" (Paris: International Energy Agency, 2002), 54.

130. Eduard Gismatullin, "French to Swap LNG for Russian Gas," *Moscow Times,* November 24, 2005, http://www.themoscowtimes.com/sitemap/free/2005/11/article/french-to-swap-lng-for-russian-gas/208368.html.

131. Sotiriou, *Russian Energy Strategy,* 146.

132. Gazprom, "Gazprom and Gaz de France Extend Contracts for Russian Natural Gas Supply to France up to 2030," December 19, 2006, http://www.gazprom.com/press/news/2006/december/article63695/.

133. Sotiriou, *Russian Energy Strategy,* 147.

134. "Stable Russia Ties, New Suppliers Key to EU Energy Security," S&P Global Platts, May 22, 2014, http://www.platts.com/latest-news/natural-gas/stpetersburg/stable-russia-ties-new-suppliers-key-to-eu-energy-26793980.

135. BP, "Working in Russia," n.d., http://www.bp.com/en_ru/russia/about-bp-in-russia/business.html.

136. Tom Bawden, "Gas Imports from Russia's Gazprom Giant to Soar after New Centrica Deal," *Independent,* May 14, 2015, http://www.independent.co.uk/news/business/news/gas-imports-from-russias-gazprom-giant-to-soar-after-new-centrica-deal-10248692.html.

137. James Henderson and Tatiana Mitrova, "The Political and Commercial Dynamics of Russia's Gas Export Strategy," OIES Paper NG 102, Oxford Institute for Energy Studies, September 2015, https://www.oxfordenergy.org/wpcms/wp-content/uploads/2015/09/NG-102.pdf.

138. Ministry of Energy of Russian Federation, "Osnovnye polozhenija proekta Energeticheskoi Strategii Rossii na period do 2035 goda," 2015, http://minenergo.gov.ru/node/1913.

139. Ambrose Evans-Pritchard, "US Hits Russia's Oil Kingpin Igor Sechin with First Energy Sanctions," *Telegraph,* April 28, 2014, http://www.telegraph.co.uk/finance/financialcrisis/10794425/US-hits-Russias-oil-kingpin-Igor-Sechin-with-first-energy-sanctions.html.

140. US Department of the Treasury, "Treasury Sanctions Russian Officials, Members of the Russian Leadership's Inner Circle, and an Entity for Involvement in the Situation in Ukraine," March 20, 2014, http://www.treasury.gov/press-center/press-releases/Pages/jl23331.aspx.

141. US Department of the Treasury, "Announcement of Treasury Sanctions on Entities within the Financial Services and Energy Sectors of Russia, against Arms or Related Materiel Entities, and Those Undermining Ukraine's Sovereignty," July 16, 2014, http://www.treasury.gov/press-center/press-releases/Pages/jl2572.aspx.

142. See Council Regulation (EU) No. 959/2014 of September 8, 2014 and No. 960/2014 of September 8, 2014 and Council Implementing Regulation

(EU) No. 961/2014 of September 8, 2014, http://eur-lex.europa.eu/legal-content/EN/TXT/PDF/?uri=OJ:L:2014:271:FULL&from=EN.

143. "Russia Sanctions Update September 2014," Chadbourne, September 24, 2014, http://www.chadbourne.com/sites/default/files/publications/russia_sanctions_update_september_2014.pdf.

144. Council of the European Union, "Crimea and Sevastopol: Further EU Sanctions Approved," December 18, 2014, http://www.consilium.europa.eu/uedocs/cms_Data/docs/pressdata/EN/foraff/146392.pdf.

145. Vera Eckert and Oleg Vukmanovic, "WRAPUP 3—Gazprom Seals Big Gas Deals in Europe despite Ukraine Crisis," Reuters, September 4, 2015, http://www.reuters.com/article/2015/09/04/gazprom-europe-idUSL5N11A0K120150904.

146. Ambrose Evans-Pritchard and Peter Spence, "Russian Crisis Turns Systemic as Rouble Crashes 13pc," *Telegraph,* December 15, 2014, http://www.telegraph.co.uk/finance/economics/11295402/Russian-crisis-turns-systemic-as-rouble-crashes-13pc.html.

147. Ivana Kottasova, "Sanctions Will Cost Russia More than $100 billion," CNN Money, April 21, 2015, http://money.cnn.com/2015/04/21/news/economy/russia-ukraine-sanctions-price/.

148. Robin Emmott, "Sanctions Impact on Russia to Be Longer Term, U.S. Says," Reuters, January 12, 2016, http://www.reuters.com/article/us-ukraine-crisis-sanctions-idUSKCN0UQ1ML20160112.

149. "Russia Balance of Trade 1997–2015," Trading Economics, n.d., http://www.tradingeconomics.com/russia/balance-of-trade.

150. Holly Ellyatt and Matt Clinch, "Putin Heralds Bounceback in Russia's Economy," CNBC, October 12, 2016, http://www.cnbc.com/2016/10/12/putin-heralds-bounceback-in-russias-economy.html.

151. "Pribyl 'Gazproma' upala na 86 procentov," Lenta, April 29, 2015, http://lenta.ru/news/2015/04/29/gazprom/.

152. Thomas Grove and Andrey Ostroukh, "Gazprom Profit Plunges on Ruble Weakness; Shakes Off Ukraine Sanctions," *Wall Street Journal,* April 29, 2015, http://www.wsj.com/articles/gazprom-profit-plunges-19-billion-1430297805.

153. Vavilov, Kovalishina, and Trofimov, "The New Export Routes,"181.

154. Henderson and Pirani, *The Russian Gas Matrix,* 315–316.

155. Alexei Topalov, "Gazprom terjaet stranu," Gazeta.ru, June 10, 2015, http://www.gazeta.ru/business/2015/06/10/6836233.shtml.

156. Andrei V. Belyi and Andreas Goldthau, "Between a Rock and a Hard Place: International Market Dynamics, Domestic Politics and Gazprom's Strategy," European University Institute Working Paper 2015/22, Robert Schuman Centre for Advanced Studies, April 2015, http://cadmus.eui.eu/bitstream/handle/1814/35398/RSCAS_2015_22.pdf?sequence=3.

157. Henderson and Pirani, *The Russian Gas Matrix,* 260, 342, 155.

158. "Alexander Novak: Ya za razvitie konkurentsii v energetike," TASS, June 23, 2015, http://tass.ru/opinions/interviews/2063780.

159. Henderson and Pirani, *The Russian Gas Matrix,* 34.

160. Topalov, "Gazprom terjaet stranu."

161. Gazprom, " 'Gazprom' na vneshnih rynkah," 2014, http://www.gazpromquestions.ru/foreign-markets/#c503.

162. Henderson and Pirani, *The Russian Gas Matrix,* 19–22.

163. "Perehod na ravnodohodnost' tsen na gas otlozhen do 2017," Vedomosti, April 21, 2014, https://www.vedomosti.ru/business/news/2014/04/21/perehod-na-ravnodohodnost-cen-na-gaz-otlozhen-do-2017-g.

164. Agnia Grigas, "Klaipeda's LNG Terminal: A Game Changer," EurActiv.com, September 22, 2014, http://www.euractiv.com/sections/energy/klaipedas-lng-terminal-game-changer-308613.

165. Andy Tully, "Gazprom Confident in European Future Despite 'New Cold War,' " Oil Price.com, February 4, 2015, http://oilprice.com/Energy/Natural-Gas/Gazprom-Confident-In-European-Future-Despite-New-Cold-War.html.

166. Henderson and Pirani, *The Russian Gas Matrix,* 60.

167. Jack Farchy, "Global Gas Market Braced for Price War," *Financial Times,* February 3, 2016, http://www.ft.com/intl/cms/s/2/c9c44750-ca50-11e5-a8ef-ea66e967dd44.html#axzz3zEx1aAOZ.

168. "Austria OMV and Russia's Gazprom Agree to Long-Term Gas Deal," *Moscow Times,* January 28, 2015, http://www.themoscowtimes.com/business/article/austria-omv-and-russia-s-gazprom-agree-to-long-term-gas-deal/515037.html.

169. "Russia Offers Ukraine Gas Price Discount ahead of Talks," EUbusiness, September 25, 2015, http://www.eubusiness.com/news-eu/ukraine-russia-gas.14hr; "Russia's Gas Discount Unlikely to Benefit Armenian Economy," *Asbarez,* April 12, 2016, http://asbarez.com/148649/russias-gas-discount-unlikely-to-benefit-armenian-economy/.

170. Mark Smedley, "Engie Breaks Oil Price Link in Gazprom Contract," *Natural Gas World*, April 12, 2016, http://www.naturalgasworld.com/engie-breaks-oil-price-link-in-gazprom-contract-29028.

171. Howard V. Rogers, "The Impact of Lower Gas and Oil Prices on Global Gas and LNG Markets," OIES Paper NG 99, Oxford Institute for Energy Studies, July 2015, 49, http://www.oxfordenergy.org/wpcms/wp-content/uploads/2015/07/NG-99.pdf.

172. Michael Birnbaum, "Russia Used to Have a Powerful Weapon in Its Energy Sector. Not Anymore," *Washington Post*, August 18, 2015, http://www.rfa.org/english/commentaries/energy_watch/russia-presses-china-for-new-gas-deal-08312015105116.html.

173. Elena Mazneva and Stepan Kravchenko, "Russia, China Sign $400B Gas Deal after Decade of Talks," Bloomberg, May 21, 2014, http://www.bloomberg.com/news/articles/2014-05-21/russia-signs-china-gas-deal-after-decade-of-talks.

174. "Russia Signs 30-Year Gas Deal with China," BBC News, May 21, 2014, http://www.bbc.com/news/business-27503017.

175. Igor Rozin, "Russia's Gas Industry Is Not Out of Gas," *Russia Direct*, November 6, 2013, http://www.russia-direct.org/qa/russias-gas-industry-not-out-gas.

176. Peter Roberts and Igor Makarov, "Russia and the Growth of LNG," *Russian Petroleum Investor*, September 2005.

177. Gazprom, "First Russian LNG Plant Launched in Sakhalin," February 18, 2009, http://www.gazprom.com/press/news/2009/february/article64569/.

178. "Sakhalin Energy Hits 1000 LNG Cargoes," *LNG World News*, August 26, 2015, http://www.lngworldnews.com/sakhalin-energy-hits-1000-lng-cargoes-milestone/.

179. Marina Zvonareva, "Russian LNG: A Five Year Window—And It's Closing," Natural Gas Europe, April 20, 2015, http://www.naturalgaseurope.com/russian-lng-market-20791.

180. "Russia Offers India Stake in Yamal Project, Petronet Mulling over the Offer," Natural Gas Asia, July 10, 2016, http://www.naturalgasasia.com/russia-offers-india-stake-in-yamal-lng-project-19004.

181. Gazprom, "Baltic LNG Project to Be Executed in Ust-Luga," January 22, 2015, http://www.gazprom.com/press/news/2015/january/article213255/.

182. Nick Cunningham, "Russia's Natural Gas Plans May Be Little More Than Hype," Oil Price.com, August 13, 2015, http://oilprice.com/Energy/Natural-Gas/Russias-Natural-Gas-Plans-May-Be-Little-More-Than-Hype.html.

183. Brian Spegele and Andrew Peaple, "Total Taps Chinese Banks to Fund Russian Project," *Wall Street Journal,* March 23, 2015, http://www.wsj.com/articles/total-seeks-10-billion-to-15-billion-in-chinese-financing-for-russian-project-1427093833.

184. EIA, "World Shale Resource Assessments," September 24, 2015, http://www.eia.gov/analysis/studies/worldshalegas/.

185. Helen Robertson, "Russia Won't Develop Shale Gas for a Decade," *Petroleum Economist,* April 19, 2013, http://www.petroleum-economist.com/Article/3194240/Russia-wont-develop-shale-gas-for-a-decade.html.

186. Kenneth J. Bird et al., "Circum-Arctic Resource Appraisal: Estimates of Undiscovered Oil and Gas North of the Arctic Circle," US Geological Survey, 2008, http://pubs.usgs.gov/fs/2008/3049/.

187. "Russia Will Defend Interests outside Its Territory—Weapons Industry Chief," *Russia Today,* October 5, 2015, https://www.rt.com/politics/317667-russian-will-defend-its-interests/.

188. Klavs A. Holm, "The Arctic—Regional Challenges with Global Consequences," *Baltic Rim Economies,* March 28, 2013, https://www.utu.fi/fi/yksikot/tse/yksikot/PEI/BRE/Documents/2013/BRE%202-2013%20web.pdf.

189. Joel K. Bourne Jr., "In the Arctic's Cold Rush, There Are No Easy Profits," *National Geographic,* March 2016, http://www.nationalgeographic.com/magazine/2016/03/new-arctic-thawing-rapidly-circle-work-oil/.

190. "Prirazlomnoye: Russia's First Arctic Offshore Field Begins Production," *Pipeline & Gas Journal* 241 (April 2014), http://www.pipelineandgasjournal.com/prirazlomnoye-russias-first-arctic-offshore-field-begins-production.

191. Bourne, "The Cold Rush," 62.

192. "Gazprom Neft Brings Second Arctic Well in Production," *Offshore Energy Today,* August 21, 2015, http://www.offshoreenergytoday.com/gazprom-neft-brings-second-arctic-well-in-production/.

193. Richard Milne, Christopher Adams, and Ed Crooks, "Oil Companies Put Arctic Projects into Deep Freeze," *Financial Times,* February 5, 2015, http://www.ft.com/cms/s/0/ae302d22-ad1b-11e4-a5c1-00144feab7de.html#slide0.

194. Ed Adamczyk, "Russia Prepares Militarization of Arctic," United Press International, October 21, 2014, http://www.upi.com/Top_News/World-News/2014/10/21/Russia-prepares-militarization-of-Arctic/2741413894698/.

195. Hannah Hoag, "Arctic Development Stalls with Tumbling Oil Prices," Huffington Post, January 22, 2016, http://www.huffingtonpost.com/entry/arctic-development-stalls_us_56a2b40be4b076aadcc6c444.

196. "The Prospects and Challenges for Arctic Oil Development," Oxford Institute for Energy Studies, November 2014, https://www.oxfordenergy.org/wpcms/wp-content/uploads/2014/11/WPM-56.pdf.

197. Anca Gurzu. "Economic Pain Pushes Russia to Drill in High Arctic," Politico, April 18, 2016, http://www.politico.eu/article/economic-pain-pushes-russia-to-drill-in-high-arctic-oil-energy-natural-gas/.

198. Irina Slav, "Russia Ramps Up Arctic Oil Production," Oil Price.com, July 21, 2016, http://oilprice.com/Energy/Energy-General/Russia-Ramps-Up-Arctic-Oil-Production.html.

199. "Obama Cancels Lease Sales in Arctic; Cedes Arctic to Russia's Vladimir Putin," Institute for Energy Research, October 23, 2015, http://instituteforenergyresearch.org/analysis/obama-cancels-lease-sales-in-arctic-cedes-arctic-to-russias-vladimir-putin/#_edn1.

4. The Politics of Dependence Transformed

1. Enerdata defines "Europe" as the European Union consisting of twenty-eight members and Albania, Bosnia-Herzegovina, Croatia, Iceland, Macedonia, Norway, Serbia and Montenegro, Switzerland, Turkey. Enerdata, "Global Energy Market Data Light," 2015, http://globaldata.enerdata.net/nrd_web/site/.

2. James A Caporaso, Gary Marks, Andrew Moravcsik, and Mark A. Pollack, "Does the European Union Represent an N of 1?," *ECSA Review* 10, no. 3 (Fall 1997): 1–5.

3. David G. Victor, Amy Jaffe, and Mark H. Hayes, eds., *Natural Gas and Geopolitics: From 1970 to 2040* (Cambridge: Cambridge University Press, 2006), 7, 56.

4. Jonathan P. Stern, *European Gas Markets: Challenge and Opportunity in the 1990's* (Aldershot, UK: Ashgate, 1990), 2.

5. Michael D. Tusiani and Gordon Sheare, *LNG: A Nontechnical Guide* (Tulsa, OK: PennWell, 2007), 198.

6. Gaz de France Suez, "GDF Suez: 50th Anniversary of the First LNG Commercial Chain between Algeria and France," April 2, 2015, http://www.gdfsuez.com/en/journalists/press-releases/50th-anniversary-first-lng-commercial-chain-algeria-france/.

7. David G. Victor and Mark H. Hayes, "Introduction to the Historical Case Studies: Research Questions, Methods, and Case Selection," in *Natural Gas and Geopolitics: From 1970 to 2040,* ed. David G. Victor, Amy Jaffe, and Mark H. Hayes (Cambridge: Cambridge University Press, 2006), 44–47.

8. Energy Sector Management Assistance Programme, "Border Oil and Gas Pipelines: Problems and Prospects," Joint UNDP / World Bank Energy Sector Management Assistance Programme, June 2003, 53–54, http://siteresources.worldbank.org/INTOGMC/Resources/crossborderoilandgaspipelines.pdf.

9. European Commission (henceforth EC), "Concerning Natural Gas," Communication from the Commission to the Council, Brussels, October 1, 1981, 1, http://aei.pitt.edu/5050/1/5050.pdf.

10. Eurostat database, "Energy Production and Imports," http://ec.europa.eu/eurostat/data/database; Eurostat database, "Imports (by Country of Origin)—Gas—Annual Data."

11. Eurostat database, "Energy Production and Imports."

12. Eurostat, "Main Origin of Primary Energy Imports, EU-28, 2003–14."

13. Eurostat database, "Simplified Energy Balances—Annual Data," author's calculations.

14. Enerdata, "Global Energy Market Data Light."

15. British Petroleum (henceforth BP), "Statistical Review of World Energy 2015," http://www.bp.com/content/dam/bp/excel/energy-economics/statistical-review-2015/bp-statistical-review-of-world-energy-2015-workbook.xlsx.

16. Enerdata, "Global Energy Market Data Light."

17. Karen Ayat, "The Zohr Field May Be Cypriot, Too," Natural Gas Europe, September 17, 2015, http://www.naturalgaseurope.com/the-zohr-field-may-be-cypriot-too-25465; Constantine Levoyannis and Mathieu Labrèche, "New Gas Discoveries in Southern Periphery May Transform European Energy Landscape," Energypost.eu, December 13, 2013, http://www.energypost.eu/new-gas-discoveries-europes-southern-periphery-may-transform-european-energy-landscape/.

18. Enerdata, "Global Energy Market Data Light."

19. Time Boersma, "The End of Dutch Natural Gas Production as We Know It," Brookings, August 4, 2016, https://www.brookings.edu/blog/order-from-chaos/2016/08/04/the-end-of-dutch-natural-gas-production-as-we-know-it/.

20. Gazprom, *Gazprom Annual Report 2014,* 2015, 22, http://www.gazprom.com/f/posts/55/477129/gazprom-annual-report-2014-en.pdf.

21. Gas Infrastructure Europe, "GLE Investment Database," April 2015, http://www.gie.eu/index.php/maps-data/lng-investment-database; author's calculations.

22. Ibid.

23. Kecse Zsuzsanna Réka, "The Role of Liquefied Natural Gas in Europe," *International Relations Quarterly* 1, no. 4 (Winter 2010/4 Tél): 8–9.

24. Eurogas, "New Eurogas Data Confirms Dynamic EU Gas Market," March 25, 2015, http://www.eurogas.org/uploads/media/Eurogas_Press_Release_-_New_Eurogas_data_confirms_dynamic_EU_gas_market.pdf.

25. Wim Groenendijk and Jacques Rottenberg, "The European LNG Terminal Infrastructure 2015: Status and Outlook," Gas Infrastructure Europe, June 17, 2015, http://www.gie.eu/index.php/maps-data/lng-map.

26. Ibid.

27. Agnia Grigas, "Klaipeda's LNG Terminal: A Game Changer," EurActiv.com, September 22, 2014, http://www.euractiv.com/sections/energy/klaipedas-lng-terminal-game-changer-308613.

28. Gas Infrastructure Europe, "GLE Investment Database."

29. Interview with Nikos Tsafos, Washington, DC, December 2015.

30. International Group of Liquefied Natural Gas Importers, "The LNG Industry," 2016, 5, http://www.giignl.org/sites/default/files/PUBLIC_AREA/Publications/giignl_2016_annual_report.pdf.

31. Groenendijk and Rottenberg, "The European LNG Terminal Infrastructure 2015."

32. Ibid.

33. "Conscious Uncoupling," *Economist,* April 5, 2014, http://www.economist.com/news/briefing/21600111-reducing-europes-dependence-russian-gas-possiblebut-it-will-take-time-money-and-sustained.

34. Gas Infrastructure Europe, "GLE Investment Database," author's calculations.

35. Eurostat, "Main Origin of Primary Energy Imports, EU-28, 2003–14."

36. Jonathan Stern, "The Russian-Ukrainian Gas Crisis of January 2006" (Oxford: Oxford Institute for Energy Studies, January 16, 2016), 7–8.

37. "Kiev and Moscow Reach Gas Agreement—Despite Limited Impact of Gas Conflict on Union, Austrian Presidency and Commission Stress Need for Common Energy Approach," Agence Europe, January 4, 2006.

38. Agnia Grigas, *The Politics of Energy and Memory between the Baltic States and Russia* (Farnham, UK: Ashgate, 2013), 50–60.

39. Giovani Caldioli, "Belarus-Russia Energy Disputes—Political and Economic Comparative Analysis," University of Bologna, January 1, 2012.

40. Tom Parfitt, "Belarus Cuts Off Russian Pipeline in Bitter Gas War," *Guardian,* January 9, 2007, http://www.theguardian.com/business/2007/jan/09/oilandpetrol.russia.

41. "Russian Gas Still Not Flowing to Europe," Agence Europe, January 13, 2009.

42. Luke Harding and Dan McLaughlin, "Deal to Resume Russian Gas Eludes EU as 11 People Die in Big Freeze-Up," *Guardian,* January 11, 2009, http://www.theguardian.com/world/2009/jan/11/russia-ukraine-gas-supplies-dispute.

43. EC, "Joint statement by the Czech EU Presidency and the European Commission," January 6, 2009, http://europa.eu/rapid/press-release_IP-09-4_en.htm?locale=en.

44. "New EU Presidency Presents Priorities and Work Programme," Agence Europe, January 6, 2009.

45. "Baltarusija apribojo rusiškų dujų tiekimą Lietuvai," *Verslo žinios,* June 23, 2010, http://vz.lt/archive/straipsnis/2010/06/23/Baltarusija_apribojo_rusisku_duju_tiekima_Lietuvai3.

46. "EU Wants New Gas Agreement by Summer," Agence Europe, March 19, 2015; "Ukraine Crisis: Russia Halts Gas Supplies to Kiev," BBC News, June 16, 2014, http://www.bbc.com/news/world-europe-27862849.

47. "Gazprom Resumes Gas Supplies from Russia to Ukraine," Euronews, December 9, 2014, http://www.euronews.com/2014/12/09/gazprom-resumes-gas-supplies-from-russia-to-ukraine/.

48. Földgázszállító Zártkörűen Működő Részvénytársaság, "FGSZ Ltd Temporarily Suspends the Gas Transmission to Ukraine," September 25,

2014, http://fgsz.hu/en/content/fgsz-ltd-temporarily-suspends-gas-transmission-ukraine.

49. Barbara Lewis, "Ukraine Appeals to EU over 'Illegal' Gazprom-Slovak Pipeline Contract," Reuters UK, June 23, 2015, http://uk.reuters.com/article/2015/06/23/uk-ukraine-gas-slovakia-idUKKBN0P31W620150623.

50. Neil Buckley, "Hungary Halts Flow of Gas to Ukraine," *Financial Times,* September 26, 2014, http://www.ft.com/intl/cms/s/0/7c5d2bf0-4552-11e4-ab86-00144feabdc0.html#axzz3mCLbMlC9.

51. Dorota Gajewska, "Reduced Natural Gas Supplies from Countries East of Poland," Polskie Górnictwo Naftowe i Gazownictwo, September 10, 2014, http://en.pgnig.pl/news/-/news-list/changeYear/2014?_newslistportlet_WAR_newslistportlet_urlTitle=reduced-natural-gas-supplies-from-countries-east-of-poland&_newslistportlet_WAR_newslistportlet_newsGroupId=18252&_newslistportlet_WAR_newslistportlet_action=newsDetails&_newslistportlet_WAR_newslistportlet_currentPage=2.

52. EC, "Green Paper—Towards a European Strategy for the Security of Energy Supply," November 29, 2000, http://eur-lex.europa.eu/legal-content/EN/TXT/?uri=CELEX%3A52000DC0769.

53. Giandomenico Majone, "The Rise of the Regulatory State in Europe," *West European Politics* 17, no. 3 (July 1, 1994): 77–101.

54. EC, "Europe 2020—EU-Wide Headline Targets for Economic Growth," June 22, 2015, http://ec.europa.eu/europe2020/europe-2020-in-a-nutshell/targets/index_en.htm.

55. Agora Energiewende, "12 Insights on Germany's Energiewende," discussion paper, Berlin, February 2013.

56. Craig Morris, "Renewables Briefly Covered 78 Percent of German Electricity," German Energy Transition, July 25, 2015, http://energytransition.de/2015/07/renewables-covered-78percent-of-german-electricity/.

57. EC, "EU on Track to Meeting 20% Renewable Energy Target," June 16, 2015, https://ec.europa.eu/energy/en/news/eu-track-meeting-20-renewable-energy-target.

58. EC, "EU Leaders Agree 2030 Climate and Energy Goals," October 24, 2014, http://ec.europa.eu/clima/news/articles/news_2014102401_en.htm.

59. Morena Skalamera and Andreas Goldthau, "Russia: Playing Hardball or Bidding Farewell to Europe?," Belfer Center for Science and International Affairs, June 2016, 29, http://belfercenter.hks.harvard.edu/files/Russia%20Hardball%20-%20Web%20Final.pdf.

60. EC, "Commission Unveils Key Energy Infrastructure Projects to Integrate Europe's Energy Markets and Diversify Sources," November 18, 2015, https://ec.europa.eu/energy/en/news/commission-unveils-list-195-key-energy-infrastructure-projects.

61. Tim Boersma, *Energy Security and Natural Gas Markets in Europe* (New York: Routledge, 2015).

62. EC, Energy, "North-South East—Energy," 2015, https://ec.europa.eu/energy/en/topics/infrastructure/north-south-east.

63. European Council, "European Council 4 February 2011 Conclusions," 2.

64. Lietuvos Energija, "Syderiai Underground Gas Storage," 2015, http://gamyba.le.lt/en/activities/projects/syderiai-underground-gas-storage-/.

65. EC, "First Gas Interconnector Poland—Lithuania Ends Energy Isolation of the Baltic States," 2015, http://europa.eu/rapid/press-release_IP-15-5844_en.htm.

66. EC, "Gas Interconnection Romania—Hungary (RO-HU)," 2015, http://ec.europa.eu/energy/eepr/projects/files/gas-interconnections-and-reverse-flow/romania-hungary-ro-hu_en.pdf.

67. Vija Pakalkaite, "Catalysts for Reforms in the EU Natural Gas Markets: Cases of Hungary, Lithuania and Romania" (PhD dissertation, Central European University, 2017).

68. European Parliament, Council of the European Union, "Concerning Common Rules for the Internal Market in Natural Gas and Repealing Directive 2003/55 / EC," July 13, 2009, http://eur-lex.europa.eu/LexUriServ/LexUriServ.do?uri=OJ:L:2009:211:0094:0136:en:PDF.

69. EC, Ausnahmegenehmigung der bundesnetzagentur für die OPAL gasleitung gemä13 art. 22 der Richtlinie 2003/55, 2009; EC, Decision on the Exemption of the "Gazelle" Interconnector According to Article 36 of Directive 2009/73/EC, 2011; Jonas Grätz, "The Impact of EU Law on Gazprom and Its Implications," Center for Security Studies (CSS), Platts 7th European Gas Summit, Vienna, September 20, 2013, http://www.platts.com/IM.Platts.Content/ProductsServices/ConferenceandEvents/2013/pc378/presentations/16.%20Dr.%20Jonas%20Gratz_ETH%20ZURICH.pdf.

70. Natural Gas World, "Brussels Allows Gazprom Greater Access to OPAL: Report," October 26, 2016, http://www.naturalgasworld.com/brussels-allows-to-extend-gazprom-access-to-opal-pipeline-report-34127.

71. European Parliament, Council of the European Union, "Concerning Common Rules for the Internal Market," art. 9.

72. EC Directorate-General Competition, "DG Competition Report on Energy Sector Inquiry," Energy, Basic Industries, Chemicals and Pharmaceuticals (Brussels: European Commission, January 10, 2007), http://ec.europa.eu/competition/sectors/energy/inquiry/index.html.

73. EC, "Commission Staff Working Paper Interpretative Note on Directive 2009/72/EC Concerning Common Rules for the Internal Market in Electricity and Directive 2009/73/EC Concerning Common Rules for the Internal Market in Natural Gas: The Unbundling Regime," January 22, 2010, https://ec.europa.eu/energy/sites/ener/files/documents/2010_01_21_the_unbundling_regime.pdf.

74. European Parliament, Council of the European Union, "Concerning Common Rules for the Internal Market," art. 49.

75. Richard Milne, "Lithuania Claims Gas Price Victory in Battle with Gazprom," *Financial Times,* May 13, 2014, http://www.ft.com/intl/cms/s/0/2b6f3ef0-dab2-11e3-9a27-00144feabdc0.html#axzz3kPTG8O3g.

76. Gleb Bryanski, "Putin Says EU Energy Laws Are Uncivilised 'Robbery,'" Reuters India, February 10, 2010, http://in.reuters.com/article/2010/11/26/idINIndia-53174220101126.

77. Andrius Sytas, "Gazprom Sells Lithuania Assets after Antitrust Fine," Reuters, June 12, 2014, http://uk.reuters.com/article/2014/06/12/uk-lithuania-gazprom-idUKKBN0EN1IF20140612.

78. EC, "Country Reports: Estonia," October 13, 2014, 64, https://ec.europa.eu/energy/sites/ener/files/documents/2014_countryreports_estonia.pdf.

79. "Natgas Tariffs Will Inevitably Climb after Latvijas Gaze Is Unbundled—Itera Latvija CEO," Latvian News Service, August 22, 2016, http://www.latviannewsservice.lv/en/topic/1905/news/51705527/.

80. "'Gazprom Clause' Issues Russia Ultimatum for Energy Co-Operation," EurActiv.com, September 20, 2007, http://www.euractiv.com/energy/gazprom-clause-issues-russia-ult-news-218748.

81. Simon Taylor, "EU Struggles to Agree on Anti-Gazprom Clause," Politico, May 21, 2008, http://www.politico.eu/article/eu-struggles-to-agree-on-anti-gazprom-clause/.

82. European Parliament, Council of the European Union, "Concerning Common Rules for the Internal Market," para. 21.

83. Ibid., art. 11.

84. Roman Kupchinsky, "Bulgaria's 'Overgas,' a Russian Spy in Canada, and Gazprom," *Eurasia Daily Monitor* 6, no. 30 (February 13, 2009), http://www.jamestown.org/single/?tx_ttnews%5Btt_news%5D=34511.

85. Grigas, *The Politics of Energy and Memory,* 95–99.

86. Group Dmitry Firtash, "Mabofi, a Group DF Company, Finally Registered as the Sole Shareholder of Emfesz Ltd.," February 13, 2013, http://groupdf.com/en/press-center/news/mabofi-a-group-df-company-finally-registered-as-the-sole-shareholder-of-emfesz-ltd/.

87. "EU Recommends Member States Renegotiate South Stream Pipeline," Radio Free Europe / Radio Liberty, December 5, 2013, http://www.rferl.org/content/eu-renegotiate-south-stream/25191193.html.

88. Nikolay Jeliazkov, "EC: We Have No Police in Bulgaria and Serbia to Monitor South Stream," September 17, 2014, http://www.bta.bg/en/c/DF/id/912669.

89. Ministry of Energy of the Republic of Lithuania, "Lithuanian Ministry of Energy Launched a Complaint to the European Commission Regarding Abuse of Dominant Position by the Russian Gas Supplier Gazprom," January 25, 2011, http://www.enmin.lt/en/news/detail.php?ID=1198.

90. EC, "Antitrust: Commission Sends Statement of Objections to Gazprom—Factsheet," April 22, 2015, http://europa.eu/rapid/press-release_MEMO-15-4829_en.htm.

91. Alan Riley, "Commission v. Gazprom: The Antitrust Clash of the Decade?," Regulatory Policy, Centre for European Policy Studies Policy Briefs, October 31, 2012, 1, http://www.ceps.eu/book/commission-v-gazprom-antitrust-clash-decade.

92. Tom Fairless and Gabriele Steinhauser, "EU Files Formal Charges against Gazprom for Abuse of Dominant Position," *Wall Street Journal,* April 22, 2015, http://www.wsj.com/articles/eu-files-formal-charges-against-gazprom-for-abuse-of-dominant-position-1429697186.

93. Gabriele Steinhauser and Tom Fairless, "Gazprom Attempts to Settle EU Antitrust Case," *Wall Street Journal,* September 21, 2015, http://www.wsj.com/articles/gazprom-attempts-to-settle-eu-antitrust-case-1442852375.

94. Katya Golubkova and Foo Yun Chee, "Gazprom Aims for Amicable Solution in EU Antitrust Case," Reuters, March 9, 2016, http://uk.reuters.com/article/uk-russia-gazprom-eu-antitrust-idUKKCN0WB2BH.

95. Rochelle Toplensky, Jack Farchy, and Henry Foy, "Brussels Nears Settlement with Gazprom After 5-Year Probe," *Financial Times*, October 24, 2016, https://www.ft.com/content/be17134e-991f-11e6-8f9b-70e3cabccfae.

96. Anna Galkina, "European Energy Security and Russian Export Strategy," Energy Research Institute of the Russian Academy of Sciences, Budapest, September 15, 2015, 8, http://ineiran.ru/articles/2015/20150915-Budapest-ERI-RAS.pdf.

97. Nektaria Stamouli, "Construction Begins on Trans Adriatic Pipeline," *Wall Street Journal,* May 17, 2016, http://www.wsj.com/articles/construction-begins-on-trans-adriatic-pipeline-1463508079.

98. "Trans Anatolian Natural Gas Pipeline project (TANAP), Turkey," Hydrocarbons-technology.com, 2016, http://www.hydrocarbons-technology.com/projects/trans-anatolian-natural-gas-pipeline-project-tanap.

99. Elmar Baghirov, "In-Depth Analysis of a Possible Nabucco West Revival," Natural Gas Europe, May 6, 2015, http://www.naturalgaseurope.com/the-revival-of-nabucco-west-myth-or-reality-23537.

100. EC, "Commission Decision of 16.5.2013 on a Prolongation of the Effects of the Ex Emption Decision of NABUCCO Gas Pipeline International GmbH from Third Party Access and Tariff Regulation Granted under Directive 2003/55/EC," May 16, 2013.

101. Mansour Kashfi, "Iranian Gas and the Nabucco Pipeline Realities," Oilprice.com, January 3, 2013, http://oilprice.com/Energy/Natural-Gas/Iranian-Gas-and-the-Nabucco-Pipeline-Realities.html.

102. "Austria Agrees to Join South Stream Project," *Pipeline & Gas Journal,* June 2010, http://pipelineandgasjournal.com/austria-agrees-join-south-stream-project.

103. "Is Nabucco-West Revivable?," Natural Gas Europe, March 10, 2015, http://www.naturalgaseurope.com/viability-nabucco-west-revival-26549.

104. Skalamera and Goldthau, "Russia," 3.

105. Ibid, 10.

106. Ralf Dickel, Elham Hassanzadeh, James Henderson, Anouk Honoré, Laura El-Katiri, Simon Pirani, Howard Rogers, Jonathan Stern, and Katja Yafimava, "Reducing European Dependence on Russian Gas—Distinguish ing Natural Gas Security from Geopolitics" (Oxford: Oxford Institute for Energy Studies, 2014), http://www.oxfordenergy.org/2014/10/reducing-european-dependence-on-russian-gas-distinguishing-natural-gas-security-from-geopolitics/.

107. Sami Andoura, Leigh Hancher, and Marc Van der Woude, "Towards a European Energy Community: A Policy Proposal," Jacques Delors Institute, 2010, http://www.delorsinstitute.eu/011-2155-Towards-a-European-Energy-Community-A-Policy-Proposal.html; Jerzy Buzek and Jacques Delors, "Full Text of the Buzek and Delors Declaration on the Creation of a European Energy Community," May 5, 2010, http://www.europarl.europa.eu/president/en/press/press_release/2010/2010-May/press_release-2010-May-4.html.

108. Donald Tusk, "A United Europe Can End Russia's Energy Stranglehold," *Financial Times,* April 21, 2014, http://www.ft.com/cms/s/0/91508464-c661-11e3-ba0e-00144feabdc0.html#axzz3SwpaEllB.

109. Sonja van Renssen, "Brussels Tests Limits of Its Powers with Energy Union," Energy Post, February 27, 2015, http://www.energypost.eu/brussels-tests-limits-powers-energy-union/.

110. "Hungary Will Resist IGA Transparency, Warns Orban," Agence Europe, February 19, 2015.

111. EC, "Communication from the Commission to the European Parliament, the Council, the European Economic and Social Committee, the Committee of the Regions and the European Investment Bank," February 25, 2015, 6, http://ec.europa.eu/priorities/energy-union/docs/energyunion_en.pdf.

112. IHS Cambridge Energy Research Associates, "Caspian Development Corporation," Final implementation report, Cambridge, UK, December 2010.

113. Interview with Nikos Tsafos, Washington, DC, January 2016.

114. EC, "Towards Energy Union: The Commission Presents Sustainable Energy Security Package," February 16, 2016, http://europa.eu/rapid/press-release_IP-16-307_en.htm.

115. "EU's Sefcovic Seeks Prior Approval of Member State Energy Deals," Reuters, January 26, 2016, http://www.reuters.com/article/eu-energy-russia-idUSL8N15A31Q.

116. Michel Derdevet, Philipp Fink, Antoine Guillou, Instytut Spraw Publicznych, Robert Schachtschneider, Daniel Scholten, and Christophe Schramm, "New and Ambitious or Just More of the Same? The Energy Union at a Crossroads," Friedrich-Ebert-Stiftung, November 23, 2015, https://www.fes.de/de/fileadmin/redaktion/FES/FESweite_Projekte/Politik_fuer_Europa/Publikationen/23112015_-_The_Energy_Union_at_a_crossroads.pdf.

117. Tim Boersma, "What's Next for Europe's Natural Gas Market?," Brookings, March 15, 2016, http://www.brookings.edu/blogs/order-from-chaos/posts/2016/03/15-europe-natural-gas-markets-boersma?rssid=energy+and+environment&utm_source=feedblitz&utm_medium=FeedBlitzRss&utm_campaign=FeedBlitzRss&utm_content=What%E2%80%99s+next+for+Europe%E2%80%99s+natural+gas+market%3f.

118. BP, "Statistical Review of World Energy 2015"; Isis Almeida, "Iran Is No Qatar, Even with World's Second-Biggest Gas Reserves," Bloomberg.com, July 15, 2015, http://www.bloomberg.com/news/articles/2015-07-14/iran-no-qatar-even-with-the-world-s-second-biggest-gas-reserves.

119. Brenda Shaffer, *Energy Politics* (Philadelphia: University of Pennsylvania Press, 2011), 46.

120. Agnia Grigas and Amir Handjani, "Big Loser in Any Nuclear Deal with Iran May Be Russia," *Reuters Blogs,* July 10, 2015, http://blogs.reuters.com/great-debate/2015/07/09/big-loser-in-any-nuclear-deal-with-iran-may-be-russia/.

121. "Iran Wants to Resume LNG Projects," Natural Gas Europe, May 7, 2015, http://www.naturalgaseurope.com/iran-eyes-resuming-huge-lng-project-with-germany-23582.

122. Cristian Kanovits and Leonela Lenes, "Middle East and North Africa: Iran and the Geopolitics behind the Different Options of Transport for Oil and Gas," Middle East Political and Economic Institute, September 13, 2016, http://mepei.com/in-focus/5463-middle-east-and-north-africa-iran-and-the-geopolitics-behind-the-different-options-of-transport-for-oil-and-gas.

123. Fereidun Fesharaki, "Opportunities in Iran Oil and Gas Sector in the Post-Sanction Era," Institute of Energy Economics in Japan (IEEJ), Tokyo, February 3, 2016; "Iran to Take Shortcut to EU with FLNG," Natural Gas Europe, April 7, 2016, http://www.naturalgaseurope.com/iranian-gas-to-take-shortcut-to-eu-with-flng-28977; Benoit Faucon, "Iran Seeks Rapid Reboot for Natural Gas Exports," *Wall Street Journal,* January 26, 2016, http://www.wsj.com/articles/iran-seeks-rapid-reboot-for-natural-gas-exports-1453821547; "Minister Skourletis: Greece Mulling LNG Cooperation with Iran," *LNG World News,* February 26, 2016, https://www.lngworldnews.com/minister-skourletis-greece-mulling-lng-cooperation-with-iran/; Danuta Slusarska and Fabio Orlando, "Iran's Energy Come-

back," Friends of Europe, March 14, 2016, http://www.friendsofeurope.org/greener-europe/irans-energy-comeback/.

124. Noble Energy, "Eastern Mediterranean," 2016, http://www.nobleenergyinc.com/Operations/International/Eastern-Mediterranean-128.html.

125. Off-record meetings with industry representatives, Tel Aviv, Israel, May 2016; Michael Ratner, "Natural Gas Discoveries in the Eastern Mediterranean," Congressional Research Service Report, August 16, 2016.

126. Government of the Netherlands, "Groningen Gas Extraction Further Reduced to 30 Billion Cubic Meters in 2015," June 23, 2015, https://www.government.nl/latest/news/2015/06/23/groningen-gas-extraction-further-reduced-to-30-billion-cubic-metres-in-2015; Will Dalrymple, "Blue Gold," Machinery.co.uk, September 1, 2015, http://www.machinery.co.uk/machinery-features/natural-gas-future-market-prospects/88927/.

127. US Department of Energy, "Technically Recoverable Shale Oil and Shale Gas Resources: An Assessment of 137 Shale Formations in 41 Countries Outside the United States," Independent Statistics & Analysis, June 2013, 1–7, http://www.eia.gov/analysis/studies/worldshalegas/pdf/fullreport.pdf.

128. Amaranta Herrero, "Fracking Ban Expanding in Spanish Regions," Environmental Justice Organizations, Liabilities and Trade, February 10, 2014, http://www.ejolt.org/2014/02/fracking-ban-expanding-in-spanish-regions/; Institute for Advanced Sustainability Studies, "Shale Gas and Fracking in Europe," June 2015, 5, http://www.iass-potsdam.de/sites/default/files/files/shale_gas_and_fracking_in_europe.pdf; Gregor Erbach, "Shale Gas and EU Energy Security," European Parliamentary Research Service, December 2014, 4–5, http://www.europarl.europa.eu/RegData/etudes/BRIE/2014/542167/EPRS_BRI%282014%29542167_REV1_EN.pdf.

129. "Lithuania's Shale Gas Potential among Highest in Central-Eastern Europe," 15min.lt, November 7, 2012, http://www.15min.lt/en/article/business/lithuania-s-shale-gas-potential-among-highest-in-central-eastern-europe-527-276743.

130. Stanley Reed, "Chevron to Abandon Shale Natural Gas Venture in Poland," *New York Times,* January 30, 2015, http://www.nytimes.com/2015/01/31/business/international/chevron-to-abandon-shale-venture-in-poland-a-setback-to-fracking-europe.html?_r=0; "Lithuania Abandons Shale

Exploration," Shale Gas International, April 10, 2015, http://www.shalegas.international/2015/04/10/lithuania-abandons-shale-exploration/; "UK Fracking Go-Ahead Boosts Shale Gas Industry," *Financial Times*, October 6, 2016, https://www.ft.com/content/09dea1e6-8af4-11e6-8cb7-e7ada1d123b1.

131. "Guidance on Fracking: Developing Shale Oil and Gas in the UK," Gov.uk, April 11, 2016, https://www.gov.uk/government/publications/about-shale-gas-and-hydraulic-fracturing-fracking/developing-shale-oil-and-gas-in-the-uk; Jon Mainwaring, "Shale Gas UK: A Pivotal Year Ahead," Rigzone, January 8, 2016, http://www.rigzone.com/news/oil_gas/a/142402/Shale_Gas_UK_A_Pivotal_Year_Ahead.

132. "Shale Gas: Controversy Just Beginning," Agence Europe, September 23, 2011; "Germany Shuts Door to Fracking," Phys.org, June 24, 2016, http://phys.org/news/2016-06-germany-door-fracking.html.

133. "Europe Acting Separately on Shale Gas," *Agence Europe,* July 12, 2013; Inés Benítez, "Growing Mobilisation against Introduction of Fracking in Spain," Inter Press Service, June 2, 2015, http://www.ipsnews.net/2015/06/growing-mobilisation-against-introduction-of-fracking-in-spain/.

134. Tomasz Dąborowski and Jakub Groszkowski, "OSW: Shale Gas in Bulgaria, the Czech Republic and Romania: Political Context—Legal Status—Outlook," Ośrodek Studiów Wschodnich, November 7, 2014; "Impact Assessment Directive Strengthened but Shale Excluded," Agence Europe, March 31, 2014.

135. James Burges, "France Discusses Ban of Imported U.S. Shale Gas," Oil Price.com, May 10, 2016, http://oilprice.com/Latest-Energy-News/World-News/France-Discusses-Ban-Of-Imported-US-Shale-Gas.html.

136. Andrius Sytas, "Chevron Leaves Lithuania as Shale Gas Prospects Remain Uncertain," Reuters, July 8, 2014, http://www.reuters.com/article/2014/07/08/us-chevron-lithuania-idUSKBN0FD1GH20140708.

137. "Chevron Withdraws from Shale Gas Activities in Romania," *Romania Insider,* February 21, 2015; Agnieszka Barteczko, "Conoco Stops Its Shale Gas Exploration in Poland," Reuters UK, June 5, 2015, http://uk.reuters.com/article/2015/06/05/conoco-poland-shalegas-idUKL5N0YR2MY20150605.

138. Dickel et al., "Reducing European Dependence on Russian Gas," 15.

139. Leonardo Maugeri, "The Coming Global Gas-Market Bust," *The National Interest,* February 24, 2015, http://nationalinterest.org/feature/the-coming-global-gas-market-bust-12316; Mariya Petkova, "Anti-Fracking

Campaigns Freaking Out Europe," *Aljazeera*, July 5, 2014, http://www.aljazeera.com/indepth/features/2014/06/anti-fracking-campaigns-freaking-out-europe-201462773614979427.html.

140. Stephen Tindale, "Europe Needs More Gas, So Should Develop Shale Gas," Interfax Global Energy, August 20, 2015, http://interfaxenergy.com/analytics/article/442/europe-needs-more-gas-so-should-develop-shale-gas.

141. United States Mission to the European Union, "Remarks at the European Parliament's LNG Roundtable," 2015, http://useu.usmission.gov/rm-05292015.html.

142. "Second US LNG Cargo Lands in Europe as Tanker Docks in Spain," S&P Global Platts, July 22, 2016, http://www.platts.com/latest-news/natural-gas/london/second-us-lng-cargo-lands-in-europe-as-tanker-26498748.

143. Centrica, "Centrica Signs Long-Term US LNG Export Deal with Cheniere," March 25, 2013, https://www.centrica.com/news/centrica-signs-long-term-us-lng-export-deal-cheniere; Enel, "Enel Signs Two 20-Year Contracts for the Supply of LNG from United States," August 4, 2014, https://www.enel.com/en-GB/media/press_releases/enel-signs-two-20-year-contracts-for-the-supply-of-lng-from-united-states/r/1661736; Iberdrola, "Iberdrola Agrees to LNG Supply Deal with Cheniere Over 20 Years Valued at Around €4.1 Billion," May 30, 2014, http://www.iberdrola.es/press-room/press-releases/national-international/2014/detail/press-release/140530_NP_01_Cheniere.html; "LNG Supply with Cheniere," Gas Natural Fenosa, June 2, 2014, http://www.prensa.gasnaturalfenosa.com/en/new-agreement-lng-cheniere-supply/; "Cheniere, Endesa Sign Corpus Christi LNG Deal," *LNG World News,* April 2, 2014, http://www.lngworldnews.com/cheniere-endesa-sign-corpus-christi-lng-deal/; "Cheniere Inks LNG Supply Deal with EDF," *LNG World News,* August 12, 2015, http://www.lngworldnews.com/cheniere-inks-lng-supply-deal-with-edf/; LITGAS, "LITGAS and Cheniere Marketing Sign Master Trade Agreement," February 28, 2015, http://www.litgas.lt/en/litgas-and-cheniere-marketing-sign-master-trade-agreement/.

144. Galkina, "European Energy Security and Russian Export Strategy."

145. "EU Gas Use Grows 4% in 2015: Eurogas," Natural Gas Europe, March 31, 2016, http://www.naturalgaseurope.com/eu-gas-use-rises-4-in-2015-eurogas-28875.

146. Eurostat database, "Supply, Transformation, Consumption—Gas—Annual Data."

147. Eurostat database, "Real GDP Growth Rate—Volume."

148. Enerdata, "Global Energy Market Data Light."

149. Eurostat database, "Supply, Transformation, Consumption—Gas—Annual Data."

150. Nikos Tsafos, "Europe's Dangerous Distraction: Pipelines," *The National Interest,* July 2, 2015, http://nationalinterest.org/feature/europes-dangerous-distraction-pipelines-13242; "Europe's Declining Gas Demand," Third Generation Environmentalism, June 11, 2015, http://e3g.org/news/media-room/europes-declining-gas-demand.

151. IEA, "Medium-Term Gas Market Report 2016," Executive Summary, 2016, http://www.iea.org/Textbase/npsum/MTGMR2016SUM.pdf.

152. Kalina Oroschakoff and Sara Stefanini, "Fiery Blowback to Brussels' Climate Change Plan," Politico, July 20, 2016, http://www.politico.eu/article/fiery-blowback-to-brussels-climate-change-plan-paris-carbon-emissions/; Jilles van den Beukel, "Why It's So Difficult to Reduce CO2 Emissions," Energy Post, February 4, 2016, http://www.energypost.eu/difficult-reduce-co2-emissions/; MCC Mercator Research Institute on Global Commons and Climate Change, "Policy Brief: Carbon Price Floor to Reform EU Emissions Trading," June 2016, 1, https://www.mcc-berlin.net/fileadmin/data/B2.3_Publications/Kurzdossiers/Emissionshandel/MCC-policy-brief_EU-emissions-trading_status_June-2016.pdf.

5. The Politics of Transit

1. Vladimir Soldatkin, "Factbox: Russia's Energy Disputes with Ukraine and Belarus," Reuters, December 21, 2012, http://www.reuters.com/article/us-russia-gas-disputes-idUSBRE8BK11T20121221.

2. Katja Yafimava, *The Transit Dimension of EU Energy Security: Russian Gas Transit across Ukraine, Belarus, and Moldova* (Oxford: Oxford University Press for the Oxford Institute for Energy Studies, 2011), 5, 319–320.

3. Margarita M. Balmaceda, *The Politics of Energy Dependency: Ukraine, Belarus, and Lithuania between Domestic Oligarchs and Russian Pressure* (Toronto: University of Toronto Press, 2013).

4. Taras Kuzio, "Ukrainian Politics, Energy, and Corruption under Kuchma and Yushchenko," lecture at Harvard University, March 7, 2008, http://www.taraskuzio.net/conferences2_files/Ukrainian_Politics_Energy.pdf.

5. Naftogaz, "Company," n.d., http://www.naftogaz.com/www/3/nakweben.nsf/0/3A25D65C2606A6C9C22570D800318869?Open

Document; Marc Champion, "Ukraine Can Kneel or Starve," Bloomberg, May 21, 2014, http://www.bloombergview.com/articles/2014-05-21/ukraine-can-kneel-or-starve.

6. "Bypassing Ukraine Will Be Costly for Russia's Gazprom," *Moscow Times*, July 19, 2015, http://www.themoscowtimes.com/business/article/bypassing-ukraine-will-be-costly-for-russia-s-gazprom-analysts/525806.html.

7. Ukrtransgaz, "Ukrtransgaz Today," last modified 2014, http://utg.ua/en/utg/company/ukrtransgaz-today.html.

8. Ukrnafta, "Company Profile," last modified 2015, http://www.ukrnafta.com/en/about/profile.

9. Naftogaz, "Naftogaz of Ukraine Joins AGSI+ Transparency Platform of Gas Infrastructure Europe," May 6, 2014, http://naftogaz-europe.com/article/en/NaftogazofUkrainejoinsAGSI.

10. In addition, oil accounted for 8.5 percent of the energy mix; and hydropower, solar, and geothermal power nearly 3 percent. Ukrstat, "Energy Balance of Ukraine 2013," December 21, 2015, https://ukrstat.org/en/operativ/operativ2012/energ/en_bal/arh_2012_e.htm.

11. Ministry of Energy and Mines, "U 2014 roci Ukraina skorotyla spozhyvannia pryrodnogo gazu na 16%," January 15, 2015, http://www.kmu.gov.ua/control/publish/article?art_id=247872221; Naftogaz Ukrainy, "Gas Consumption in Ukraine Declined by 21% in 2015," February 2, 2016, http://naftogaz.com/www/3/nakweben.nsf/0/DAA0FB1F385DF293C2257F4C00581262?OpenDocument.

12. Georg Zachmann and Dmytro Naumenko, "Ukraine Energy Update 2015/16," German Advisory Group to Ukraine, March 2016, http://www.beratergruppe-ukraine.de/wordpress/wp-content/uploads/2016/05/TN_01_2016_en.pdf.

13. "Ukraina za rik skorotyla import gazu z Rosii na 15%," Ukrainian Independent Information Agency of News (henceforth UNIAN), January 10, 2014, http://www.unian.ua/society/871180-ukrajina-za-rik-skorotila-import-gazu-z-rosiji-na-15.html.

14. Naftogaz, "Ukraine Purchased 63% of Its Imported Gas in Europe in 2015," February 2, 2016, http://www.naftogaz.com/www/3/nakweben.nsf/0/8FD7A9A348A0844DC2257F4C005802FD?OpenDocument/.

15. US Energy Information Administration (henceforth EIA), "Technically Recoverable Shale: Oil and Shale Gas Resources," September 2015, x-2,

https://www.eia.gov/analysis/studies/worldshalegas/pdf/Eastern_Europe_BULGARIA_ROMANIA_UKRAINE_2013.pdf.

16. "Shale Gas Reserves and Major Fields of Ukraine," Unconventional Gas in Ukraine, June 14, 2013, http://shalegas.in.ua/en/shale-gas-resources-in-ukraine/.

17. John McGraw, "Ukraine: On the Road toward Energy Independence?," Center for Strategic and International Studies, June 21, 2013, http://csis.org/blog/ukraine-road-toward-energy-independence.

18. "Chevron vyrishila vyty z proekty po osvojeniu Oleskoi ploshi—dzeherelo," *Ukrainskaya Pravda,* December 15, 2014, http://www.epravda.com.ua/news/2014/12/15/512959/.

19. "Eni nachnet bureniye slantsevoi skvazhiny v Ukrainye v kontse goda—Minekologii," *RIA Novosti Ukraina,* January 21 2015, http://rian.com.ua/economy/20150121/362368605.html.

20. *Natural Gas and Ukraine's Energy Future,* special report, Ukraine Policy Dialogue, February 2012, http://s05.static-shell.com/content/dam/shell-new/local/country/zaf/downloads/pdf/research-reports/Ukraine-Policy-Dialogue-report.pdf.

21. "Shell likvidue dvi gazovi sverdlovyny na Kharkivshini," *Ukrainskaya Pravda,* March 12, 2015, http://www.pravda.com.ua/news/2015/03/12/7061235/; Roman Olearchyk, "Shell to Withdraw from Shale Gas Exploration in Eastern Ukraine," *Forbes,* June 11, 2015, http://www.ft.com/intl/cms/s/0/0c66011e-104a-11e5-bd70-00144feabdc0.html#axzz3iU41KbIb.

22. Unconventional Gas in Ukraine, "Gas Hydrates in Ukraine," May 2, 2014, http://shalegas.in.ua/en/gazogidraty-v-ukrayini/; "Ukraine to Tap on Black, Azov Sea Shelf." *Oil and Gas Journal,* November 27, 2000, http://www.ogj.com/articles/print/volume-98/issue-48/exploration-development/ukraine-to-tap-gas-on-black-azov-sea-shelf.html.

23. "Ukraine Stolen Gas Claim Raises Stakes in Dispute with Russia," Voice of America, May 27, 2014, http://www.voanews.com/content/ukraine-stolen-gas-claim-raises-stakes-in-dispute-with-russia-/1923315.html.

24. "Russia Illegally Explores Natural Resources in Ukraine's Maritime Economic Zone," *Ukraine Today,* July 17, 2016, http://uatoday.tv/business/russia-continues-to-explore-natural-resources-in-ukraine-s-marine-economic-zone-698306.html.

25. "Europe" here includes EU members, Turkey, Norway, Switzerland, and the Balkan states. Alexander Metelitsa, "16% of Natural Gas Consumed

in Europe Flows through Ukraine," EIA, March 14, 2014, http://www.eia.gov/todayinenergy/detail.cfm?id=15411.

26. Simon Pirani and Katja Yafimava, "Russian Gas Transit across Ukraine Post 2019: Pipeline Scenarios, Gas Line Consequences, and Regulatory Constraints," OIES Paper NG 105, Oxford Institute for Energy Studies, February 2016, https://www.oxfordenergy.org/wpcms/wp-content/uploads/2016/02/Russian-Gas-Transit-Across-Ukraine-Post-2019-NG-105.pdf.

27. Interview with Oksana Ishchuk, Kyiv, Ukraine, December 2015; Ralf Dickel, Elham Hassanzadeh, James Henderson, Anouk Honoré, Laura El-Katiri, Simon Pirani, Howard Rogers, Jonathan Stern, and Katja Yafimava, "Reducing European Dependence on Russian Gas: Distinguishing Natural Gas Security from Geopolitics," OIES Paper NG 92, Oxford Institute for Energy Studies, October 2014, 34, http://www.oxfordenergy.org/wpcms/wp-content/uploads/2014/10/NG-92.pdf.

28. Yafimava, *The Transit Dimension,* 5, 319.

29. Rawi Abdelal, "Razlichnoe ponimanie vzaimozavisimosti: Nacionalnaja bezopasnost' I torgovlia energoresursami mezhdu Rosiej, Ukrainoj I Belarusijej," in *Swords and Sustenance* (Cambridge, MA: MIT Press), 101; "European Union Natural Gas Import Price Chart," Y Charts, http://ycharts.com/indicators/europe_natural_gas_price/chart/#/?securities=include:true,id:I:EUNGIP,,&calcs=&correlations=&zoom=custom&startDate=09%2F13%2F1994&endDate=09%2F15%2F2015&format=real&recessions=false&chartView=&splitType=single&scaleType=linear&securitylistName=&securitylistSecurityId=&securityGroup.

30. "Gazovie Kompleksi Ukrainii i Possii: Ot vzaimovygodnogo sotrudnichestva: K strategicheskomu partnerstvu," *Segodnya,* July 7, 1999, http://www.segodnya.ua/oldarchive/8fe81c48cc069863c22567a60045ae2a.html; Randall Newnham, "Oil, Carrots, and Sticks: Russia's Energy Resources as a Policy Tool," *Journal of Eurasian Studies* 2, no. 2 (July 2011): 134–143, http://www.sciencedirect.com/science/article/pii/S187936651100011X.

31. "Gazprom zvynuvachuye Yushchenka v tomu, shcho vin nichogo ne zrozumiv," *Ukrainskaya Pravda,* December 20, 2005, http://www.pravda.com.ua/news/2005/12/20/3036556/.

32. Daniel Treisman, *The Return: Russia's Journey from Gorbachev to Medvedev* (New York: Free Press, 2011), 115.

33. "Gazprom Cuts Off Ukraine's Gas Supplies," CNN, January 1, 2009, http://www.cnn.com/2009/WORLD/europe/01/01/russia.ukraine.gas/index.html?eref=rss_latest.

34. Jonathan Stern, "The Russian-Ukrainian Gas Crisis of 2006," Oxford Institute for Energy Studies, January 16, 2006, http://www.oxfordenergy.org/wpcms/wp-content/uploads/2011/01/Jan2006-RussiaUkraineGasCrisis-JonathanStern.pdf; Balmaceda, *The Politics of Energy Dependency,* 127.

35. "Gazovaja promyshlenost," *Kommersant,* November 27, 2001, http://www.kommersant.ru/doc/295883.

36. Alla Eremenko, "Rynok Gaza Ukrainiy: V Litsah, tsifrah i Faktah," ZN, UA, May 6, 1996, http://gazeta.zn.ua/ECONOMICS/rynok_gaza_ukrainy_v_litsah,_tsifrah_i_faktah.html.

37. Andrew E. Kramer, "Gazprom Cuts Off Gas to Ukraine," *New York Times,* January 1, 2009, http://www.nytimes.com/2009/01/01/business/worldbusiness/01iht-gas.2.19026546.html; Natalia Grib and Oleg Gavrish, "Dolgie gazoprovody," *Kommersant,* January 15, 2009, http://www.kommersant.ru/doc/1102775.

38. Gazprom, "Gazprom 'na vneshnih rynkah,'" last modified 2015, http://www.gazpromquestions.ru/foreign-markets/#c503.

39. Simon Pirani, *Russian and CIS Gas Markets and Their Impact on Europe* (Oxford: Oxford Institute for Energy Studies, 2009), 8; Dickel et al., "Reducing European Dependence," 185.

40. Anders Aslund, *How Ukraine Became a Market Economy and Democracy* (Washington, DC: Peterson Institute for International Economics, 2009).

41. Balmaceda, *The Politics of Energy Dependency,* 134.

42. Edward C. Chow, "Ukraine and Russian Gas: The Never Ending Crisis," Center for Strategic and International Studies, December 19, 2013, http://csis.org/publication/ukraine-and-russian-gas-never-ending-crisis; Henderson and Pirani, "Reducing European Dependence," 187.

43. Margarita M. Balmaceda, *Energy Dependency, Politics and Corruption in the Former Soviet Union: Russia's Power, Oligarchs' Profits and Ukraine's Missing Energy Policy, 1995–2006* (New York: Routledge, 2007), 89.

44. Balmaceda, *The Politics of Energy Dependency,* 113.

45. Ibid., 151.

46. Aslund, *How Ukraine Became a Market Economy,* 107.

47. Chi-Kong Chyong, "The Role of Russian Gas in Ukraine," European Council on Foreign Relations, April 16, 2014, http://www.ecfr.eu/article

/commentary_the_role_of_russian_gas_in_ukraine248; Robert Legvold and Celeste A. Wallander, eds., *Swords and Sustenance* (Cambridge, MA: MIT Press, 2004), 93.

48. Aslund, *How Ukraine Became a Market Economy,* 170.

49. Alla Burakovskaya, "A Ukraine Mystery: Who Did Russian Gas Sales Benefit?," McClatchy DC, February 11, 2009, http://www.mcclatchydc.com/news/nation-world/world/article24524623.html; Stern, "The Russian-Ukrainian Gas Crisis of January 2006; Mikhail Korchemkin, "Rosukrenergo Wins, Gazprom and NAK Lose," Eastern European Gas Analysis, March 6, 2006, http://www.eegas.com/ukrtran6.htm.

50. Burakovskaya, "A Ukraine Mystery"; Mikhailo Gonchar, Alexander Duleba, and Oleksandr Malynovskyi, "Ukraine and Slovakia in a Post-Crisis Architecture of European Energy Security," Research Center of the Slovak Foreign Policy Association, 2011, http://gpf-europe.com/upload/iblock/70a/ua_sk_ch-book_eng.pdf.

51. Balmaceda, *The Politics of Energy Dependency.*

52. Yafimava, *The Transit Dimension of EU Energy Security*, 5, 319–320.

53. Michael R. Gordon, "Russia and Ukraine Finally Reach Accord on Black Sea Fleet," *New York Times*, May 29, 1997, http://www.nytimes.com/1997/05/29/world/russia-and-ukraine-finally-reach-accord-on-black-sea-fleet.html.

54. Agnia Grigas, *Beyond Crimea: The New Russian Empire* (New Haven, CT: Yale University Press, 2016).

55. "Soglashenie mezhdu Ukrainoj i Rossijskoj Federacijej po voprosam prebyvania Chernomoskogo flota Rossijskoj Federaciji na teritorii Ukrainy," *Verkhovna Rada,* April 27, 2010, http://zakon4.rada.gov.ua/laws/show/643_359; "Tymoshenko hoche noviy Maydan?," UNIAN, April 27, 2010, http://www.unian.ua/politics/352065-timoshenko-hoche-noviy-maydan.html.

56. "Ukraina ne sobiraetsia denonsirovat no odno iz soglashnij otnositelno CF RF," UNIAN, April 4, 2014, http://www.unian.net/politics/902933-ukraina-ne-sobiraetsya-denonsirovat-ni-odno-iz-soglasheniy-otnositelno-chf-rf.html.

57. "Russia Offers Ukraine Major Economic Assistance," BBC, December 17, 2013, http://www.bbc.com/news/world-europe-25411118.

58. "Gazovi peregovori z Roseiuy," Zaxid, http://zaxid.net/news/showList.do?gazov&tagId=51186.

59. "Ukrainian police detains Naftogaz Chairman of Board Yevhen Bakulin," TASS, March 21, 2014, http://tass.ru/en/world/724838; "Austria

Refuses US Bid to Extradite Dmytro Firtash," BBC, May 1, 2015, http://www.bbc.com/news/world-europe-32544699.

60. Richard Balmforth, "Ukrainian Oligarch under Fire after Night Raid on State Oil Firm," Reuters, March 20, 2015, http://www.reuters.com/article/2015/03/20/us-ukraine-crisis-kolomoisky-idUSKBN0MG2A320150320; Oleg Sukhov and Olena Goncharova, "Lawmakers Want Kolomoisky Fired after Outburst Involving His Attempts to Hold on to Oil Firm," *Kyiv Post,* March 20, 2015, http://www.kyivpost.com/content/ukraine/lawmakers-want-kolomoisky-fired-after-he-snaps-at-journalist-384042.html.

61. Inna Koval, "Kachestvo gaza za chto my platim—nyuansy naduvatelstva," *Argument,* April 6, 2015, http://argumentua.com/stati/kachestvo-gaza-za-chto-my-platim-nyuansy-naduvatelstva.

62. "Postanova Kabinetu Ministriv Ukrainy Pro vnesenia zmin do norm spozhyvania pryrodnogo gazy u razi vidsutnosti lichilnykiv," *Verkhovna Rada,* April 29, 2015, http://zakon0.rada.gov.ua/laws/show/237-2015-%D0%BF.

63. Margarita Balmaceda and Peter Rutland, "Ukraine's Gas Politics," openDemocracy, May 8, 2014, https://www.opendemocracy.net/od-russia/margarita-balmaceda-peter-rutland/ukraines-gas-politics; Pami Aalton, ed., *Russia's Energy Policies: National Interregional and Global Levels* (Surrey, UK: Edward Elgar, 2012), 140.

64. Edward Chow, "Edward Chow: Energy Reform—The Clock Has Struck Midnight," *Kyiv Post,* September 25, 2015, http://www.kyivpost.com/opinion/op-ed/edward-chow-energy-reform-the-clock-has-struck-midnight-398688.html.

65. Reza Moghadam, "Ukraine Unveils Reform Program with IMF Support," International Monetary Fund, April 30, 2014, http://www.imf.org/external/pubs/ft/survey/so/2014/NEW043014A.htm.

66. International Monetary Fund, "IMF Executive Board Completes First Review of Ukraine's EFF and Approves US$1.7 Billion Disbursement," July 31, 2015, https://www.imf.org/external/np/sec/pr/2015/pr15364.htm; European Commission, "Market Legislation," last modified January 31, 2016, https://ec.europa.eu/energy/en/topics/markets-and-consumers/market-legislation.

67. Alexander Bor, "World Bank Approves $500 Mil Loan Guarantees for Ukraine Gas Imports," S&P Global Platts, October 19, 2016, http://www.platts.com/latest-news/natural-gas/kiev/world-bank-approves-500-mil-loan-guarantees-for-26573770.

68. Roman Olearchyk, "Ukraine Opens Gas Sector to Investors with Anti-Monopoly Law," *Financial Times*, April 10, 2015, http://www.ft.com/intl/cms/s/0/db4e4afa-ded5-11e4-852b-00144feab7de.html#axzz3jLFbXfKy.

69. Wojciech Konończuk, "Why Ukraine Has to Reform Its Gas Sector," Centre for Eastern Studies (OSW) *Commentary*, September 2, 2015, http://www.osw.waw.pl/en/publikacje/osw-commentary/2015-09-02/reform-1-why-ukraine-has-to-reform-its-gas-sector; Janez Kopač, "Energy Com munity Country Brief: Spotlight on Ukraine," April 10, 2015, https://www.energy-community.org/portal/page/portal/ENC_HOME/DOCS/3672149/13AAC2CF97597F39E053C92FA8C0DA2E.PDF.

70. Janez Kopač "Energy Community Country Brief"; "Ukraine's Government Approves Naftogaz Unbundling Plan," Naftogaz, July 4, 2016, http://www.naftogaz.com/www/3/nakweben.nsf/0/471E4A2222A20B92C2257FE6003174D0?OpenDocument&year=2016&month=07&nt=News&.

71. "Pavlo Rozenko, "Subsidy Will Be Extended Automatically to the Next Heating Period," Ukrainian Government Portal, July 4, 2016, http://www.kmu.gov.ua/control/en/publish/article?art_id=248948340.

72. Naja Bentzen, "Ukraine's Economic Challenges: From Ailing to Failing?," European Parliament, June 2015, http://www.europarl.europa.eu/RegData/etudes/IDAN/2015/559497/EPRS_IDA(2015)559497_EN.pdf.

73. Chow, "Edward Chow."

74. "Russia Halts Gas Supplies to Ukraine after Talks Breakdown," BBC, July 1, 2015, http://www.bbc.com/news/world-europe-33341322.

75. Katya Golubovka, Denis Pinchuk, and Jan Lopatka, "Bypassing Ukraine Will be Costly for Russia's Gazprom—Analysts," Reuters, July 16, 2015, http://www.reuters.com/article/2015/07/16/ukraine-crisis-gazprom-idUSL5N0ZV1KM20150716; "Gazprom: S 2019 goda rossiyskiy gaz poydot v Evropu v obhod Ukrainiy," NTV, January 1, 2015, http://www.ntv.ru/novosti/1295722/.

76. Nana Chornaya, "Ukraine May Respond to the Hysteria of Gazprom by Refusing to Buy Russian Gas," UNIAN, June 2, 2015, http://www.unian.info/politics/1084631-ukraine-may-respond-to-the-hysteria-of-gazprom-by-refusing-to-buy-russian-gas.html.

77. Naftogaz, *Naftogaz Annual Report 2014*, September 29, 2015, 82, http://www.naftogaz.com/files/Zvity/Naftogaz_Annual_Report_2014_engl.pdf.

78. Anna Shiryaevskaya, "Ukraine to Seek European Gas Imports as Russia Offers Price Cut," Bloomberg, December 18, 2013, http://www.bloomberg.com/news/articles/2013-12-18/ukraine-to-seek-european-gas-imports-as-russia-offers-price-cut.

79. Carolina Novak, "The Implications of Russia's Use of Gas as a Political Weapon," Energy Policy Group, December 2014, http://enpg.ro/details-125-Rusia_rsquo_s_Use_of_Gas_as_a_Political_Weapon_in_the_Ukrainian_crisis_Carolina_Novac.html; "Ukraine Imports Record Volume of Reverse-Flow Gas from EU," UNIAN, January 30, 2015, http://www.unian.info/economics/1038322-ukraine-imports-record-volume-of-reverse-flow-gas-from-eu.html.

80. Naftogaz, "Ukraine Purchased 63% of Its Imported Gas in Europe in 2015"; "Romania, Bulgaria, Ukraine to Test Gas Flows by End of 2016," ICIS, July 20, 2016, http://www.icis.com/resources/news/2016/07/20/10018112/updated-romania-bulgaria-ukraine-to-test-gas-flows-by-end-of-2016/; "French Gas for Ukraine," UNIAN, November 2, 2016, https://www.kyivpost.com/ukraine-politics/unian-french-gas-ukraine.html.

81. Dag Mjaaland, "Facilitating Reverse Flow Capacity from West to East," Naftogaz, June 2, 2015, http://www.naftogaz.com/files/Information/2015-06-02%20-%20Facilitating%20Reverse%20Flow%20Capacity%20from%20West%20to%20East.pdf; Naftogaz, "Ukraine Purchased 63% of Its Imported Gas in Europe in 2015."

82. Svitlana Pyrkalo, "EBRD, Ukraine Agree Naftogaz Reform, Sign US $300 Million Loan for Winter Gas Purchases," European Bank for Reconstruction and Development (EBRD), October 23, 2015, http://www.ebrd.com/news/2015/ebrd-ukraine-agree-naftogaz-reform-sign-us-300-million-loan-for-winter-gas-purchases.html.

83. Pasquale de Micco, "A Cold Winter to Come? The EU Seeks Alternatives to Russian Gas," European Parliament, October 2014, http://www.europarl.europa.eu/RegData/etudes/STUD/2014/536413/EXPO_STU(2014)536413_EN.pdf.

84. "In Russia's Crosshairs, Ukraine Revives LNG Plan," OilPrice.com, July 1, 2014, http://oilprice.com/Energy/Energy-General/In-Russias-Crosshairs-Ukraine-Revives-LNG-Plans.html.

85. "Ukraine to Build LNG Plant with US Investment," *Ukraine Today,* July 17, 2015, http://uatoday.tv/news/ukraine-signs-memo-with-us-frontera-on-construction-of-lng-terminal-457763.html.

86. Pavlo Zagarodnuyk, "Pryrodniy Gaz v Ukraini: Tradycijni dzherela," Nadga Group, http://ua-energy.org/upload/files/Pavlo_Zagorodniuk _Presentation_Ua.pdf.

87. Elena Struk, "Vytik gazu," Kontrakty, March 2009, http://archive .kontrakty.ua/gc/2009/12/13-vitik-gazu.html?lang=ua.

88. "V vanco schitayut chto reshenie o razryive dogovora narushaet zakon," UNIAN, June 19, 2008, http://www.unian.net/society/124315-v -vanco-schitayut-chto-reshenie-o-razryive-dogovora-narushaet-zakon .html.

89. Alla Eremenko, "Renta zabuksovala," ZN, UA, July 3, 2015, http:// gazeta.zn.ua/energy_market/renta-zabuksovala-_.html; "Ukraine Cancels Gas Price Benefits for Households Starting May 1," *Ukraine Today,* May 2, 2016, http://uatoday.tv/news/ukraine-cancels-gas-price-benefits-for -households-as-of-may-1-641951.html.

90. Tatiana Manenok, "Belarus-Russia: Dependency or Addiction?," Heinrich-Böll-Stiftung Warsaw, January 2014, https://pl.boell.org/sites /default/files/downloads/Belarus_Russia_energy.pdf.

91. "Belarus Natural Gas Consumption," Y Charts, last modified July 8, 2015, https://ycharts.com/indicators/belarus_natural_gas _consumption.

92. EIA, "International Energy Statistics," last modified 2014, http://www.eia.gov/cfapps/ipdbproject/IEDIndex3.cfm?tid=79&pid =79&aid=2#.

93. World Bank, "Energy Imports, Net (%-Energy Use)," last modified 2015, http://data.worldbank.org/indicator/EG.IMP.CONS.ZS; Alena Rakava, "Energy Security of Belarus: Stereotypes, Threats, and Trends," Office for a Democratic Belarus, 2011, http://odb-office.eu/files/docs/Energy%20 Security%20of%20Belarus%20eng.pdf; "V minuvshem rodu Belarus' snizila import rossiiskogo gaza pochti na 700 mil. dollarov," *Neft Rossii,* March 1, 2016, http://www.oilru.com/news/504120/.

94. International Energy Agency, "Belarus: Indicators for 2012," updated 2015, http://www.iea.org/statistics/statisticssearch/report/?year=2012&country =Belarus&product=Indicators; "Natural Gas—Proved Resources (Cubic Meters)," *Index Mundi,* June 30, 2015, http://www.indexmundi.com/g/g.aspx ?c=bo&v=98.

95. Belorusneft, "Exploration," last modified 2013, http://www .belorusneft.by/sitebeloil/en/center/exploration/.

96. Mikhail Korchemkin, "Major Gas Pipelines of Belarus and Comments to Russia-Belarus Gas Dispute," East European Gas Analysis, December 30, 2006, http://www.eegas.com/belarus1.htm.

97. EIA, "Russia," July 28, 2015, http://www.eia.gov/beta/international/analysis.cfm?iso=RUS; Gazprom, "Yamal-Europe," http://www.gazprom.com/about/production/projects/pipelines/yamal-evropa/.

98. Gazprom, "Underground Storage," n.d., http://www.gazprom.com/about/production/transportation/underground-storage/.

99. Margarita Balmaceda, *Living the High Life in Minsk: Russian Energy Rents, Domestic Populism and Belarus' Impending Crisis* (Budapest: Central European University Press, 2014), 39; Gazprom Transgaz Belarus, "Tranzit rossi'skogo gaza cherez Belarus' v 2014 rodu sostavit 45.4 mlrd kubometrov," November 19, 2014, http://belarus-tr.gazprom.ru/press/about-company/2014/11/44/.

100. Balmaceda, *The Politics of Energy Dependency,* 38.

101. "Belarus-Russia Energy Disputes: Political and Economic Comparative Analysis," *Pecob's Energy Policy Studies,* January 2012, http://www.pecob.eu/flex/cm/pages/ServeAttachment.php/L/EN/D/4%252Ff%252Fd%252FD.3945ee2563c52557b221/P/BLOB%3AID%3D3450.

102. Balmaceda, *Living the High Life in Minsk,* 43.

103. "Russia Oil Row Hits Europe Supply," BBC, January 8, 2007, http://news.bbc.co.uk/2/hi/business/6240473.stm.

104. "Russia-Belarus Gas Deal Reached," BBC, January 31, 2007, http://news.bbc.co.uk/2/hi/europe/6221835.stm.

105. "Belarus Buys One Week's Respite in Russian Gas Row," Reuters, August 3, 2007, http://www.reuters.com/article/2007/08/03/gazprom-belarus-settlement-idUSL0347959120070803.

106. Balmaceda, *The Politics of Energy Dependency,* 188; Balmaceda, *Living the High Life in Minsk,* 43; "Russia, Belarus: The Kremlin Turns the Screws," Stratfor, July 31, 2007, https://www.stratfor.com/analysis/russia-belarus-kremlin-turns-screws.

107. "Factbox: Russia's Energy Disputes with Ukraine and Belarus," Reuters, December 21, 2012, http://www.reuters.com/article/2012/12/21/us-russia-gas-disputes-idUSBRE8BK11T20121221; "Lukashenko Said Medvedev Humiliated Belarus with Sausages, Butter, and Pancakes," UNIAN, June 23, 2010, http://www.unian.info/world/372146-lukashenko-said-medvedev-humiliated-belarus-with-sausages-butter-and-pancakes.html.

108. "Belarus Clears Gas Debt with Russia," BBC, June 23, 2010, http://www.bbc.com/news/10395370.

109. Andrew Kramer, "Gas Deal with Belarus Gives Control of Pipeline to Russia," *New York Times,* November 25, 2011, http://www.nytimes.com/2011/11/26/world/europe/in-deal-with-belarus-russia-gets-control-of-yamal-europe-pipeline.html?_r=0; Alan Cullison, "Russia Tightens Its Grip in Belarus," *Wall Street Journal,* November 26, 2011, http://www.wsj.com/articles/SB10001424052970203764804577060312640356138.

110. Interview with Tatiana Manenok, Minsk, Belarus, January, 4, 2016.

111. Jeff Mower, "Twenty Four Months Later, a New High In Oil Prices," S&P Global Platts, November 8, 2010, http://blogs.platts.com/2010/11/08/twenty_four_mon/.

112. Alex Juric, "Tatyana Manenok Razryv zavisimosti belarusi ot rossiyskih energoresurov iz oblasti utopii," January 8, 2014, http://eurobelarus.info/news/economy/2014/08/01/tat-yana-manenok-razryv-zavisimosti-belarusi-ot-rossiyskih-energoresursov-iz-oblasti-utopii.html.

113. Manenok, "Belarus-Russia."

114. Balmaceda, *Living the High Life in Minsk,* 135.

115. Zuzanna Nowak, Jarosław Ćwiek-Karpowicz, and Jakub Godzimirski, "The Power to Influence Europe? Russia's Grand Gas Strategy," Polish Institute of International Affairs (PISM), March 2015, https://www.pism.pl/files/?id_plik=19367.

116. Balmaceda, *The Politics of Energy Dependency,* 206.

117. Legvold and Wallander, *Swords and Sustenance,* 103; Simon Pirani and Katja Yafimava, "CIS Gas Markets and Transit," in *The Russian Gas Matrix: How Markets Are Driving Change,* ed. James Henderson and Simon Pirani (Oxford: Oxford Institute for Energy Studies, 2014), 201.

118. Lucy Zubrute, "Kijevas praso baltarusijos uztikrinti duju tranzita is lietuvos," Verslo Zinios, July 24, 2015, http://vz.lt/sektoriai/energetika/2015/07/24/kijevas-praso-baltarusijos-uztikrinti-duju-tranzita-is-lietuvos#ixzz3iXrULHNG.

119. "Lithuania's Capacity to Supply Gas to Belarus Limited, Former Energy Minister Says," Delfi, November 5, 2014, http://en.delfi.lt/lithuania/energy/lithuanias-capacity-to-supply-gas-to-belarus-limited-former-energy-minister-says.d?id=66306276.

120. "Lukashenko: Belarus Is Ready to Offer a Larger Gas Transit Capacity to Gazprom," *Belarus News,* August 21, 2015, http://eng.belta.by

/president/view/lukashenko-belarus-is-ready-to-offer-a-larger-gas-transit-capacity-to-gazprom-84287-2015.

121. Aliaksandr Kudrytski and Elena Mazneva, "Gazprom to Build Tallest Tower in Belarus as Ukraine Ties Fray," Bloomberg, August 21, 2015, http://www.bloomberg.com/news/articles/2015-08-21/gazprom-to-build-tallest-tower-in-belarus-as-ukrainian-ties-fray.

122. "Belarus President Lukashenko in His Own Words," Reuters, November 27, 2012, http://www.reuters.com/article/2012/11/27/us-belarus-lukashenko-extracts-idUSBRE8AQ0V520121127.

123. EIA, "Russia."

124. Gazprom, "Rossiya I belarussiya opredelili noviye usloviya postavki I transportirovki gaza. 'Beltransgaz' perexodit k Gazpromu," November 25, 2011, http://www.gazprom.ru/press/news/2011/november/article124284/.

125. Balmaceda, *Living the High Life in Minsk,* 175.

126. Matthew Rojansky, Balázs Jarábik, Tatiana Kouzina, Sergej Satsuk, Olga Stuzhinskaya, and Rodger Potock, "Belarus: Stable Instability?," Carnegie Endowment for International Peace, June 18, 2012, http://carnegieendowment.org/2012/06/18/belarus-stable-instability/b0wi.

127. Alexei Pikulik and Dzianis Meyantsou, "Belarus," Freedom House, 2012, https://freedomhouse.org/report/nations-transit/2012/belarus; Balmaceda, *Living the High Life in Minsk,* 55.

128. Balmaceda, *The Politics of Energy Dependency,* 195.

129. Ibid., 162; Balmaceda, *Living the High Life in Minsk,* 183.

6. The Politics of Isolated Suppliers

1. Sabit Bagirov, "Azerbaijan Oil: Glimpses of a Long History," January 1, 2012, http://sam.gov.tr/wp-content/uploads/2012/01/1.-AZERBAIJANI-OIL-GLIMPSES-OF-A-LONG-HISTORY.pdf.

2. "The History of Oil in Azerbaijan," *Azerbaijan International,* Summer 1994, http://azer.com/aiweb/categories/magazine/22_folder/22_articles/22_historyofoil.html; Aitor Ciarreta and Shahriyar Nasirov, "Analysis of Azerbaijan Oil and Gas Sector," United States Association for Energy Economics, 2011, 2, http://www.usaee.org/usaee2011/submissions/OnlineProceedings/Ciarreta_Nasirov-Article1.pdf.

3. Pinar Batur-VanderLippe and Stephen Simmons, "Oil and Regional Relations in the Caucasus and Central Asia in the Post-Soviet Period," in *Oil in the New World Order*, ed. Kate Gillespie and Clement M. Henry (Gainesville: University Press of Florida, 1995), 164.

4. The companies included Amoco, BP, McDermott, Unocal, SOCAR, Lukoil, Statoil, Exxon, Turkish Petrolium, Penzoil, Itochu, Remko, and Delta. "Contract of the Century," Azerbaijan, n.d., http://www.azerbaijans.com/content_775_en.html.

5. Interview with Richard Pomfret, January 2016.

6. "Azerbaijan: Production-Sharing Agreement," June 2002, http://www.angelfire.com/dragon/asif/neft.htm; Ciarreta and Nasirov, "Analysis of Azerbaijan Oil and Gas Sector," 12.

7. "Shah Deniz Stage 1," BP, n.d., http://www.bp.com/en_az/caspian/operationsprojects/Shahdeniz/SDstage1.html.

8. "2015 Year-End Results," BP, February 26, 2016, http://www.bp.com/en_az/caspian/press/businessupdates/2015-year-end-results.html.

9. "Shah Deniz—Project Timeline," BP, n.d., http://www.bp.com/en_az/caspian/operationsprojects/Shahdeniz/projecthistory.html; "Shaz Deniz, Azerbaijan," offshore-technology.com, n.d., http://www.offshore-technology.com/projects/shah_deniz/"Shah Deniz Stage 2," BP, n.d., http://www.bp.com/en_az/caspian/operationsprojects/Shahdeniz/SDstage2.html.

10. "Trans Anatolian Natural Gas Pipeline project (TANAP), Turkey," hyrdrocarbons-technology.com, n.d., http://www.hydrocarbons-technology.com/projects/trans-anatolian-natural-gas-pipeline-project-tanap/.

11. "South Caucasus Pipeline," BP, n.d., http://www.bp.com/en_az/caspian/operationsprojects/pipelines/SCP.html.

12. US Energy Information Administration (henceforth EIA), "Azerbaijan: International Energy Data and Analysis," June 24, 2016, http://www.eia.gov/beta/international/analysis.cfm?iso=AZE.

13. Isabel Gorst, "Construction of TANAP Pipeline Begins in Turkey as EU and Russia Spar for Upper Hand," *Financial Times*, March 3, 2015, http://blogs.ft.com/beyond-brics/2015/03/18/construction-of-tanap-pipeline-begins-in-turkey-as-eu-and-russia-spar-for-upper-hand/.

14. "TANAP Construction to Commence in April," Natural Gas World, March 16, 2015, http://www.naturalgasworld.com/tanap-construction-to-commence-in-april-22735.

15. "TAP's Shareholders," Trans-Adriatic Pipeline, March 7, 2015, http://www.tap-ag.com/about-us/our-shareholders.

16. "TAP Begins Construction of Access Roads and Bridges in Albania," Trans-Adriatic Pipeline, March 7, 2015, http://www.tap-ag.com/news-and-events/2015/07/03/tap-begins-construction-of-access-roads-and-bridges-in-albania.

17. "TAP Project Milestones," Trans-Adriatic Pipeline, n.d., http://www.tap-ag.com/the-pipeline/project-timeline/tap-project-milestones.

18. Vladimir Socor, "SCP, TANAP, TAP: Segments of the Southern Gas Corridor to Europe," *Eurasia Daily Monitor* 11, no. 8 (January 15, 2014), http://www.jamestown.org/single/?tx_ttnews%5Bany_of_the_words%5D=Statoil&tx_ttnews%5Btt_news%5D=41821&tx_ttnews%5BbackPid%5D=7#.VXiEXFzq9Ro.

19. BP, "BP Statistical Review of World Energy June 2015," n.d., 23, 28, http://www.bp.com/content/dam/bp/pdf/energy-economics/statistical-review-2015/bp-statistical-review-of-world-energy-2015-full-report.pdf.

20. EIA, "Azerbaijan."

21. "Azerbaijan May Develop Its Shale Deposits—SOCAR Says," Shale Gas International, January 1, 2015, http://www.shalegas.international/2015/05/01/azerbaijan-may-develop-its-shale-deposits-socar-says/.

22. "Gazprom Resumes Gas Deliveries to Georgia and Armenia," Gazprom, January 29, 2006, http://www.gazprom.com/press/news/2006/january/article88122/.

23. Elizabeth Stonor, "Georgian North Caucasus—Transcaucasus Pipeline Restored after Explosion," Independent Chemical Information Service, January 30, 2006, http://www.icis.com/resources/news/2006/01/30/9285095/georgian-north-caucasus-transcaucasus-pipeline-restored-after-explosion/.

24. C.J. Chivers, "Explosions in Southern Russia Sever Gas Lines to Georgia," *New York Times*, January 23, 2006, http://www.nytimes.com/2006/01/23/international/europe/23georgia.html.

25. Ian Jeffries, *Economic Developments in Contemporary Russia* (New York: Routledge, 2011), 162.

26. "Iranuli Gazi Sakartveloshi Shemovida," March 1, 2006, http://www.civil.ge/geo/article.php?id=11592.

27. "South Caucasus Pipeline Project."

28. Georgian Ministry of Energy, official data request, November 2015. See also "Gazpromi da Sokari Sakartvelostvis oms itskeben—Azerbaijanuli media," NewPosts.ge, October 9, 2015, http://newposts.ge/?l=G&id=88795.

29. "Sakartveloshi 'gazpromis' shemosvlis ekonomikuri safrtkheebi," *Radiotavisupleba*, October 17, 2015, http://www.tavisupleba.mobi/a/gazpromi-da-riskebi/27311581.html.

30. Sakartvelos tkmit, azerbaijanidan damatebit gazs miigebs da meti moculobis gazi 'gazpromidan' aghar daschirdeba," Civil.ge, March 4, 2016, http://www.civil.ge/geo/article.php?id=30167.

31. Ariel Cohen, "Developing a Western Energy Strategy for the Baltic Sea Region and Beyond," Atlantic Council, November 10, 2015, http://www.atlanticcouncil.org/images/publications/Black_Sea_Energy.pdf.

32. "Kaladze 'fronteras' sakhelmtsipos tsinashe angarishvaldebulebis tavidan aridebis mcdelobashi adanashaulebs," *PolitCommersant*, February 16, 2016, http://politcommersant.ge/detalpage?newsid=4174.

33. "Remittance Man," *Economist*, September 7, 2013, http://www.economist.com/news/asia/21584999-russia-attempts-draw-tajikistan-and-kyrgyzstan-back-its-orbit-remittance-man.

34. Alexander Cooley, *Great Games, Local Rules: The New Great Power Contest in Central Asia* (Oxford: Oxford University Press, 2012), 160.

35. Sebastien Peyrouse, *Turkmenistan: Strategies of Power, Dilemmas of Development* (Armonk, NY: M. E. Sharpe, 2012), 170, 222.

36. Luca Anceschi, *Turkmenistan's Foreign Policy: Positive Neutrality and the Consolidation of the Turkmen Regime* (London: Routledge, 2009), 155.

37. Joanna Lillis, "Kazakhstan, Turkmenistan and Iran Launch Railroad to Get Trade on Track," EurasiaNet.org, December 3, 2014, http://www.eurasianet.org/node/71166.

38. Cooley, *Great Games, Local Rules,* 204.

39. Eni, "World Oil and Gas Review for 2014," September 2014, https://www.eni.com/world-oil-gas-review-2014/sfogliabile/O-G-2014.pdf; BP, "BP Statistical Review of World Energy June 2015."

40. "Economy," Ministry of Foreign Affairs of Turkmenistan, n.d., http://www.mfa.gov.tm/en/tukrmenistan/economy.

41. The Eni's World Oil and Gas Review, however, places Russia at the top of gas rich countries in the world with an amount of 49.5 TCM. BP, "BP Statistical Review of World Energy June 2015."

42. Abdelghani Henni, "Gas for Cash: The Future of Turkmenistan," Society of Petroleum Engineers, November 11, 2014, http://www.spe.org/news/article/Turkmenistan-Gas-for-Cash.

43. "Centralnaya Aziya: Geopolitika i Ekonomika Regiona," Institut Strategichekyh Otsenok i Analiza, Moscow, 2010, 50, http://www.isoa.ru/docs/central_asia-book.pdf.

44. Nazar Alaolmolki, *Life after the Soviet Union: The Newly Independent Republics of the Transcaucasus and Central Asia* (Albany: State University of New York Press, 2001), 102.

45. Richard Pomfret, *The Central Asian Economies since Independence* (Princeton, NJ: Princeton University Press, 2006), 13.

46. "Galkynysh Gas Field, Turkmenistan," hydrocarbons-technology.com, n.d., http://www.hydrocarbons-technology.com/projects/-galkynysh-gas-field-turkmenistan/; Karine M. Renaud, "2013 Minerals Yearbook: Turkmenistan," United States Geology Survey, May 2015, http://minerals.usgs.gov/minerals/pubs/country/2013/myb3-2013-tx.pdf; Vladimir Komarov, "Galkynysh Field: Stage of Development," *Turkmenistan: Ihe Golden Age Online Newspaper,* March 5, 2014, http://www.turkmenistan.gov.tm/_eng/?id=3297.

47. BP, "BP Statistical Review of World Energy June 2015"; "Potencial Turkmenistana po narashivaniyu eksporta prirodnogo gaza," Trend News Agency, February 8, 2015, http://www.trend.az/casia/turkmenistan/2361962.html.

48. Ibid.

49. "Obespechenie globalnoy energobezopasnosti—factor ustoychivogo razvitiya," Asgabat.net, May 30, 2015, http://asgabat.net/stati/yekonomika/obespechenie-globalnoi-yenergobezopasnosti-faktor-ustoichivogo-razvitija.html.

50. Henni, "Gas for Cash."

51. "European Gas Demand Fell 11% on Year in 2014 to 409 BCM: Eurogas," S&P Global Platts, March 25, 2015, http://www.platts.com/latest-news/natural-gas/london/european-gas-demand-fell-11-on-year-in-2014-to-26049067; Eurogas, "Gas Demand in EU Rises for the First Time in Four Years, According to New Eurogas Data," March, 30, 2016, http://www.eurogas.org/uploads/media/Eurogas_Press_Release_-_Gas_demand_in_EU_rises_for_the_first_time_in_four_years_according_to_new_Eurogas_data_01.pdf.

52. International Energy Agency (henceforth IEA), "Statistics: Balances for 2012 in Thousand Tons of Oil Equivalent (Ktoe) on a Net Calorific Value Basis," n.d., http://www.iea.org/statistics/statisticssearch/report/?country=KAZAKHstAN&product=balances&year=2012.

53. "GDP per Capita 2013," World Bank, n.d., http://www.data.worldbank.org.

54. "GDP per Capita Ranking 2015: Data and Charts," Knoema, January 11, 2016, http://knoema.com/sijweyg/gdp-per-capita-ranking-2015-data-and-charts.

55. Interview with Nygmet Ibadildin, Kazakhstan, January 2016; Edward Schatz and Elena Maltseva, "Kazakhstan's Authoritarian 'Persuasion,'" *Post-Soviet Affairs* 28, no. 1 (2012): 45–65.

56. "Nazarbayev: "Economy First, then Politics," *Euronews,* October 15, 2010, http://www.euronews.com/2010/01/15/nazarbayev-economy-first-then-politics/.

57. Bruce Pannier, "A New Wave of Ethnic Russians Leaving Kazakhstan," *Radio Free Europe / Radio Liberty,* February 9, 2016, http://www.rferl.org/a/qishloq-ovozi-kazakhstan-ethnic-russians-leaving/27541817.html.

58. EIA, "World Shale Resource Assessments," September 2015, http://www.eia.gov/analysis/studies/worldshalegas/; BP, "BP Statistical Review of World Energy June 2015"; Eni, "World Oil and Gas Review for 2014"; Ministry of Environmental Protection of the Republic of Kazakhstan, *Sostavlenie ekologo-energeticheskix reytingov predpriyatiy Kazaxstana na 2013,* no. 07-1-223 (Astana, n.d.), 5.

59. Kulpash Konirova, "Kaspiy stanet odnim iz klyuchevix regionov mirovogo neftyanogo rinka," *Kazenergy* 4, no. 59 (2013), http://kazenergy.com/en/4-59-2013/11549-2013-10-30-08-47-58.html.

60. Asian Development Bank, *Energy Outlook for Asia and the Pacific* (Mandaluyong City: Asian Development Bank, 2013).

61. The numbers are estimated based on the map of distribution of European population in Kazakhstan in 2012. "Zhdut li Russkie v Kazakhstane 'vezhlivyh' ludei," *ArgumentUA,* October 23, 2014, http://argumentua.com/stati/zhdut-li-russkie-v-kazakhstane-vezhlivykh-lyude; Agnia Grigas, *Beyond Crimea: The New Russian Empire* (New Haven, CT: Yale University Press, 2016).

62. Shamil Yenikeyeff, "Kazakhstan's Gas: Export Markets and Export Routes," Oxford Institute for Energy Studies, November 2008, 3, http://www

.oxfordenergy.org/wpcms/wp-content/uploads/2010/11/NG25-Kazakhstans gasExportMarketsandExportRoutes-ShamilYenikeyeff-2008.pdf.

63. Elena Kosolapova, "Kazakhstan Increases Gas Production," Trend News Agency, January 19, 2016, http://en.trend.az/casia/kazakhstan /2482426.html; Government of the Republic of Kazakhstan, *Koncepciya razvitiya gazovogo sektora Respubliki Kazaxstan to 2030 goda,* online.zakon. kz, December 5, 2014, http://online.zakon.kz/Document/?doc_id=31641775; BP, "BP Statistical Review of World Energy June 2015."

64. Simon Pirani, "Central Asian and Caspian Gas Production and the Constraints on Export," OIES Paper NG 69, Oxford Institute for Energy Studies, 2012, 45, http://www.oxfordenergy.org/wpcms/wp-content/uploads /2012/12/NG_69.pdf.

65. Nygmet Ibadildin, "Role of the Old and New Institutional Framework in Combating the Resource Curse in Kazakhstan" (PhD dissertation, University of Tampere, April 29, 2011), https://tampub.uta.fi/bitstream /handle/10024/66726/978-951-44-8391-2.pdf?sequence=1.

66. James Perkins, "Kazakhstan Plans to Develop Shale Programme," *Shale Energy Insider,* December 1, 2014, http://www.shaleenergyinsider.com /2014/12/01/kazakhstan-will-develop-shale-programme/.

67. A. A. Bulekbay, "Unconventional Tight, Coal Bed and Shale Gas: Perspectives in Kazakhstan?," *Kazenergy* 56, no. 1 (2013): 50–59.

68. Karim Massimov, "Shale Gas to Be Exported—Karabalin," The Prime Minister of Kazakhstan, December 3, 2014, http://www.primeminister.kz /news/show/24/slantsevyj-gaz-v-rk-mozhet-byt-napravlen-na-eksport -ukarabalin-/03-12-2014?lang=en; "Shale Gas in Kazakhstan Can Go for Export," *Caspian Energy News,* December 4, 2014, http://www.caspianenergy .net/en/oil-and-gas/17151-shale-gas-in-kazakhstan-can-go-for-export.

69. Marlene Laruelle and Sebastien Peyrouse, *Globalizing Central Asia: Geopolitics and the Challenges of Economic Development* (Armonk, NY: M. E. Sharpe, 2013), 128–130.

70. Erica Marat, "Uzbekistan after Karimov: Its Grim Prospects," *Foreign Affairs,* September 8, 2016, https://www.foreignaffairs.com/articles /2016-09-07/uzbekistan-after-karimov.

71. BP, "BP Statistical Review of World Energy June 2015"; Eni, "World Oil and Gas Review for 2014."

72. "Osnovnie pokazateli," Uzbekneftegaz, n.d., http://www.ung.uz/ru /business/indicators.

73. Pirani, "Central Asian and Caspian Gas Production," 35.

74. "Perviy gaz Uzbekistana," Uzbekneftegaz, n.d., http://www.ung.uz/press_center/smi/pervyj_gaz_uzbekistana.

75. "Oil and Gas Sector," Uzinfoinvest, n.d., http://www.uzinfoinvest.uz/eng/investment_opportunities/by_industry/oil_and_gas_sector/.

76. BP, "BP Statistical Review of World Energy June 2015."

77. IEA, "Statistics."

78. "Uzbekistan: The Economics of Efficiency," World Bank, April 30, 2013, http://www.worldbank.org/en/results/2013/04/30/uzbekistan-the-economics-of-efficiency.

79. "Uzbekistan Oil and Gas Report Includes 10-Year Forecasts to 2023," *Business Monitor International,* November 2014, 17.

80. James Fishelson, "From the Silk Road to Chevron: The Geopolitics of Oil Pipeline in Central Asia," *Journal of Russian and Asian Studies,* December 12, 2012, http://www.sras.org/geopolitics_of_oil_pipelines_in_central_asia.

81. "Uzbekistan's First Gas; EIA, Kazakhstan: Overview," Uzbekneftegaz, January 14, 2015, http://www.eia.gov/beta/international/analysis.cfm?iso=KAZ; Aleksandra Jarosiewicz, "Will Tajik Gas Change the Balance of Power in Central Asia?," *Ośrodek Studiów Wschodnich,* June 26, 2013, http://www.osw.waw.pl/en/publikacje/analyses/2013-06-26/will-tajik-gas-change-balance-power-central-asia.

82. "International Cooperation in the Turkmenistan's Oil and Gas Sector: Achievements and Prospects," State News Agency, November 20, 2014, http://tdh.gov.tm/en/official-news-2/12015-2014-11-20-16-17-00.

83. Natalia Konarzewska, "Turkmenistan Advances Westward Natural Gas Export," *Central Asia-Caucasus Analyst,* March 26, 2016, http://www.cacianalyst.org/publications/analytical-articles/item/13345-turkmenistan-advances-westward-natural-gas-export.html.

84. Sabina Amangeldi, "KazTransGas Secures $1.8bn from China to Build a Gas Pipeline," *Halykfinance,* December 14, 2012, http://www.halykfinance.kz/en/site/index/research/news:81750.

85. Martha Brill Olcott, "Central Asia's Catapult to Independence," *Foreign Affairs* 71, no. 3 (1992): 118–128, http://www.foreignaffairs.com/articles/47979/martha-brill-olcott/central-asias-catapult-to-independence.

86. Jeffrey Mankoff, *Russian Foreign Policy: The Return of Great Power Politics,* 2nd ed. (Lanham, MD: Rowman & Littlefield, 2012), 245.

87. EIA, "International Energy Statistics—Turkmenistan," http://www.eia.gov/cfapps/ipdbproject/iedindex3.cfm?tid=3&pid=26&aid=1&cid=TX,&syid=1985&eyid=2013&unit=BCF.

88. Alaolmolki, *Life after the Soviet Union*, 102.

89. Vladimir Paramonov, Oleg Stolpovsky, and Alexey Strokov, "Russia and Uzbekistan: Oil and Gas Cooperation," *International Affairs*, July 20, 2010, http://en.interaffairs.ru/events/249-russia-and-uzbekistan-oil-and-gas-cooperation.html.

90. EIA, "International Energy Statistics—Turkmenistan."

91. Marat Gurt, "Turkmenistan Stifles U.S. Gas Plans," *Moscow Times*, June 3, 2000, http://www.themoscowtimes.com/news/article/turkmenistan-stifles-us-gas-plans/262464.html.

92. Global Witness, "It's a Gas: Funny Business in the Turkmen-Ukrainian Gas Trade," April 2006, 26, https://www.globalwitness.org/sites/default/files/pdfs/its_a_gas_april_2006_lowres.pdf.

93. "Turkmenistan Boosting Gas Exports to China," Energy Global, February 11, 2015, http://www.energyglobal.com/downstream/gas-processing/11022015/Turkmenistan-boosting-gas-exports-to-China-230/; Imogen Bell, ed., *Eastern Europe, Russia and Central Asia 2003* (London: Europa, 2002), 454.

94. Ariel Cohen, "The Putin-Turkmenbashi Deal of the Century: Towards a Eurasian Gas OPEC?," *Central Asia–Caucasus Analyst*, May 7, 2003.

95. IEA, "Optimizing Russian Natural Gas. Reform and Climate Policy," 2006, http://www.iea.org/publications/freepublications/publication/russiangas2006.pdf.

96. Vladimir Socor, "Russia Resuming Gas Imports from Turkmenistan on a Small Scale," *Eurasia Daily Monitor* 7, no. 1 (January 2010), http://www.jamestown.org/single/?tx_ttnews%5Btt_news%5D=35866&no_cache=1#.VaVAecZViko; Sergey S. Zhiltsov, "Russia's Policy toward the Pipeline Transport in the Caspian Region: Results and Prospects," in *Oil and Gas Pipelines in the Black–Caspian Seas Region*, ed. Sergey S. Zhiltsov (New York: Springer Berlin Heidelberg, 2016), 92.

97. Tim Gould, Isabel Murray, Jonathan Sinton, Dagmar Graczyk, Christopher Segar, and Audun Wiig, "Perspectives on Caspian Oil and Gas Development," IEA, 2008, 11, http://www.asiacentral.es/docs/caspian_perspectives_iea_dec08.pdf.

98. Gawdat Bahgat, "Europe's Energy Security: Challenges and Opportunities," *International Affairs* 82, no. 5 (2006): 961.

99. Creelea Henderson, "Shifting Sands in Central Asia: Geopolitics of Natural Gas Flows," *Perspective* 20, no. 2 (February 2010), http://www.bu.edu/iscip/Vol20/henderson.html.

100. Howard V. Rogers, "The Impact of Lower Gas and Oil Prices on Global Gas and LNG Markets," OIES Paper NG 99, Oxford Institute for Energy Studies, July 2015, 3, http://www.oxfordenergy.org/wpcms/wp-content/uploads/2015/07/NG-99.pdf; Vadim Pokhlebkin, "Why Are Crude Oil Prices Dropping? Not Why Many People Think," Elliotwave International, April 16, 2013, https://www.elliottwave.com/freeupdates/archives/2013/04/16/Why-Are-Crude-Oil-Prices-Dropping-Not-Why-Many-People-Think.aspx.

101. "Central Asia Considers Alternatives to Gazprom for Exporting Gas: Will Silk Road Producers Actually Prefer Europe to Russia?," *Central Asia Online,* November 18, 2011, http://centralasiaonline.com/en_GB/articles/caii/features/main/2011/11/18/feature-01?change_locale=true.

102. Boris Barkanov, "The Geo-Economic of Eurasian Gas: The Evolution of Russia-Turkmen Relations in Natural Gas (1992–2010)," in *Export Pipelines from the CIS Region,* ed. Andreas Heinrich and Heiko Pleines (Stuttgart: Ibidem, 2014), 165–167.

103. "Turkmens Blame Gazprom for Pipeline Blast," *Kyiv Post,* April 10, 2009, https://www.kyivpost.com/article/content/world/turkmens-blame-gazprom-for-pipeline-blast-39377.html.

104. "Gas Purchases," Gazprom, n.d., http://www.gazprom.com/about/production/central-asia/; Sevinj Mamadova, "Changing Market Dynamics in Central Asia: Declining Russian Interests and Emerging Chinese Presence—Analysis," Caspian Center for Energy and Environment, March 6, 2015, http://www.eurasiareview.com/06032015-changing-market-dynamics-in-central-asia-declining-russian-interests-and-emerging-chinese-presence-analysis/.

105. Ilya Kuvakin, "Gazprom to Suspend Gas Supplies from Turkmenistan," RosBusinessKonsalting, January 4, 2016, http://www.rbc.ru/business/04/01/2016/568a60d89a79477ae3bf88fc; "Turkmenistan Says Russia's Gazprom Has Not Paid for Any Gas This Year," Reuters, July 8, 2015, http://uk.reuters.com/article/2015/07/08/gas-turkmenistan-gazprom-idUKL8N0ZO30Q20150708; Catherine Putz, "Russia Takes Turkmenistan to Court over the Price of Gas,"

Diplomat, July 28, 2015, http://thediplomat.com/2015/07/russia-takes-turkmenistan-to-court-over-the-price-of-gas/.

106. Vladimir Milov, "Ups and Downs of the Russia-Turkmenistan Relations," in *Russian Energy Security and Foreign Policy,* ed. Adrian Dellecker and Thomas Gomart (London: Routledge, 2011), 100.

107. Stephen Blank, "The Strategic Implications of the Turkmenistan-China Pipeline Project," *China Brief* 10, no. 3 (February 4, 2010): 10, http://www.jamestown.org/uploads/media/cb_010_14.pdf.

108. Alexander Vershinin, "China Loans Turkmens $4bln in Exchange for Gas," *Bloomberg Businessweek,* April 26, 2011, http://www.businessweek.com/ap/financialnews/D9MRG3CO0.htm.

109. Payrav Chorshanbiyev, "Tajikistan Seeks Consultant for Construction of Its Section of Turkmen-Chinese Gas Pipeline," *Asia Plus,* February 24, 2015, http://news.tj/en/news/tajikistan-seeks-consultant-construction-its-section-turkmen-chinese-gas-pipeline; China National Petroleum Corporation, "Flow of Natural Gas from Central Asia," n.d., http://www.cnpc.com.cn/en/FlowofnaturalgasfromCentralAsia/FlowofnaturalgasfromCentralAsia2.shtml.

110. Andrew C. Kuchins, Jeffrey Mankoff, and Oliver Backes, "Central Asia in a Reconnecting Eurasia: Turkmenistan's Evolving Foreign, Economic and Security Interests" (Washington, DC: Center for Strategic and International Studies, May 2015), 12.

111. "India, Pakistan Edge Closer to Joining SCO Security," *The Express Tribune,* June 24, 2016, http://tribune.com.pk/story/1129533/india-pakistan-edge-closer-joining-sco-security-bloc/.

112. Interview with Farkhod Aminjonov, April 2016.

113. "Potencial Turkmenistana po Narashivaniyu Eksporta Prirodnogo Gaza."

114. "Is Turkmenistan Losing Iran as a Gas Customer?," Radio Free Europe, August 14, 2014, http://www.rferl.org/content/qishloq-ovozi-turkmenistan-iran-gas/26530894.html.

115. "Saparmurat Niyazov Inaugurates Gas Compressor Station at Korpeje Natural Gas Field," Turkmenistan.ru, September 14, 2005, http://www.turkmenistan.ru/?page_id=3&lang_id=en&elem_id=7108&type=event&sort=date_desc; "Turkmenistan, Iran Launch Gas Pipeline," *Pipeline and Gas Journal* 238, no. 1 (January 2011), http://www.pipelineandgasjournal.com/turkmenistan-iran-launch-gas-pipeline.

116. Danila Bochkarev, "EU Investment in Turkmenistan's Energy Upstream: Mission Impossible?," *European Energy Review,* April 30, 2015, http://europeanenergyreview.com/site/pagina.php?id=4363; "Turkmenistan Boosting Gas Exports to China."

117. Gas Exports to Reach 40 bcm per Year," *Iran Daily,* August 3, 2016, http://www.iran-daily.com/News/156181.html.

118. Farkhod Aminjonov, "Is the Iran–Turkmenistan Gas Trade Coming to the End?," Eurasian Research Institute, February 2, 2016, http://eurasian-research.org/en/research/comments/energy/iran%E2%80%93turkmenistan-gas-trade-coming-end-0.

119. Huseyn Hasanov, "Negotiations on Trans-Caspian Gas Pipeline Held Successfully in Brussels," Trend News Agency, March 10, 2012, http://en.trend.az/capital/energy/2001834.html; Vladimir Socor, "Aliyev, Erdogan Sign Inter-Governmental Agreement on Trans-Anatolia Gas Pipeline to Europe," *Eurasia Daily Monitor* 9, no. 122 (June 2012), http://www.jamestown.org/programs/edm/single/?tx_ttnews%5Btt_news%5D=39545&tx_ttnews%5BbackPid%5D=27&cHash=2e9f386bf569ef7ea670cde5a5c3784c#.VaBs6sZViko.

120. Jason Bordoff and Trevor Houser, "American Gas to the Rescue? The Impact of US LNG Exports on European Security and Russian Foreign Policy," Center for Global Energy Policy, September 2014, 29, http://energypolicy.columbia.edu/sites/default/files/energy/CGEP_American%20Gas%20to%20the%20Rescue%3F.pdf.

121. "Ashgabat Declaration," European Commission, May 1, 2015, https://ec.europa.eu/commission/2014-2019/sefcovic/announcements/ashgabat-declaration_en.

122. Paolo Sorbello, "Europe and Turkmenistan Make Nice," *Diplomat,* May 17, 2015, http://thediplomat.com/2015/05/europe-and-turkmenistan-make-nice/.

123. "Berdymukhamedov i Merkel Obsudili Transcaspiiski Gazoprovod," Day.az, August 30, 2016, http://news.day.az/economy/816023.html.

124. Altay Alti, "Turkey, Turkmenistan, Azerbaijan Aim to Boost Connectivity," *Asia Times,* September 29, 2016, http://www.atimes.com/turkey-turkmenistan-azerbaijan-aim-boost-connectivity/.

125. Huseyn Hasanov, "Turkmenistan Intends to Host Trilateral Summit on Trans-Caspian Gas Pipeline Project," Trend News Agency, July 15, 2016, http://en.trend.az/business/economy/2562186.html.

126. Barbara Janusz, "The Caspian Sea Legal Status and Regime Problems," Chatham House, August 2005, https://www.chathamhouse.org/sites/files/chathamhouse/public/Research/Russia%20and%20Eurasia/bp0805caspian.pdf.

127. "Turkmen Gas Could Reach Europe through Iran: Commissioner," EUbusiness, May 1, 2015, http://www.eubusiness.com/news-eu/turkmenistan-gas.10zl.

128. "Turkmenistan-Afghanistan-Pakistan-India [TAPI] Pipeline," Ministry of Mines and Petroleum Islamic Republic of Afghanistan, n.d., http://mom.gov.af/en/page/4717.

129. Zafar Bhutta, "Four Countries Ink Deal for $10 Billion TAPI Gas Pipeline Project," *The Express Tribune,* March 4, 2016, http://tribune.com.pk/story/1058949/tapi-gas-pipeline-four-countries-ink-deal-for-10-billion-project/.

130. Ibid.

131. "Turkmenistan Accelerated the Construction of the TAPI Pipeline," *Nebit-Gaz,* October 10, 2016, http://www.oilgas.gov.tm/en/blog/506/in-turkmenistan-accelerated-the-construction-of-the-tapi-pipeline.

132. "Turkmenistan Speeds up Tapi, Eyes EU," *Natural Gas World,* August 1, 2016, http://www.naturalgasworld.com/turkmenistan-speeds-up-tapi-eyes-eu-30878.

133. Dipanjan Chaudhury, "Afghanistan to Form 7,000 Member Security Force to Guard $10 Billion Tapi Pipeline," *The Economic Times*, December 30, 2015, http://economictimes.indiatimes.com/news/defence/afghanistan-to-form-7000-member-security-force-to-guard-10-billion-tapi-pipeline/articleshow/50373647.cms.

134. Ralf Dickel, Elham Hassanzadeh, James Henderson, Anouk Honoré, Laura El-Katiri, Simon Pirani, Howard Rogers, Jonathan Stern, and Katja Yafimava, "Reducing European Dependence on Russian Gas: Distinguishing Natural Gas Security from Geopolitics," OIES Paper NG 92, Oxford Institute for Energy Studies, October 2014, 15, https://www.oxfordenergy.org/wpcms/wp-content/uploads/2014/10/NG-92.pdf; Iana Dreyer and Gerald Stang, "The Shale Gas 'Revolution': Challenges and Implications for the EU," European Union Institute for Security Studies, February 2013, 2.

135. Anouk Honoré, "The Outlook for Natural Gas Demand in Europe," OIES Paper NG 87, Oxford Institute for Energy Studies, June 2014, 71, http://www.oxfordenergy.org/wpcms/wp-content/uploads/2014/06/NG-87.pdf.

136. Alex Forbes, "China Forecast to Consume 420 BCM of Gas in 2020," *Gastechnews,* May 6, 2014, http://www.gastechnews.com/lng/china-forecast-to-consume-420-BCM-of-gas-in-2020/.

7. The Politics of Demand

1. International Energy Agency (henceforth IEA), "World Energy Outlook 2015," November 10, 2015, 7, http://www.worldenergyoutlook.org/media/weowebsite/2015/151110_WEO2015_presentation.pdf.

2. Naureen Malik, "Half of America's Gas Exports Claimed by Asia in Next 3 Years," Bloomberg, July 21, 2016, http://www.bloomberg.com/news/articles/2016-07-21/half-of-america-s-gas-exports-claimed-by-asia-in-next-3-years.

3. US Energy Information Administration (henceforth EIA), "China: International Energy Data and Analysis," May 14, 2015, http://www.eia.gov/beta/international/analysis.cfm?iso=CHN.

4. National Development and Reform Commission (hereafter NDRC), "U.S.-China Joint Statement on Climate Change," February 15, 2014, http://en.ndrc.gov.cn/newsrelease/201402/t20140218_579304.html.

5. NDRC, "China's National Plan for Addressing Climate Change (2015–2020)," September 19, 2014, http://www.sdpc.gov.cn/gzdt/201411/W020141104591413713551.pdf.

6. United Nations Economic and Social Commission for Asia and the Pacific, "COP21: Statement at the China Low Carbon Development Strategy Workshop," December 9, 2015, http://www.unescap.org/speeches/cop21-statement-china-low-carbon-development-strategy-workshop; "COP21 Climate Change Summit Reaches Deal in Paris," BBC News, December 13, 2015, http://www.bbc.com/news/science-environment-35084374.

7. Not including Hong Kong (an additional 3.2 billion cubic meters). British Petroleum (henceforth BP), "BP Statistical Review of World Energy 2015," June 2015, http://www.bp.com/content/dam/bp/pdf/energy-economics/statistical-review-2016/bp-statistical-review-of-world-energy-2015-full-report.pdf; BP, "BP Statistical Review of World Energy June 2016," June 2016, http://www.bp.com/content/dam/bp/pdf/energy-economics/statistical-review-2016/bp-statistical-review-of-world-energy-2016-full-report.pdf, 23.

8. EIA, "China"; BP, "BP Statistical Review of World Energy June 2016," 23.

9. BP, "BP Statistical Review of World Energy June 2016," 22, 32.

10. Anjli Raval, "Slowing China Demand Stalls 'Golden Age of Gas': Subdued Industrialised Nation Growth Slows Global Energy Demand," *Financial Times,* June 8, 2016, https://next.ft.com/content/fe83a638-2c8f-11e6-bf8d-26294ad519fc.

11. Keun-Wook Paik, *Sino-Russian Oil and Gas Cooperation: The Reality and Implications* (Oxford: Oxford University Press for the Oxford Institute for Energy Studies, 2012), 205–220.

12. "China's Wholesale Natural Gas Prices Cut in Major Reform," CNTV, April 2, 2015, http://english.cntv.cn/2015/04/02/VIDE1427915397363397.shtml.

13. Oleg Vukmanovic and Henning Gloystein, "LNG Boom over as China Looks to Sell Out of Long-Term Deals," Reuters, December 9, 2014, http://www.reuters.com/article/2014/12/09/china-gas-imports-idUSL6N0TN1HM20141209.

14. Michael Ratner, Gabriel M. Nelson, and Susan V. Lawrence, "China's Natural Gas: Uncertainty for Markets," Congressional Research Service Report, May 2, 2016.

15. Ibid., 1, 11, 23–25.

16. Raval, "Slowing China Demand Stalls 'Golden Age of Gas' "; IEA, "IEA Sees Major Shifts in Global Gas Trade over next Five Years," June 8, 2016, https://www.iea.org/newsroomandevents/pressreleases/2016/june/iea-sees-major-shifts-in-global-gas-trade-over-next-five-years.html; BP, "BP Statistical Review of World Energy June 2016," 1, 4; IEA, "IEA Medium-Term Market Report 2016: Market Analysis and Forecasts to 2021," June 7, 2016, https://www.iea.org/newsroomandevents/speeches/160608_MTGMR2016_presentation.pdf, 1, 3; Jude Clemente, "China's Rising Natural Gas Demand, Pipelines, and LNG," *Forbes,* April 24, 2016, http://www.forbes.com/sites/judeclemente/2016/04/24/chinas-rising-natural-gas-demand-pipelines-and-lng/#2fd624ca6a38.

17. BP, "BP Statistical Review of World Energy 2015."

18. EIA, "China"; International Gas Union, *2016 World LNG Report* (Oslo, Norway: International Gas Union, June 2016), http://www.igu.org/publications/2016-world-lng-report, 12, 68–69.

19. International Gas Union, *2016 World LNG Report,* 12; interview with Aloulou Faouzi, Project Manager of Global Hydrocarbon Supply Modeling at the Energy Information Administration, December 31, 2015.

20. *The Report, Qatar 2010* (London: Oxford Business Group, 2010), 117; Wan Zhihong, "CNOOC Buys More LNG from Qatar," *China Daily,*

November 14, 2009, http://www.chinadaily.com.cn/cndy/2009-11/14/content_8971029.htm.

21. Australia Pacific LNG, "About the Project," n.d., http://www.aplng.com.au/about-project/about-project; EIA, "Country Analysis—China," May 14, 2015, https://www.eia.gov/beta/international/analysis.cfm?iso=CHN; Nobuyuki Higashi, "Natural Gas in China: Market Evolution and Strategy," IEA Working Paper Series, Energy Markets and Security (Paris: IEA / OECD, June 2009), 18, https://www.iea.org/publications/freepublications/publication/nat_gas_china.pdf.

22. Zheng Yao, "Sino-Australian LNG Projects Today and Challenges," *Sino-Global Energy,* April 2014, http://en.cnki.com.cn/Article_en/CJFDTotal-SYZW201504002.htm.

23. Merrick White and David Phua, "Natural Gas Imports into China—Prospects for Growth," Energy Law Exchange, September 10, 2014, http://www.energylawexchange.com/natural-gas-imports-china-prospects-growth/; "Stroitelstvo gazoprovoda Uzebkistan-KNR nachnetsja letom 2016 goda," *Sputnik Uzbekistan,* December 18, 2015, http://ru.sputniknews-uz.com/economy/20151218/1360294.html.

24. BP, "BP Statistical Review of World Energy 2016."

25. "Controversial Pipeline Now Fully Operational," *Myanmar Times,* October 2013, http://www.mmtimes.com/index.php/business/8583-controversial-pipeline-now-fully-operational.html.

26. "Iran Eyes Chinese Gas Market," Natural Gas Asia, November 14, 2015, http://www.naturalgasasia.com/iran-eyes-chinese-gas-market-17003.

27. Morteza Soleymannezhad, "Iran's Oil Sector: An Analysis from the Current Risk Standpoint," *Forbes,* August 20, 2015, http://www.forbes.com/sites/mortezasoleymannezhad/2015/08/20/irans-oil-sector-an-analysis-from-the-current-risk-standpoint/; Syd Nejad, "The Resurgence of Iran Upstream: Will Rewards Outweigh Risks?," Society of Petroleum Engineers, December 18, 2015, http://www.spe.org/news/article/the-resurgence-of-iran-upstream-will-rewards-outweigh-risks; Najmeh Bozorgmehr, "Iran Offers Flexible Oil Contracts to Attract Foreign Investors," *Financial Times,* November 28, 2015, http://www.ft.com/cms/s/0/b2f6bf58-95b2-11e5-95c7-d47aa298f769.html#axzz3wUhdhDqq.

28. BP, "BP Statistical Review of World Energy June 2016," 20; EIA, "China"; Ministry of Land and Resources, "Chinese Mineral Resources,"

October 2014, http://www.mlr.gov.cn/zwgk/qwsj/201501/P02015012053162 5261319.pdf.

29. Ministry of Land and Resources, "National Reserves Evaluation of Oil and Gas," May 6, 2015, http://www.mlr.gov.cn/xwdt/jrxw/201505 /t20150506_1349846.htm.

30. China National Petroleum Corporation (hereafter CNPC), "CNPC Discovers Largest Monomer Marine Uncompartmentalized Gas Reservoir in China," February 10, 2014, http://www.cnpc.com.cn/en/nr2014/201402/4f 0bf0203d374acb9fb3bba7b0d86d3e.shtml.

31. China National Offshore Oil Corporation, "Oil and Gas Exploration and Development," July 2015, http://www.cnooc.com.cn/col/col6301/index.html.

32. BP, "BP Statistical Review of World Energy June 2016," 22.

33. Shardul Sharma, "WellDog to Form China JV to Develop CBM Resources," *Natural Gas World,* November 13, 2016, http://www.naturalgas world.com/welldog-to-form-chinese-jv-to-develop-cbm-resources-34403.

34. Paik, *Sino-Russian Oil and Gas Cooperation,* 194–195.

35. Faouzi Aloulou, "Shale Gas Development in China Aided by Government Investment and Decreasing Well Cost," EIA, September 30, 2015, http://www.eia.gov/todayinenergy/detail.cfm?id=23152; Philip Andrews-Speed and Christopher Len, "China Coalbed Methane: Slow Start and Still Work in Progress," National University Singapore Energy Studies Institute, policy brief no. 4, December 5, 2014, 4, http://esi.nus.edu.sg/docs /default-source/esi-policy-briefs/china-cbm-slow-start-and-still-work-in -progress.pdf?sfvrsn=2; Qin Yong and Ye Jianping, "A Review on Development of CBM Industry in China," American Association of Petroleum Geologists, Asia Pacific Region, Geoscience Technology Workshop, "Opportunities and Advancements in Coal Bed Methane in the Asia Pacific," Brisbane, Queensland, Australia, February 12, 2015, 1, http://www.searchanddiscovery.com/documents/2015/80454yong/ndx _yong.pdf.

36. EIA, "China; Fan Gao, "Will There Be a Shale Gas Revolution in China by 2020?," OIES Paper NG 59, Oxford Institute for Energy Studies, April 2012, 6, https://www.oxfordenergy.org/wpcms/wp-content/uploads /2012/04/NG-61.pdf.

37. CNPC, "China's First Shale Gas Well," January 13, 2015, http://news .cnpc.com.cn/system/2015/01/13/001524531.shtml.

38. Ministry of Land and Resources, "Summary of Shale Gas Development," September 18, 2014, http://www.mlr.gov.cn/xwdt/zsdwdt/201409/t20140918_1330243.htm.

39. Sinopec, "200th Well in Fulin Shale Gas Field," February 25, 2015, http://www.sinopecgroup.com/group/xwzx/gsyw/20150225/news_20150225_336253964241.shtm.

40. "New Shale Gas Discovery in South-West China Adds to Already Prolific Fueling Field," Shale Gas International, July 12, 2016, http://www.shalegas.international/2016/07/12/new-shale-gas-discovery-in-south-west-china-adds-to-already-prolific-fuling-field/.

41. Ministry of Land and Resources, "Summary of Shale Gas Development"; EIA, "Shale Gas Gross Withdrawals," July 31, 2015, http://www.eia.gov/dnav/ng/NG_PROD_SUM_A_EPG0_FGS_MMCF_A.htm.

42. "New Shale Gas Discovery in South-West China Adds to Already Prolific Fueling Field."

43. Zou Caineng, "Development of China Shale Gas for the Last Decade," Sixth Annual Conference of China Petroleum and Geology, June 16, 2015–June 18, 2015, *China Science Daily*, July 24, 2015 (in Chinese), http://www.sinopecgroup.com/group/xwzx/hgzc/20150724/news_20150724_326719936199.shtml.

44. NDRC, "National Shale Gas Development Plan (2011–2015)," May 16, 2012 (in Chinese), http://www.mlr.gov.cn/xwdt/jrxw/201203/P020120316624004741256.pdf.

45. Chen Aizhu, Judy Hua, and Charlie Zhu, "China Finds Shale Gas Challenging, Halves 2020 Output Target," Reuters, August 7, 2014, http://www.reuters.com/article/2014/08/07/us-china-shale-target-idUSKBN0G71FX20140807; Song Yen Ling, "China Cuts 2020 Shale Gas Output Target as Challenges Persist," S&P Global Platts, September 18, 2014, http://www.platts.com/latest-news/natural-gas/singapore/china-cuts-2020-shale-gas-output-target-as-challenges-27641138.

46. Chen Aizhu. "China Back to Drawing Board as Shale Gas Fails to Flow," Reuters, September 5, 2013. http://www.reuters.com/article/2013/09/05/china-shale-idUSL4N0GY23420130905; Energy Study Institute, "China Shale Gas: Can the Pace Be Sustained?" January 5, 2015, http://www.esi.nus.edu.sg/docs/default-source/esi-policy-briefs/china-shale-gas-can-the-pace-be-sustained.pdf?sfvrsn=2.

47. Paul Stevens, *The "Shale Gas Revolution": Hype and Reality* (London: Chatham House, September 2010); Edward C. Chow, "The Future of Natural Gas in China," Center for Strategic and International Studies, December 18, 2013, https://www.csis.org/analysis/future-natural-gas-china-why-world-cares; Constanza Jacazio, On Chinese Gas Demand—Perspective of a Senior Gas Analyst at the IEA, interview by Agnia Grigas, June 28, 2016.

48. Song Yen Ling, "Sinopec, Petrochina Persevere with Shale Gas as Drilling Costs Fall," S&P Global Platts, March 30, 2015, http://www.platts.com/latest-news/natural-gas/singapore/sinopec-petrochina-persevere-with-shale-gas-as-27259583; Aloulou, "Shale Gas Development in China."

49. Florence Tan, "China's Sinopec Hunting U.S. Shale Deals, but Prices High," Reuters, June 16, 2015, http://www.reuters.com/article/us-sinopec-usa-oil-idUSKBN0OW1ZW20150616; Melanie Hart, "Current Patterns and Future Opportunities; Mapping Chinese Direct Investment in the U.S. Energy Economy" (Washington, DC: Center for American Progress, July 2015), 8, https://cdn.americanprogress.org/wp-content/uploads/2015/07/30051604/ChinaEnergyFDI-Final.pdf; Aloulou, "Shale Gas Development in China."

50. "ConocoPhillips Halts Sichuan Shale Gas Talks with PetroChina," Bloomberg, July 22, 2015, http://www.bloomberg.com/news/articles/2015-07-22/conocophillips-halts-shale-gas-talks-with-cnpc-for-sichuan-field.

51. Interview with Aloulou Faouzi.

52. BP, "BP Statistical Review of World Energy June 2016," 22–23.

53. National Energy Administration, "Discovery of Natural Gas in South China Sea," March 20, 2014, http://www.nea.gov.cn/2014-03/20/c_133200146.htm.

54. Six trillion cubic feet (169.9 billiob cbic meters) of LNG passes through the South China Sea. EIA, "International Energy Data and Analysis of the South China Sea," February 7, 2013, http://www.eia.gov/beta/international/regions-topics.cfm?RegionTopicID=SCS.

55. United States Geological Survey, "An Estimate of Undiscovered Conventional Oil and Gas Resources of the World," April 18, 2012, http://pubs.usgs.gov/fs/2012/3042/fs2012-3042.pdf.

56. EIA, "International Energy Data and Analysis of the South China Sea."

57. Ibid.

58. Ronald D. Ripple, "The Geopolitics of Australian Natural Gas Development" (Cambridge, MA, and Houston, TX: Harvard University's

Belfer Center and Rice University's Baker Institute Center for Energy Studies, January 14, 2014), 10, http://belfercenter.hks.harvard.edu/files/CES-pub-GeoGasAustralia-011414.pdf.

59. Shannon Tiezzi, "Why China Is Stopping Its South China Sea Island-Building," *Diplomat,* June 16, 2015, http://thediplomat.com/2015/06/why-china-is-stopping-its-south-china-sea-island-building-for-now/.

60. "The Stockholm International Peace Research Institute Military Expenditure Database," 2015, http://www.sipri.org/research/armaments/milex/milex_database/milex-data-1988-2014.

61. Republic of the Philippines, "On the Enhanced Defense Cooperation Agreement," April 28, 2014, http://www.gov.ph/2014/04/28/qna-on-the-enhanced-defense-cooperation-agreement/; "A Philippines Defense Equipment Wish-list Submitted in Japan," Manila Livewire, May 27, 2015, http://www.manilalivewire.com/2015/05/a-philippines-defense-equipment-wish-list-submitted-in-japan/.

62. Chris Blake and Manirajan Ramasamy, "ASEAN 'Differences' on South China Sea Threaten Meeting Message," Bloomberg, August 6, 2015, http://www.bloomberg.com/news/articles/2015-08-06/asean-differences-on-south-china-sea-threaten-meeting-message; Charlie Campbell, "After Days of Deadlock, ASEAN Releases Statement on South China Sea Dispute," *Time,* July 25, 2016, http://time.com/4421293/asean-beijing-south-china-sea-cambodia-philippines-laos/.

63. Ben Blanchard and Benjamin Kang Lim, " 'Give Them a Bloody Nose': Xi Pressed for Stronger South China Sea Response," Reuters, July 31, 2016, http://www.reuters.com/article/us-southchinasea-ruling-china-insight-idUSKCN10B10G.

64. Ibid.; Council on Foreign Relations, "Territorial Disputes in the South China Sea," Global Conflict Tracker, August 1, 2016, http://www.cfr.org/global/global-conflict-tracker/p32137#!/conflict/territorial-disputes-in-the-south-china-sea; John Ruwitch and Brenda Goh, "China Says to Hold Drills with Russia in South China Sea," Reuters, July 28, 2016, http://www.reuters.com/article/us-southchinasea-china-drills-idUSKCN1080O8.

65. Martin Fackler, "China and Japan in Deal over Contested Gas Fields" *New York Times,* June 19, 2008, http://www.nytimes.com/2008/06/19/world/asia/19sea.html?_r=0; Zhou Shan, "China and Japan Agree on Joint Gas Exploration in East China Sea," *Epoch Times,* June 27, 2008, http://www

.theepochtimes.com/n3/1530765-joing-gas-exploration-china-japan/; "Japan Warns China Making New Platform in East China Sea," *Sydney Morning Herald,* July 12, 2015, http://www.smh.com.au/world/japan-warns-china-making-new-platform-in-east-china-sea-20150712-giad2b.html.

66. Jonathan Stern, ed., *Natural Gas in Asia: The Challenges of Growth in China, India, Japan, and Korea* (Oxford: Oxford Institute for Energy Studies, 2008), 1.

67. BP, "BP Statistical Review of World Energy June 2016," 23; Central Intelligence Agency (henceforth CIA), "The World Fact Book (Japan)," September 4, 2015, https://www.cia.gov/library/publications/the-world-factbook/geos/ja.html.

68. EIA, "Country Analysis—Japan," January 30, 2015, https://www.eia.gov/beta/international/analysis.cfm?iso=JPN.

69. BP, "BP Statistical Review of World Energy June 2016," 28.

70. "Will Japan Receive Russian Pipeline Gas?," *Russia beyond the Headlines,* May 26, 2015.

71. Eriko Amaha, "Japan LNG Demand Expected to Fall by 2020 on Nuclear Restarts, Renewables," S&P Global Platts, December 15, 2015, http://www.platts.com/latest-news/natural-gas/tokyo/japan-lng-demand-expected-to-fall-by-2020-on-27051779.

72. Congressional Research Service, "U.S. Natural Gas Exports," January 28, 2015, https://www.fas.org/sgp/crs/misc/R42074.pdf; EIA, "Natural Gas Weekly Update, for Week Ending August 17, 2016," August 18, 2016, http://www.eia.gov/naturalgas/weekly/archive/2016/08_18/index.cfm.

73. National Bureau of Asian Research, "The Trans-Pacific Partnership as a Pathway for US Energy Exports to Japan," January 2015. http://www.nbr.org/downloads/pdfs/eta/ES_essay_cutler_012815.pdf.

74. BP, "BP Statistical Review of World Energy June 2016," 28.

75. CIA, "The World Fact Book (South Korea)," July 11, 2016, https://www.cia.gov/library/publications/the-world-factbook/geos/ks.html.

76. EIA, "International Energy Data and Analysis (South Korea)," last modified April 1, 2014, http://www.eia.gov/beta/international/analysis.cfm?iso=KOR.

77. BP, "BP Statistical Review of World Energy June 2016," 28.

78. "South Korea's Kogas to Cut LNG Imports in Response to Weaker Local Demand," S&P Global Platts, November 20, 2014, http://www.platts

.com/latest-news/natural-gas/seoul/south-koreas-kogas-to-cut-lng-imports-in-response-27856421.

79. "Korea Gas Formalizes Deal to Resell Part of US Sabine Pass LNG to Total," S&P Global Platts, January 7, 2014, http://www.platts.com/latest-news/natural-gas/seoul/korea-gas-formalizes-deal-to-resell-part-of-us-27797547.

80. US Department of Energy, "LNG Monthly Report," July 2015, http://energy.gov/fe/downloads/lng-monthly-report-2015.

81. "GDF Suez to Deliver US Shale Gas to Taiwan," Shale World, March 31, 2014, http://www.shale-world.com/2014/03/31/gdf-suez-deliver-us-shale-gas-taiwan/.

82. LNG for Shipping, "Bunkering," 2016, https://lngforshipping.eu/lng-for/bunkering; Charles Lee, "South Korea Looks to Position as New LNG Bunkering Hub to Boost Bargaining Power," S&P Global Platts, July 15, 2016, http://www.platts.com/latest-news/natural-gas/seoul/south-korea-looks-to-position-as-new-lng-bunkering-26493191.

83. President Xi, "Korea-China Relations Better than Ever," Korea.net, July 4, 2014, http://www.korea.net/NewsFocus/Society/view?articleId=120421.

84. Lally Weymouth, "President Park's Talk with the Washington Post," *Washington Post,* June 11, 2015, https://www.washingtonpost.com/opinions/an-interview-with-south-korean-president-park-geun-hye/2015/06/11/15abee3e-1039-11e5-9726-49d6fa26a8c6_story.html.

85. Ministry of Foreign Affairs of the People's Republic of China, "The Joint Declaration of the Government of the People's Republic of China and the Republic of Korea," July 3, 2014, http://www.mfa.gov.cn/chn//pds/ziliao/1179/t1171410.htm.

86. Alexander's Gas and Oil Connections, "China's Hanas Signs Gas Deal with SK Group," August 8, 2014, http://www.gasandoil.com/news/2014/08/china2019s-hanas-signs-gas-deal-with-sk-group.

87. EIA, "Medium-Term Market Report 2016," 11.

88. Anupama Sen, "Gas Pricing Reform in India: Implications for the Indian Gas Landscape," OIES Paper NG 96, Oxford Institute for Energy Studies, April 2015, 5, http://www.oxfordenergy.org/wpcms/wp-content/uploads/2015/04/NG-96.pdf.

89. BP, "Country Insights: India," 2016, http://www.bp.com/en/global/corporate/energy-economics/energy-outlook-2035/country-and-regional-insights/india-insights.html.

90. BP, "BP Statistical Review of World Energy June 2016," 40.

91. Mike Mellish, "China and India Drive Recent Changes in World Coal Trade," EIA, November 20, 2015, http://www.eia.gov/todayinenergy/detail.cfm?id=23852&src=email; BP, "BP Statistical Review of World Energy June 2016," 23, 33.

92. Suhasini Raj and Ellen Barry, "Delhi Closes Over 1,800 Schools in Response to Dangerous Smog," *New York Times*, November 4, 2016, http://www.nytimes.com/2016/11/05/world/asia/delhi-closes-over-1800-schools-in-response-to-dangerous-smog.html?_r=1.

93. BP, "BP Statistical Review of World Energy June 2016," 22, 23, 28.

94. "In a First, Natural Gas Hydrates Discovered in the Indian Ocean," *Hindu,* July 26, 2016, http://www.thehindu.com/sci-tech/energy-and-environment/in-a-first-natural-gas-hydrates-discovered-in-the-indian-ocean/article8902112.ece.

95. Gaurav Agnihotri, "India Is Becoming a Massive Importer of LNG," Business Insider, June 23, 2015, http://www.businessinsider.com/india-is-becoming-a-huge-importer-of-lng-2015-6; "Qatar Slashes Price of LNG It Sells to India," Natural Gas Asia, December 31, 2015, http://www.naturalgasasia.com/qatar-slashes-price-of-lng-it-sells-to-india-17368; "Report: GAIL Seeking LNG Contract Revision with Gazprom," *LNG World News,* July 25, 2016, http://www.lngworldnews.com/report-gail-seeking-lng-contract-revision-with-gazprom/?utm_source=emark&utm_medium=email&utm_campaign=daily-update-lng-world-news-2016-07-26&uid=27940.

96. EIA, "U.S. Natural Gas Re-Exports by Country," August 31, 2015, http://www.eia.gov/dnav/ng/ng_move_expc_s1_a.htm; Anna Shiryaevskaya, "India Demand Surge Sucks Up LNG Otherwise Meant for Europe," Bloomberg, July 4, 2016, http://www.bloomberg.com/news/articles/2016-07-04/india-demand-surge-sucks-up-lng-otherwise-destined-for-europe; Office of Fossil Energy, U.S. Department of Energy, "LNG Monthly Report—October 2016," 2–3, http://energy.gov/sites/prod/files/2016/10/f33/LNG%202016_0.pdf.

97. Sunjoy Joshi and Najeed Jung, "Natural Gas in India," in *Natural Gas in Asia: The Challenges of Growth in China, India, Japan, and Korea,* ed. Jonathan Stern (Oxford: Oxford Institute for Energy Studies, 2008), 83–89; "Bangladesh Contracts Accelerate for Its First LNG Terminal," *LNG World News,* July 25, 2016, http://www.lngworldnews.com/bangladesh-contracts

-excelerate-for-its-first-lng-terminal/; "GAIL India Postpones $7-Billion LNG Ship Tender by One Month," *Economic Times India,* March 2, 2016, http://economictimes.indiatimes.com/industry/energy/oil-gas/gail-india-postpones-7-billion-lng-ship-tender-by-one-month/articleshow/51220786.cms.

98. Ministry of Petroleum and Natural Gas of India, "Natural Gas Scenario in India," March, 2015, http://petroleum.nic.in/docs/abtng.pdf.

99. "ONGC, Gail to Take 12.5% Stake in Chinese Gas Pipeline," *The Economic Times,* December 29, 2009, http://articles.economictimes.indiatimes.com/2009-12-29/news/28469400_1_korea-gas-corp-870-km-pipeline-gas-pipeline.

100. Catherine Putz, "Narendra Modi's Agenda in Central Asia," *Diplomat,* July 2, 2015, http://thediplomat.com/2015/07/narendra-modis-agenda-in-central-asia-energy-terrorism-and-china/.

101. Abdujalil Abdurasulov, "Is Turkmenistan's Gas Line a Pipe Dream?," BBC News, July 16, 2015, http://www.bbc.com/news/world-asia-32981469.

102. IEA, "World Energy Outlook 2015," 3.

103. Bhupendra Kumar Singh, "Energy Security and India-China Cooperation," International Association for Energy Economics, 2010, http://www.iaee.org/en/publications/newsletterdl.aspx?id=92.

104. Tao Wang, "Shared Energy Interests an Opportunity for Sino-Indian Cooperation," *Diplomat,* May 23, 2014, http://thediplomat.com/2014/05/shared-energy-interests-an-opportunity-for-sino-indian-cooperation/.

105. James Henderson and Simon Pirani, "The Changing Balance of the Russian Gas Matrix and the Role of the Russian State," in *The Russian Gas Matrix: How Markets Are Driving Change,* ed. James Henderson and Simon Pirani (Oxford: Oxford Institute for Energy Studies, 2014), 388.

106. Sakhalin Energy, "20th Anniversary of Sakhalin Energy," 2014, http://www.sakhalinenergy.ru/en/media-centre/press_releases/item.wbp?article_id=59b7d662-8a65-4eff-9a1a-c83b83bf45ff.

107. Morena Skalamera and Andreas Goldthau, "Russia: Playing Hardball or Bidding Farewell to Europe? Debunking the Myths of Eurasia's New Geopolitics of Gas," discussion paper, Geopolitics of Energy Project (Cambridge, MA: Harvard Kennedy School, Belfer Center for Science and International Affairs, June 2016), 16–17, http://belfercenter.hks.harvard.edu/files/Russia%20Hardball%20-%20Web%20Final.pdf.

108. Paik, *Sino-Russian Oil and Gas Cooperation,* 1–19.

109. Xiangdong Geng, *Interpretation of Chinese Diplomacy with Illustrations* (Hong Kong: Zhonghua Book Company, 2010).

110. Jeremy Page, "Russian Oil Route Will Open to China," *Wall Street Journal,* September 26, 2010, http://www.wsj.com/articles/SB10001424052748704082104575515543164948682; "China-Russia Crude Oil Pipeline Cumulative Transportation 68.42 Million Tons," *China News,* June 20, 2015, http://www.chinanews.com/ny/2015/06-20/7357538.shtml; Elizabeth Wishnick, "The 'Power of Siberia': No Longer a Pipe Dream," Program on New Approaches to Research and Security in Eurasia, August 2014, http://www.ponarseurasia.org/memo/"power-siberia"-no-longer-pipe-dream.

111. Gazprom, "Altai Project, 2011," n.d., http://www.gazprom.com/about/production/projects/pipelines/altai/.

112. "Oil and Gas Cooperation Events between China and Russia," International Petroleum Economics, June 2013, http://mall.cnki.net/magazine/magadetail/GJJJ201506.htm.

113. See also Paik, *Sino-Russian Oil and Gas Cooperation,* 358–362, 397.

114. Gazprom, "Altai Project."

115. E. C. Chow, "Russia-China Gas Deal and Redeal," Center for Strategic and International Studies, May 11, 2015, http://csis.org/publication/russia-china-gas-deal-and-redeal.

116. "Gazprom to Sign Monumental Gas Deal with China," *Russia Today,* May 19, 2014, http://www.rt.com/business/159880-gazprom-china-russia-cnpc.

117. Interfax, "Putin in Yakutsk to Inaugurate Construction of Pipeline to China," September 1, 2014, http://www.rferl.org/content/russia-china-gas-pipeline-yakutsk/26559938.html; CNPC, "Construction of Chinese Part of the Eastern Route of Russia-China Gas Pipeline Commenced," June 30, 2015, http://www.cnpc.com.cn/en/nr2015/201506/e95714f5bcf94ebeacfa2060a996adf0.shtml.

118. Zhang Gaoli, "China-Russia Natural Gas Pipeline in China Territory Officially Started," Xinhua Economic Information, July 6, 2015 (in Chinese), http://jjckb.xinhuanet.com/2015-07/06/content_553520.htm.

119. Gavin Hanrahan, "What Is a Heads of Agreement and Are They Legally Binding?," Turnbull Hill Lawyers, April 22, 2015, http://www.turnbullhill.com.au/articles/what-is-a-heads-of-agreement-and-are-they-legally-binding-.html.

120. Gazprom, "Russia and China Consider State of Negotiations on Gas Supply via Western Route," June 2015, http://www.gazprom.com/press/news/2015/june/article228565/.

121. "Russia's Gazprom Says Gas Supplies to China May Start Next Decade," Reuters, August 26, 2015, http://www.reuters.com/article/2015/08/26/russia-china-gas-idUSL5N1113OT20150826; "The Reborn Variable to Natural Gas, the Price Is the Focus of the China-Russia Game," *China Economic Herald,* August 5, 2015, http://www.ceh.com.cn/cjpd/2015/08/866120.shtml; "Gazprom Cuts Power of Siberia Spending for 2016," Natural Gas Asia, June 29, 2016, http://www.naturalgasasia.com/gazprom-cuts-power-of-siberia-spending-for-2016-18923; Jacazio, On Chinese Gas Demand, interview by Agnia Grigas.

122. Paik, *Sino-Russian Oil and Gas Cooperation,* 398, 391.

123. Jack Farchy, "Gazprom's China Contract Offers No Protection," *Financial Times,* August 10, 2015, http://www.ft.com/cms/s/0/4ac1cdd6-3f79-11e5-9abe-5b335da3a90e.html#axzz3k7k88eLP.

124. "Kremlin Pivot to China Slowed as Projects Delayed," Reuters, August 27, 2015, http://www.reuters.com/article/2015/08/27/russia-china-projects-idUSL5N11034S20150827.

125. Henderson and Pirani, "The Changing Balance of the Russian Gas Matrix," 388.

126. Rosneft, "Rosneft and Sinopec to Jointly Control Udmurtneft," November 17, 2006, http://www.rigzone.com/news/oil_gas/a/38309/Rosneft_and_Sinopec_to_Jointly_Control_Udmurtneft.

127. Rosneft, "Rosneft and CNPC Sign Memorandum to Expand Cooperation in Upstream Projects in East Siberia," October 18, 2013, http://www.rosneft.com/news/pressrelease/18102013.html.

128. Tim Daiss, "U.S.-Led Sanctions Squeeze Massive Russian Gas Project, But Chinese Funds May Hold The Answer," *Forbes,* November 18, 2015, http://www.forbes.com/sites/timdaiss/2015/11/18/us-led-sanctions-squeeze-massive-russian-gas-project-but-chinese-funds-may-hold-the-answer/; Katya Golubkova, "RIA: Russia's Rosneft in Talks to Allow China into Offshore Arctic Projects," Rigzone, November 16, 2015, http://www.rigzone.com/news/oil_gas/a/141631/RIA_Russias_Rosneft_in_Talks_to_Allow_China_into_Offshore_Arctic_Projects?utm_source=DailyNewsletter&utm_medium=email&utm_term=2015-11-16&utm_content=&utm_campaign=Exploration_2.

129. Keun-Wook Paik, "Natural Gas in Korea," in *Natural Gas in Asia: The Challenges of Growth in China, India, Japan, and Korea,* ed. Jonathan Stern (Oxford: Oxford Institute for Energy Studies, 2008), 201–208; Edward C. Chow and Zachary D. Cuyler, "New Russian Gas Export Projects-From Pipe Dreams to Pipelines" (Washington, DC: Center for Strategic and International Studies, July 22, 2015).

130. Skalamera and Goldthau, "Russia," 17.

131. "Russia Is Capable of Meeting All Japan's Demands for LNG—Rosneft CEO," TASS, November 6, 2015, http://tass.ru/en/economy/834320.

132. "Tokyo Gas Wants to Build a Natural Gas Pipeline to Russia," Oil & Gas 360, May 27, 2015, http://www.oilandgas360.com/tokyo-gas-wants-to-build-a-natural-gas-pipeline-to-russia/; Anna Kuchma, "Will Japan Receive Russian Pipeline Gas?," *Russia beyond the Headlines,* May 26, 2015, http://rbth.com/business/2015/05/26/will_japan_receive_russian_pipeline_gas_46345.html.

133. Anne-Sophie Corbeau, Sammy Six, and Rami Shabaneh, "Natural Gas: Entering the New Dark Age?," KAPSARC Energy Workshop Series (Riyadh: King Abdullah Petroleum Studies and Research Center [KAPSARC], August 2015), 8, http://www.globallnghub.com/custom/domain_4/extra_files/attach_107.pdf.

134. Henderson and Pirani, "The Changing Balance of the Russian Gas Matrix," 388.

135. "In China-Russia Gas Deal, Why China Wins More," *Fortune,* June 20, 2014, http://fortune.com/2014/06/20/in-china-russia-gas-deal-why-china-wins-more/.

136. Bobo Lo, *Axis of Convenience: Moscow, Beijing, and the New Geopolitics* (Washington, DC: Brookings Institution Press, 2008); Bobo Lo, "Russian Foreign Policy" (Washington, DC: Brookings Institution, February 8, 2016).

137. "Putin Pivots to the East," *Economist,* May 24, 2014, http://www.economist.com/news/china/21602727-natural-gas-deal-boosts-vladimir-putin-and-sends-message-america-putin-pivots-east.

138. Richard Weitz, "The Russia-China Gas Deal: Implications and Ramifications," *World Affairs,* September / October 2014, http://www.worldaffairsjournal.org/article/russia-china-gas-deal-implications-and-ramifications.

139. Jonathan Stern, “Conclusion,” in *Natural Gas in Asia: The Challenges of Growth in China, India, Japan, and Korea,* ed. Jonathan Stern (Oxford: Oxford Institute for Energy Studies, 2008), 386.

140. BP, “BP Statistical Review of World Energy June 2016,” 28; International Gas Union, *2016 World LNG Report,* 6.

141. “Long Term Applications Received by DOE / FE to Export Domestically Produced LNG from the Lower-48 States (as of March 18, 2016),” Energy .gov, March 18, 2016, http://energy.gov/sites/prod/files/2016/03/f30 /Summary%20of%20LNG%20Export%20Applications.pdf; Michael Levi, “What the TPP Means for LNG,” Council on Foreign Relations, November 17, 2015, http://blogs.cfr.org/sivaram/2015/11/17/what-the-tpp-means-for-lng/.

142. Mark Agerton, “Global LNG Pricing Terms and Revisions: An Empirical Analysis,” 2014, https://bakerinstitute.netfu.rice.edu/media/files /files/3a17c1c4/CES-pub-GlobalLNG-092414.pdf.

143. Victoria Zaretskaya and Scott Bradley, “Natural Gas Prices in Asia Mainly Linked to Crude Oil, but Use of Spot Indexes Increases,” EIA, September 29, 2015, https://www.eia.gov/todayinenergy/detail.cfm?id =23132; Dale Nijoka, Barry Munro, and Foster Mellen, “Competing for LNG Demand: The Pricing Structure Debate” (Toronto: Ernst & Young, 2014), 6, http://www.ey.com/Publication/vwLUAssets/Competing-for-LNG-demand /$FILE/Competing-for-LNG-demand-pricing-structure-debate.pdf.

144. Howard V. Rogers, “The Impact of Lower Gas and Oil Prices on Global Gas and LNG Markets,” Oxford Institute for Energy Studies, July 2015, 36–37, http://www.oxfordenergy.org/wpcms/wp-content/uploads /2015/07/NG-99.pdf.

145. Ibid.; OECD / IEA, “The Asian Quest for LNG in a Globalising Market,” Partner Country Series (Paris, 2014), 117–119, http://www.iea.org /publications/freepublications/publication/PartnerCountrySeriesTheAsian QuestforLNGinaGlobalisingMarket.pdf; interview with Holly Morrow, New York, January 22, 2016.

146. Eriko Amaha, “Japan Looks at Boosting LNG Spot Market Trading, Creating Trading Hub,” S&P Global Platts, April 13, 2016, http://www.platts .com/latest-news/natural-gas/tokyo/japan-looks-at-boosting-lng-spot -market-trading-27458588.

147. Tsuyoshi Inajima and Emi Urabe, “Tokyo Gas in Talks with European Firms to Swap LNG Cargoes,” Bloomberg, August 1, 2016, http://www

.bloomberg.com/news/articles/2016-07-31/tokyo-gas-in-talks-with-european-firms-to-swap-u-s-lng-cargoes.

148. Jason Bordoff, "How Exporting U.S. Liquefied Natural Gas Will Transform the Politics of Global Energy," *Wall Street Journal,* November 17, 2015, http://blogs.wsj.com/experts/2015/11/17/how-exporting-u-s-liquefied-natural-gas-will-transform-the-politics-of-global-energy/.

149. Ministry of Economy, Trade and Industry, Japan, "Japan's LNG Policy and Potential Issues on LNG," July 15, 2015, 8, http://www.globallnghub.com/custom/domain_4/extra_files/attach_122.pdf.

150. Ron Bousso, Dmitry Zhdannikov, and Chen Aizhu, "In Shell-BG Review, China Wants Concessions on Huge Gas Deals," Reuters, November 19, 2015, http://www.reuters.com/article/2015/11/19/us-bg-m-a-shell-china-idUSKCN0T81CJ20151119#OG3EFK8dzVmIjXqu.97; Ministry of Economy, Trade and Industry, Japan, "Japan's LNG Policy and Potential Issues on LNG," 9; Promit Mukherjee, "NTPC Refuses to Buy Expensive Natural Gas from GAIL," *Live Mint,* November 13, 2015, http://www.livemint.com/Industry/3IRl9y3hW18qIEjbgmns5J/NTPC-refuses-to-buy-expensive-natural-gas-from-GAIL.html.

151. "U.S. Gas Exports: The Pipe Dream," Natural Gas Europe, August 3, 2015, http://www.naturalgaseurope.com/us-gas-exports-the-pipe-dream-lng-24879; Michel Rose and Oleg Vukmanovic, "Cheap Coal, Giant Batteries May Keep Gas Waiting for Its Golden Age," Reuters, June 4, 2015, reuters.com/article/2015/06/04/energy-gas-demand-idUSL5N0YP4O7 20150604.

152. IEA, "Despite Decline in Oil Prices, Natural Gas Demand Outlook Revised Down," June 4, 2015, https://www.iea.org/newsroomandevents/pressreleases/2015/june/despite-decline-in-oil-prices-natural-gas-demand-outlook-revised-down.html.

153. EIA, "Short-Term Energy Outlook," November 2016, http://www.eia.gov/forecasts/steo/pdf/steo_full.pdf.

154. Aibing Guo, Elena Mazneva, and Stepan Kravchenko. "Russia Fails to Sign China Gas Deal at Shanghai Meeting," Bloomberg, May 20, 2014, http://www.bloomberg.com/news/articles/2014-05-20/russian-chinese-leaders-silent-on-gas-deal-in-shanghai.

155. Allison Good, "North American LNG Likely to Weather 'New Normal' in Global Gas Markets," S&P Global Market Intelligence, July 6, 2015, https://www.snl.com/InteractiveX/ArticleAbstract.aspx?id=33161673.

Conclusion

1. David Koranyi, "A US Strategy for Sustainable Energy Security," Atlantic Council Strategy Papers, March 2016, 13.

2. Brenda Shaffer, *Energy Politics* (Philadelphia: University of Pennsylvania Press, 2011), 4, 40.

3. "U.S. Senators Call for Slowdown of LNG Export Approval," *LNG World News,* September 27, 2016, http://www.lngworldnews.com/u-s-senators-call-for-slowdown-of-lng-export-approvals/?utm_source=emark&utm_medium=email&utm_campaign=daily-update-lng-world-news-2016-09-28&uid=27940.

4. Remarks by John Kerry at the US-EU Energy Council meeting, April 2014, http://www.state.gov/secretary/remarks/2014/04/224287.htm.

5. British Petroleum, "BP Energy Outlook 2035: Country and Regional Insights—Asia Pacific, 2015," http://www.bp.com/content/dam/bp/pdf/energy-economics/energy-outlook-2015/Regional_insights_Asia_Pacific.pdf.

Acknowledgments

This work covering the natural gas markets and political developments across North America, Europe, Asia, and beyond is a product of a two-year research project that has been made possible only with the help of dedicated research assistants and the input of fellow scholars, experts, and event participants. Together they have offered their analytical skills, their regional and market expertise, and their time to this endeavor.

For contributing to this book's study of global gas markets, the historical developments of natural gas, and LNG infrastructure and trade dynamics, specifically in the United States, Europe, and Asia, I would like to acknowledge the invaluable support of Hannah H. Braun. This project also benefited tremendously from the carefully compiled research on US LNG exports by Jason Czerwiec and Lukas Trakimavičius. The Russian and European case studies would not have been possible without the dedicated contributions of Sabina Kubekė and Vija Pakalkaitė, along with Joel Hicks. Mari Dugas and Dmytro Kondratenko were enormously helpful to the examination of the Ukraine and Belarus. I must gratefully acknowledge the input and insights of Farkhod Aminjonov and Dinara Pisareva on Central Asia and of Tengiz Sultanishvili on the South Caucasus. And finally, without the careful research of Ye Tian, Lijia Long, and Daizong Wu, the passages on Asia and China would not have been as revealing and instructive.

I also would like to express my gratitude to my colleagues, and to those scholars and experts who have read and commented on parts of this book, including Robert Anderson, Margarita Balmaceda, Tim Boersma, Guy Caruso,

Aloulou Faouzi, Seth Freeman, Michael Gonchar, Douglas Hengel, Fred Hutchison, Nygmet Ibadildin, Oksana Ishchuk, Vytis Jurkonis, Suedeen Kelly, Tatiana Manenok, Holly Morrow, Alexander Motyl, Richard Pomfret, Lucian Pugliaresi, Peter Ragauss, Jonathan Stern, David Sweet, Nikos Tsafos, Warren Wilczewski, and Katja Yafimava. Likewise, my editor at Harvard University Press, Andrew John Kinney, deserves special mention for his unwavering support for this project from the start. Finally, I would like to thank my family and especially my husband, Paulius, who has supported and encouraged me during this long journey.

Index